Dairy Sector: Opportunities and Sustainability Challenges

Dairy Sector: Opportunities and Sustainability Challenges

Editor

Rajeev Bhat

MDPI • Basel • Beijing • Wuhan • Barcelona • Belgrade • Manchester • Tokyo • Cluj • Tianjin

Editor
Rajeev Bhat
ERA-Chair in VALORTECH
Estonian University of Life Sciences
Tartu
Estonia

Editorial Office
MDPI
St. Alban-Anlage 66
4052 Basel, Switzerland

This is a reprint of articles from the Special Issue published online in the open access journal *Sustainability* (ISSN 2071-1050) (available at: www.mdpi.com/journal/sustainability/special_issues/ Dairy_Opportunities_Sustainability).

For citation purposes, cite each article independently as indicated on the article page online and as indicated below:

LastName, A.A.; LastName, B.B.; LastName, C.C. Article Title. *Journal Name* **Year**, *Volume Number*, Page Range.

ISBN 978-3-0365-3870-9 (Hbk)
ISBN 978-3-0365-3869-3 (PDF)

© 2022 by the authors. Articles in this book are Open Access and distributed under the Creative Commons Attribution (CC BY) license, which allows users to download, copy and build upon published articles, as long as the author and publisher are properly credited, which ensures maximum dissemination and a wider impact of our publications.

The book as a whole is distributed by MDPI under the terms and conditions of the Creative Commons license CC BY-NC-ND.

Contents

About the Editor vii

Preface to "Dairy Sector: Opportunities and Sustainability Challenges" ix

Rajeev Bhat, Jorgelina Di Pasquale, Ferenc Istvan Bánkuti, Tiago Teixeira da Silva Siqueira, Philip Shine and Michael D. Murphy
Global Dairy Sector: Trends, Prospects, and Challenges
Reprinted from: *Sustainability* **2022**, *14*, 4193, doi:10.3390/su14074193 1

Aisha Hassan, Li Cui-Xia, Naveed Ahmad, Muzaffar Iqbal, Kramat Hussain and Muhammad Ishtiaq et al.
Safety Failure Factors Affecting Dairy Supply Chain: Insights from a Developing Economy
Reprinted from: *Sustainability* **2021**, *13*, 9500, doi:10.3390/su13179500 9

Ridha Ibidhi, Tae-Hoon Kim, Rajaraman Bharanidharan, Hyun-June Lee, Yoo-Kyung Lee and Na-Yeon Kim et al.
Developing Country-Specific Methane Emission Factors and Carbon Fluxes from Enteric Fermentation in South Korean Dairy Cattle Production
Reprinted from: *Sustainability* **2021**, *13*, 9133, doi:10.3390/su13169133 33

Susanne Hoischen-Taubner, Jonas Habel, Verena Uhlig, Eva-Marie Schwabenbauer, Theresa Rumphorst and Lara Ebert et al.
The Whole and the Parts—A New Perspective on Production Diseases and Economic Sustainability in Dairy Farming
Reprinted from: *Sustainability* **2021**, *13*, 9044, doi:10.3390/su13169044 45

Erin Percival Carter and Stephanie Welcomer
Designing and Distinguishing Meaningful Artisan Food Experiences
Reprinted from: *Sustainability* **2021**, *13*, 8569, doi:10.3390/su13158569 63

Margherita Masi, Yari Vecchio, Gregorio Pauselli, Jorgelina Di Pasquale and Felice Adinolfi
A Typological Classification for Assessing Farm Sustainability in the Italian Bovine Dairy Sector
Reprinted from: *Sustainability* **2021**, *13*, 7097, doi:10.3390/su13137097 77

Patrick S. Byrne, James G. Carton and Brian Corcoran
Investigating the Suitability of a Heat Pump Water-Heater as a Method to Reduce Agricultural Emissions in Dairy Farms
Reprinted from: *Sustainability* **2021**, *13*, 5736, doi:10.3390/su13105736 91

Elizabeth G. Ross, Carlyn B. Peterson, Yongjing Zhao, Yuee Pan and Frank M. Mitloehner
Manure Flushing vs. Scraping in Dairy Freestall Lanes Reduces Gaseous Emissions
Reprinted from: *Sustainability* **2021**, *13*, 5363, doi:10.3390/su13105363 109

Supriya Verma, Friedhelm Taube and Carsten S. Malisch
Examining the Variables Leading to Apparent Incongruity between Antimethanogenic Potential of Tannins and Their Observed Effects in Ruminants—A Review
Reprinted from: *Sustainability* **2021**, *13*, 2743, doi:10.3390/su13052743 121

Lukáš Čechura and Zdeňka Žáková Kroupová
Technical Efficiency in the European Dairy Industry: Can We Observe Systematic Failures in the Efficiency of Input Use?
Reprinted from: *Sustainability* **2021**, *13*, 1830, doi:10.3390/su13041830 145

Janis Brizga, Sirpa Kurppa and Hannele Heusala
Environmental Impacts of Milking Cows in Latvia
Reprinted from: *Sustainability* **2021**, *13*, 784, doi:10.3390/su13020784 165

Sara Zanni, Mariana Roccaro, Federica Bocedi, Angelo Peli and Alessandra Bonoli
LCA to Estimate the Environmental Impact of Dairy Farms: A Case Study
Reprinted from: *Sustainability* **2022**, *14*, 6028, doi:10.3390/su14106028 177

Tiago Teixeira da Silva Siqueira, Danielle Galliano, Geneviève Nguyen and Ferenc Istvan Bánkuti
Organizational Forms and Agri-Environmental Practices: The Case of Brazilian Dairy Farms
Reprinted from: *Sustainability* **2021**, *13*, 3762, doi:10.3390/su13073762 193

Elizabeth G. Ross, Carlyn B. Peterson, Angelica V. Carrazco, Samantha J. Werth, Yongjing Zhao and Yuee Pan et al.
Effect of SOP "STAR COW" on Enteric Gaseous Emissions and Dairy Cattle Performance
Reprinted from: *Sustainability* **2020**, *12*, 10250, doi:10.3390/su122410250 213

Philip Shine, Michael D. Murphy and John Upton
A Global Review of Monitoring, Modeling, and Analyses of Water Demand in Dairy Farming
Reprinted from: *Sustainability* **2020**, *12*, 7201, doi:10.3390/su12177201 225

About the Editor

Rajeev Bhat

Dr. Rajeev Bhat is presently working as a Professor and ERA-Chair-holder (European Research Area Chair) in VALORTECH (Food By-Products Valorisation Technologies) at the Estonian University of Life Sciences. He has extensive "Research and Teaching" experience of more than two decades (23 years) in the field of agri-food technology, with expertise focusing mainly on issues pertaining to "Sustainable Food Production" and "Food Security". Prof. Bhat has been a visiting professor in many of the recognized universities and is also a recipient of several prestigious international awards and recognitions conferred by various institutions of higher learning and research establishments.

Preface to "Dairy Sector: Opportunities and Sustainability Challenges"

The global dairy industry is experiencing rapid transitions and it is worthy to regularly share some of the updated vital information with all the relevant stakeholders. In this regard, this book, which is an outcome of the SI: "Dairy Sector: Opportunities and Sustainability Challenges" (*Sustainability* journal), covers various topics relevant to available opportunities and overcoming recurrent sustainability challenges in the entire dairy sector. Published as individual articles (original and review articles), various multidisciplinary themes have been excellently covered by renowned experts. The articles cover various interesting aspects related to production and supply chain management, application of novel processing technologies, climate change impacts, environmental issues, safety issues, production diseases, artisan food experiences, and much more. The contributing researchers have meticulously identified existing gaps and proposed inventive solutions that can be beneficial to the dependent dairy industry.

The future of the dairy sector holds lots of positive and promising features, but still needs to remain circumspect in overcoming recurring sustainability challenges. The success will definitely depend on embracing a 'state of the art' approach and dealing with major sustainability challenges with a universal approach. Meeting the demands of all stakeholders (farming community, dependent industries, consumers, policy-makers, and others) is important, specifically taking into account regional socio-economic-environmental security issues.

As the Special Issue editor, my sincere appreciation goes to the Editor-in-Chief (*Sustainability* journal), handling editors, all of the contributing authors, and members of the MDPI publishing team who were involved.

Rajeev Bhat
Editor

Editorial

Global Dairy Sector: Trends, Prospects, and Challenges

Rajeev Bhat [1,*], Jorgelina Di Pasquale [2], Ferenc Istvan Bánkuti [3], Tiago Teixeira da Silva Siqueira [4], Philip Shine [5] and Michael D. Murphy [5]

1. ERA-Chair in Food (By-) Products Valorisation Technologies (VALORTECH), Estonian University of Life Sciences, Fr. R. Kreutzwaldi 1, 51006 Tartu, Estonia
2. Faculty of Veterinary Medicine, University of Teramo, 64100 Teramo, Italy; jdipasquale@unite.it
3. Programa de Pós-graduação em Zootecnia (PPZ/UEM), Universidade Estadual de Maringá, Paraná 87020-900, Brazil; fibankuti@uem.br
4. Cirad, The Mediterranean and Tropical Livestock Systems Unit (SELMET), 34070 Montpellier, France; tiago.teixeira_da_silva_siqueira@cirad.fr
5. Department of Process, Energy and Transport Engineering, Munster Technological University, T12 P928 Cork, Ireland; philip.shine@cit.ie (P.S.); michaeld.murphy@cit.ie (M.D.M.)
* Correspondence: rajeev.bhat@emu.ee

Currently, there is a strong need to find practical solutions towards meeting the expected efficiency and overcoming recurring sustainability challenges in the global dairy sector. Improving dairy production and its supply chain explicitly depends on adopting a sustainable 'state of the art' based approach. Carefully evaluating and understanding certain key sustainability indicators through a holistic approach is highly imperative. Appropriate design and application of novel green technologies, implementation of life cycle analysis, upgradation and optimization of the entire production line are some of the most important factors to be considered. It is vital that due considerations are given to the demands of the producers, consumers, and the dependent dairy industries. Nevertheless, in the future, concerns and challenges over the socio-economic–environmental security should not be ignored.

Rapid transitions are being witnessed in the global dairy sector, which has provided not only substantial commercial opportunities, but has also brought about several sustainability challenges. Some of the challenges include global population expansion, urbanization that has taken over the rural sectors, climate change effects, emerging diseases, and safety issues, in addition to the waste management strategies adopted (on- and off-farm). According to the Food and Agriculture Organization of the United Nations (FAO), Asia registered the highest volume increase in the global milk production output in 2020, with increased international trading of whole milk powder, whey, and cheese [1]. In the European Union (EU) alone, milk production growth is expected to decelerate to nearly 0.5% annually, and by 2031 reaching up to 162 million tonnes. In addition, organic milk production is expected to grow in the EU (reaching up to 8% in 2031), leading to economic gains, imparting environmental benefits, and overall animal welfare [2].

Of late, there has been mounting pressure from global consumers to reduce environmental stress from the dairy production sector, particularly when this is directly related to climate change. Nevertheless, a remarkable positive transition towards balanced production, reduced environmental impacts, and improvements witnessed in economic efficiency and social security have changed the components of sustainability in the dairy sector. Interesting information is available recommending the significance of reducing energy consumption in dairy industries and thereby providing crucial information on energy mitigation actions [3,4].

Hence, the need of the present day is to find practical solutions to the growing pressure towards meeting the expected efficiency and overcoming recurring sustainability challenges in the global dairy industries. Of course, gaining success in the dairy production and supply

Citation: Bhat, R.; Di Pasquale, J.; Bánkuti, F.I.; Siqueira, T.T.d.S.; Shine, P.; Murphy, M.D. Global Dairy Sector: Trends, Prospects, and Challenges. *Sustainability* 2022, *14*, 4193. https://doi.org/10.3390/su14074193

Received: 23 March 2022
Accepted: 24 March 2022
Published: 1 April 2022

Publisher's Note: MDPI stays neutral with regard to jurisdictional claims in published maps and institutional affiliations.

Copyright: © 2022 by the authors. Licensee MDPI, Basel, Switzerland. This article is an open access article distributed under the terms and conditions of the Creative Commons Attribution (CC BY) license (https://creativecommons.org/licenses/by/4.0/).

chain sector explicitly depends on adopting a sustainable 'state of the art' based approach. Carefully evaluating and understanding certain key sustainability indicators with a holistic approach is highly imperative. Appropriate design and application of novel green technologies, implementation of life cycle analysis, up-gradation, and optimization of the entire production line are some of the key factors to be carefully measured and considered. Besides, it is vital that due consideration is given to the demands of the producers, consumers, and the dependent dairy industries. Nevertheless, concerns over the environment, social security, and economy of the region should not be ignored. Precise planning (both 'on and off' farm) assumes importance especially when circular economy strategies need to be considered and adopted. Furthermore, the roadmap to sustainable economies has been effectively provided via the 'European Green Deal' to ensure a sustainable economy by directing the climate/environmental challenges into opportunities.

With this as the background, this Special Issue (SI) *'Dairy Sector: Opportunities and Sustainability Challenges'* focuses on identifying present opportunities, recent advances, and options to overcome future sustainability challenges in the global dairy sector. A 'Web of Science' search with keywords such as dairy industry, dairy metabolites, innovative technologies, sustainable production, valorization of wastes and by-products, circular economy, climate change, carbon footprint modelling, regulatory and legislative issues revealed a high increase in the scientific publications over the recent years in the dairy sector. This SI comprises 13 published articles focusing on various issues pertaining to the dairy sector that are discussed meticulously. Some interesting outcomes of the published articles are discussed in the preceding text.

The global dairy industry has always faced safety issues along the supply chain, and this has tremendously increased over recent years. As the dairy industry fulfills humans' basic necessities, safety failure factors affecting the supply chain can have tremendous impacts. This aspect has been explored by Hassan et al. [5]. In this study, a total of 25 failure factors were identified via literature reviews and by considering the opinion of experts from the dairy industry and academicians. In addition, interpretive structural modelling was applied to analyze mutual interaction among the barriers. Further, in this report, '*Matrice d'Impacts Croises Multiplication Appliques a un Classement*' technique (MICMAC technique) was adopted to identify the value of safety failure factors (SFFs), which was centered on the driving and dependence power. According to the researchers, results generated from this study will assist prime decision-makers in the dairy industries to precisely design supply chain activities and thereby efficiently manage certain identified barriers.

Furthermore, the global dairy sector is influenced directly by the consequences of global warming, and there is a significant release of greenhouse gas emissions witnessed. Ibidhi et al. [6] have reported on the importance of generating knowledge and a database on country-specific emission factors (EFs) in order to assess the national enteric methane emissions, which is expected to support mitigation action assessment. In a study reported from Latvia by Brizga et al. [7], environmental impacts of milking cows with diverse management practices are discussed. The researchers identified that land use differed more in the largest farms, which used nearly 2.25 times lesser land per kg of milk compared to the smallest farms. For the mid-sized farms, the potential of global warming, terrestrial acidification, marine eutrophication, and eco-toxicity was comparatively high. The researchers opined that if the presently used domestic farm-based protein feeds are replaced with imported high-protein soy-based feed, then the environmental impacts in dairy production can significantly increase (e.g., increased land use by 18% with global warming potential by 43%). Further, the researchers opined that the environment-based policy for handling the farms needs to sensibly contemplate the complete consequences of operation size on the environmental quality, and thus facilitate the 'best practice' for each farm type and driving for the systematic changes there-off. The constraints of the study and future research to enhance data quality, allocation methods, providing farm-size-specific information, breeding, storage of manure and handling are also being discussed in this article.

In order to assess the potential influence on the milk production, Ross et al. [8] evaluated the efficacy of commercial feed additives (SOP STAR COW), which had the ability to decrease the enteric emissions from dairy cow and cattle's performances. Results revealed both control and SOP-treated cows to exhibit identical outcomes for the amount (kg) of milk fat and milk protein produced in a day. Further, no changes in enteric emission or milk parameters were noticed between control and SOP-treated cows. The researchers opined that future work should focus on understanding the effects of long-term supplementation or high dose of 'SOP STAR COW' to regulate mitigation effects on methane emissions and increase in milk production can be ascertained. In another report, Ross et al. [9], by adopting various free-stall management techniques studied the means to lessen ammonia, greenhouse gas, and some air pollutants from lactating dairy cattle wastes. From the study, it was concluded that removal of dairy manure by 'scraping' holds a high prospective to enhance gaseous emissions such as ammonia and other greenhouse gases.

On another note, Byrne et al. [10] investigated the suitability of a heat pump water-heater system to reduce agricultural emissions in dairy farms in Ireland. The energy and cost-efficiency of heat pump system were compared with five other water heaters. Results of this investigation showed high efficiency, but the economic costs and complexity of the solar-gas system were found to be a major deterrent factor. The researchers concluded that the heat pump was cost-effective, competent, and a feasible option for dairy farmers who are aiming to reduce carbon footprint as well as energy bills.

In an interesting study from Brazil, Siqueira et al. [11] investigated the connection between organizational forms and adoption of the agri-environmental based practices by undertaking a case study of six archetypes of dairy farms. From the study, the researchers proposed a systemic approach and concluded that it is important to consider dairy farms as a heterogeneous organizational form (as human capital investment, resources, market, and other information access) in policy design to fast-track agro-ecological transitions.

The technical efficiency of the European dairy processing industries by adopting selected novel methods related to productivity and efficiency analysis was investigated by Čechura and Žáková Kroupová [12]. Accordingly, the input use efficiency was evaluated in ten selected European countries. Results of the study were constructed based on the 'Amadeus dataset', which indicated dairy products manufacturing companies to highly exploit the production possibilities during 2006-2018. It was also observed that the overall technical inefficiency (OTE) to be a result of 'short-term' shocks and unsystematic failures. In addition, in the European dairy processing industries, 'meta-frontier estimates' showed a specific degree of systematic failures like permanent managerial failures as well as structural problems.

In a noteworthy work from Italy, Masi et al. [13] used the 'cluster analysis technique' to recognize dairy farms (3 types) categorized based on the indicators of environment, social, and economic sustainability (dimensions of sustainability) with emergent structural relationships that were based on a structural characteristic of dairy farms. The classification rendered it feasible to portray 'state of art' of the Italian dairy sector and further helped to comprehend how diverse farms types can answer new European trajectories.

The expectations on specialty foods vs. conventional food products that can affect the overall well-being of consumers, and how small-scale artisan producers use this information towards designing better customer experiences are reported by Percival Carter and Welcomer [14]. The study revealed causal mediation analysis in expectations to mediate the link between product types and the utility of product information. The researchers concluded by stating that as consumers' choices evolve, small-scale producers should adopt various approaches to manage their products, thus helping them towards identifying unique opportunities for differentiation and to increase the profitability.

Further on, an interesting and novel perspective to measure the effects of production diseases on economic sustainability in dairy farms is reported by Hoischen-Taubner et al. [15]. The researchers opined that changing the perception of production diseases via reflecting it as an indemnity damage and a risk to the farms' economic viability can further change the processes involved to minimize the production diseases.

On another note, this Special Issue has two interesting review articles focusing on the global review of monitoring, modeling, and analyses of water demand in dairy farming by Shine et al. [16] and on measuring variables leading to a seeming incongruity between anti-methanogenic potential of tannins and their perceived effects in ruminants byVerma et al [17]. In the review by Shine et al. [16], monitoring of water consumption in dairy farms has been documented. The review has components exploring dairy water consumption, prediction modelling, and analysis, followed by discussions highlighting some of the normal trends through dairy water literature. The authors have concluded by indicating that globally more studies are required that can focus on the consumption of water within the milking parlor. The researchers have also opined that to guarantee best practices, improving the perception of dairy water consumption via statistical analyses as well as empirical modelling can yield increased prediction confidence and improved appeal of empirical models as a substitute for 'physical metering'.

Verma et al. [17] have reviewed the variability of anti-methanogenic potential of forages and attempted to ascertain the reasons for inconsistencies in results. They have discussed options for optimization to produce comparable and reproducible results. These researchers have proposed a link between plant metabolome, physiology, and their anti-methanogenic potential that can be duly considered for improving the sustainable intensification of livestock. The review is concluded by stating that a comparison of condensed tannins (CT) fingerprints between different species can be of use to understand various factors defining their anti-methanogenic potential, thereby offering an important background to assess interactions between plant constituents and rumen microflora, thus promoting ruminant health.

The Future:

The global climate changes we are facing in recent decades and the transformations witnessed have prompted experts, politicians, and the public to question who are the main culprits and what actions need to be taken to mitigate these effects. The swift expansion in intensive farming systems has held the consideration of experts, who hypothesized large contributions to climate change by the livestock system, raising a question on sustainability among these farming techniques, specifically related to the negative impacts on the environment and animal welfare. These factors have led to enhanced social costs compared to costs of private systems, imparting negative effects on the entire society. This holds true when the livestock farming sector is of focus, wherein a larger number of inputs are used in the intensive farming systems, as well as when an enormous amount of wastes are being generated. Today, the basic requirement is to endure the social acceptability of these procedures and offer the dependent sector to be much more competent and be environmentally sustainable to overcome challenges of the future. In this perspective, the interests at stake are wide and not all 'Nations' are taking steps to improve the production performance of the supply chain with a view to sustainability. The new European trajectory like that of the 'Farm-to-Fork' strategy, places sustainability objectives of the sector in line with the demands of the planet. Indeed, in recent years, the entire sector is undergoing rapid transitions with novel sustainable production methods being developed. A factual classification of dairy farms covering all aspects of production systems can offer an outline of factors that can be demonstrative of the farm types from a structural point and at the level of intensity in the production system, which are envisaged to be useful for secondary evaluation, mainly with regard to farm sustainability performance. Further, sustainability assessment should be carefully monitored and calibrated based on the farm's structural asset such as farming area, number of animals raised, age and education of farmers as well as on the production methods employed (e.g., organic).

Moreover, accelerating the use of environmental sustainability practices in dairy systems is also of much importance. The demand for environmental sustainability practices has increased in a variety of economic sectors worldwide. In animal production, in which the environment (soil, water, atmosphere, and temperature, among others) is one of the main factors of production, such demands are even more necessary and of urgency. Small changes in the production environment can result in important negative impacts on animal production as well as on the environment and society as a whole. Although environmental sustainability practices have been adopted in many dairy production systems, such practices should be increased in the coming years. Acceleration of the adoption of environmental sustainability practices is generally promoted by two major axes. First, through laws that regulate and oversee such practices and, second, through the generation of market incentives. Between these two axes, the generation of laws and oversight mechanisms may present more weaknesses, especially considering delays in the definition of laws and the difficulty, and high costs of monitoring compliance to governmental requirements. Thus, there is a tendency that acceleration of the adoption of sustainability practices in dairy systems be driven by market mechanisms, particularly those that result in milk valorization. For instance, farmers who adopt a set of more environmentally sustainable production practices could be paid more per liter of milk. Valorization of milk produced in systems with reduced environmental impact generally results from stable commercial relations, mainly via the establishment of purchase and sale contracts with clear clauses that can be verified by the different parties involved. Such a scenario is not observed in some countries that rank among the world's largest milk producers. Studies that analyze the characteristics of different forms of organization and identify market niches that may value milk produced by systems with reduced environmental impact are also important. Furthermore, identification of constraints or incentives to adopt environmental sustainability practices from the perspective of farmers may contribute to the creation of important public-private strategies for accelerating the use of environmental sustainability practices in dairy production.

Moreover, water demand in dairy farming will most likely increase in the coming years. Freshwater, encompassing only 3% of the global water supply, represents earth's most valuable natural resource [18]. However, the majority of the freshwater is entrenched in the glaciers, polar ice-caps, or in groundwater aquifers. A meager 0.4% of the earth's total freshwater is instantly available from rivers, lakes, or/and from [19]. Nearly 70% of the global freshwater is utilized by the agricultural sector, and hence this necessitates considerable perfection in water-use efficiency. Nevertheless, agricultural production depends not only on the availability of land, but also on the free accessibility to freshwater [20]. Further, the global water-footprint generated post-production of dairy products in the dairy sector is estimated to contribute nearly 7% of the total share, and this is expected to increase in future decades the world over, coupled with a predicted augmentation in the consumption of dairy-based products [20,21]. Even though green water forms the major category of water used in the stall- as well as pasture-based dairy farms, 'on-farm' blue water can also be more vulnerable to water shortage owing to its localized supply from rivers, lakes, or ground water aquifers, especially during periods of modest rainfall [16]. Therefore, the future direction of research related to water demand in dairy farming may trend towards identifying strategies to reduce blue water usage, which is composed of parlor and animal drinking water. In particular, strategies may focus on mitigating parlor water use as reducing drinking water may negatively impact milk production. Some of the factors like farm size, farming systems, milk pre-cooling, milking systems, and washing practices can have a profound influence on parlor water use. Strategies for reducing parlor water use may include the installation and use of a closed-loop milk pre-cooling system, wherein a buffer tank can store and supply water to the farm as and when required [22]. When water is essential for the pre-cooling of milk, this can flow via the plate cooler at the set water to milk ratio, and this flows back to the buffer tank, thus ensuring excess water use is minimal. Additionally, an effective blue water mitigation strategy should be identified and fixing of leaking water systems is a necessity. In New Zealand, 26% of 'stock' drinking water was discovered to be lost due to leakage [23]. Precise calculation on the blue water use in dairy farms is often seen as a limiting factor with the number of farms being included in water foot-printing studies. This might be due to high investment and maintenance costs, and time constraints linked with metering types of equipment to monitor direct 'on-farm' water use. The majority of research investigations have focused on designing prediction models to simulate 'on-farm' water usage. The application of machine-learning algorithms has revealed improved total dairy water prediction accuracy by ~23%, which facilitates for 'coarse model' inputs exclusive of compromising on the accuracy [24]. Developing precise working models for 'on-farm' water consumption can definitely support an expansion in the dairy farm numbers and provide inputs for water foot-printing studies. This will offer an effective environment wherein non-linear impacts on dairy water use due to changed farm practices, equipment, or meteorological conditions can be quantified and be of practical use.

Finally, there is a need for more research to be undertaken, specifically in relation to the impact dairy farming has on the environment, society, and regional economies.

Author Contributions: R.B.: Conceptualization, writing, reviewing and editing; J.D.P., F.I.B., T.T.d.S.S., P.S. and M.D.M.: Writing, reviewing, and editing. All authors have read and agreed to the published version of the manuscript.

Funding: This research received no external funding.

Conflicts of Interest: The authors declare no conflict of interest.

References

1. FAO. The Food and Agriculture Organization of the United Nations, Rome, Italy. 2021. Available online: https://www.fao.org/3/cb4230en/cb4230en.pdf & https://www.fao.org/publications/card/fr/c/CB5635EN/ (accessed on 27 January 2022).
2. *The EU Agricultural Outlook Report EU Agricultural Outlook 2021–31: Consumer Behaviour to Influence Meat and Dairy Markets*; Agriculture and Rural Development: Brussels, Belgium, 2021. Available online: https://ec.europa.eu/info/news/eu-agricultural-outlook-2021-31-consumer-behaviour-influence-meat-and-dairy-markets-2021-dec-09_en (accessed on 27 January 2021).
3. Bataille, C.G.F. Physical and policy pathways to net-zero emissions industry. *Wiley Interdiscip. Rev. Clim. Chang.* **2020**, *11*, e633. [CrossRef]
4. Malliaroudaki, M.I.; Watson, N.J.; Ferrari, R.; Nchari, L.N.; Gomes, R.L. Energy management for a net zero dairy supply chain under climate change. *Trends Food Sci. Technol.* **2022**, in press. [CrossRef]
5. Hassan, A.; Cui-Xia, L.; Ahmad, N.; Iqbal, M.; Hussain, K.; Ishtiaq, M.; Abrar, M. Safety failure factors affecting dairy supply chain: Insights from a developing economy. *Sustainability* **2021**, *13*, 9500. [CrossRef]
6. Ibidhi, R.; Kim, T.H.; Bharanidharan, R.; Lee, H.J.; Lee, Y.K.; Kim, N.Y.; Kim, K.H. Developing country-specific methane emission factors and carbon fluxes from enteric fermentation in South Korean dairy cattle production. *Sustainability* **2021**, *3*, 9133. [CrossRef]
7. Brizga, J.; Kurppa, S.; Heusala, H. Environmental impacts of milking cows in Latvia. *Sustainability* **2021**, *13*, 784. [CrossRef]
8. Ross, E.G.; Peterson, C.B.; Carrazco, A.V.; Werth, S.J.; Zhao, Y.; Pan, Y.; DePeters, E.J.; Fadel, J.G.; Chiodini, M.E.; Poggianella, L.; et al. Effect of SOP "STAR COW" on enteric gaseous emissions and dairy cattle performance. *Sustainability* **2020**, *12*, 10250. [CrossRef]
9. Ross, E.G.; Peterson, C.B.; Zhao, Y.; Pan, Y.; Mitloehner, F.M. Manure flushing vs. scraping in dairy freestall lanes reduces gaseous emissions. *Sustainability* **2021**, *13*, 5363. [CrossRef]
10. Byrne, P.S.; Carton, J.G.; Corcoran, B. Investigating the suitability of a heat pump water-heater as a method to reduce agricultural emissions in dairy farms. *Sustainability* **2021**, *13*, 5736. [CrossRef]
11. Siqueira, T.T.D.S.; Galliano, D.; Nguyen, G.; Bánkuti, F.I. Organizational forms and agri-environmental practices: The case of Brazilian dairy farms. *Sustainability* **2021**, *13*, 3762. [CrossRef]
12. Čechura, L.; Žáková Kroupová, Z. Technical efficiency in the European dairy industry: Can we observe systematic failures in the efficiency of input use? *Sustainability* **2021**, *13*, 1830. [CrossRef]
13. Masi, M.; Vecchio, Y.; Pauselli, G.; Di Pasquale, J.; Adinolfi, F. A typological classification for assessing farm sustainability in the Italian bovine dairy sector. *Sustainability* **2021**, *13*, 7097. [CrossRef]
14. Percival Carter, E.; Welcomer, S. Designing and distinguishing meaningful artisan food experiences. *Sustainability* **2021**, *13*, 8569. [CrossRef]
15. Hoischen-Taubner, S.; Habel, J.; Uhlig, V.; Schwabenbauer, E.M.; Rumphorst, T.; Ebert, L.; Möller, D.; Sundrum, A. The whole and the parts—A new perspective on production diseases and economic sustainability in dairy farming. *Sustainability* **2021**, *13*, 9044. [CrossRef]
16. Shine, P.; Murphy, M.D.; Upton, J. A global review of monitoring, modeling, and analyses of water demand in dairy farming. *Sustainability* **2020**, *12*, 7201. [CrossRef]
17. Verma, S.; Taube, F.; Malisch, C.S. Examining the variables leading to the apparent incongruity between anti-methanogenic potential of tannins and their observed effects in ruminants-A Review. *Sustainability* **2021**, *13*, 2743. [CrossRef]
18. Koehler, A. Water use in LCA: Managing the planet's freshwater resources. *Int. J. Life Cycle Assess.* **2008**, *13*, 451–455. [CrossRef]

19. UN Environment. *GEO-6—Healthy Planet Healthy People*, 1st ed.; Cambridge University Press: Cambridge, UK, 2019. [CrossRef]
20. Bruinsma, J.; Alexandratos, N. World Agriculture Towards 2030/2050: The 2012 Revision. ESA Work Pap No 12-03 2012:44–44. 2012. Available online: https://www.fao.org/docrep/016/ap106e/ap106e.pdf (accessed on 20 March 2022).
21. Mekonnen, M.M.; Hoekstra, A.Y. The green, blue and grey water footprint of crops and derived crop products. *Hydrol. Earth Syst. Sci. Discuss.* **2011**, *8*, 763–809. [CrossRef]
22. Shine, P.; Scully, T.; Upton, J.; Shalloo, L.; Murphy, M.D. Electricity & direct water consumption on Irish pasture based dairy farms: A statistical analysis. *Appl. Energy* **2018**, *210*, 529–537. [CrossRef]
23. Higham, C.D.D.; Horne, D.; Singh, R.; Scarsbrook, M.R.R.; Kuhn-Sherlock, B.; Scarsbrook, M.R.R. Water use on non-irrigated pasture-based dairy farms: Combining detailed monitoring and modeling to set benchmarks. *J. Dairy Sci.* **2017**, *100*, 828–840. [CrossRef] [PubMed]
24. Shine, P.; Murphy, M.D.; Upton, J.; Scully, T. Machine-learning algorithms for predicting on-farm direct water and electricity consumption on pasture based dairy farms. *Comput. Electron. Agric.* **2018**, *150*, 74–87. [CrossRef]

Article

Safety Failure Factors Affecting Dairy Supply Chain: Insights from a Developing Economy

Aisha Hassan [1,*], Li Cui-Xia [1,*], Naveed Ahmad [2,3], Muzaffar Iqbal [3,4], Kramat Hussain [4], Muhammad Ishtiaq [5] and Maira Abrar [6]

1. College of Economics and Management, Northeast Agricultural University, Harbin 150030, China
2. School of Management, Northwestern Polytechnical University, Xi'an 710000, China; naveedahmad@mail.nwpu.edu.cn
3. Department of Business Administration, Lahore Leads University, Lahore 54000, Pakistan; m.shah@tju.edu.cn
4. College of Management and Economics, Tianjin University, Tianjin 300072, China; kramat381@tju.edu.cn
5. School of Economics, Kohat University of Science and Technology, Kohat 26000, Pakistan; m.ishtiaq@kust.edu.pk
6. Business School, Konkuk University, Seoul 100011, Korea; maya.maan90@gmail.com
* Correspondence: aishaahassan4@gmail.com (A.H.); licuixia.883@163.com (L.C.-X.)

Abstract: Safety issues in the dairy industry have attracted greater attention in recent years, and the public have showed an intensive concern regarding safety failure in the dairy supply chain. Since the dairy industry is closely associated with humans and fulfills basic necessities, it is necessary to explore safety failure factors (SFFs) affecting the supply chain of the dairy industry. This paper aims to explore the SFFs of the dairy supply chain using an interpretive structural modeling technique (ISM) and Matrice d'Impacts Croises Multiplication Appliques a un Classement (MICMAC) analysis in a Pakistani context. A total of twenty-five failure factors have been identified through literature reviews and the opinion of an expert team, including managerial and technical experts from the dairy industry, as well as academics. Interpretive structural modeling (ISM) is applied to analyze the mutual interaction among barriers and to develop a structural model. The MICMAC technique is used to identify the importance of SFFs based on their driving and dependence power. The results of this study will help decision-makers in the dairy industry to plan their supply chain activities more effectively and efficiently by managing the identified barriers.

Keywords: dairy industry; supply chain management; safety failure factors; interpretive structural modeling; MICMAC analysis

1. Introduction

Effective supply chain management seems to be a crucial concern in today's intensifying competitive business environment, and it has to be dealt with in a global business context [1]. Information and communication technology developments are essential tools for an effective supply chain. A supply chain is the chain of different activities involved in converting raw material into a final product to fulfill customers' needs [1]. By viewing it in this way, the supply chain can be improved competently.

In recent years, most researchers of supply chain networks have been focused on agri-business theory [2]. Dairy products have shaped the diets of many populations across the world [3]. Sustainability plays an important role in the dairy supply chain, minimizing unit production cost while adding flexibility to products or processes. Sustainability and efficiency can be attained through supply chain collaboration, innovation, mitigation of uncertainties, and lean and green initiatives [4–8]. Technology can also be beneficial to exploring the plausible future of the food supply chain. Moreover, competitiveness leads industries towards sustainability [9].

To compete globally and develop sustainable agri-products, the safety standards of dairy products are crucial. Increased competitiveness in dairy industry development is necessary in order to meet the safety standards by prioritizing some crucial actors, including technological transfer, research and development (R&D), trade policies, and social and political agendas [10]. However, SFFs in the dairy industry have been reported extensively in different contexts [11,12]. To improve the regulatory system of food safety in the dairy sector, many countries have formed strict regulations and established institutions [13].

Due to the investment price capping policy in Pakistan, the requirements of local demand for milk cannot be fulfilled. In 2016–2017, Pakistan imported milk and cream with the value of USD 234 million, despite the fact that the country is the fifth largest producer of milk in the world. Dairy contributes 11% to the GDP of Pakistan. In a conducive environment, this industry would have the potential to become an economic powerhouse in the country. Almost 44% of children under the age of 5 years are stunted, and almost 15% are starving. This is a national concern, and should be acknowledged by the government in order to support the dairy sector in producing quality and safe milk, and to make the milk available for every household of the country. Some other main reasons for the low animal production include the low genomic potential of cows, the lack of forage resources, the conservation of outmoded farming methods and the chaotic marketing system. Therefore, the quantification of livestock in the context of economic growth is necessary because many policymakers can use this information to identify the potential impact on the economy of dairy farms, and their benefits related to industries and societies. A lack of technological innovation, inventory management and supplier management are the major issues in developing countries, including Pakistan, contributing to dairy safety failures. Most previous researchers have not stressed the SFFs affecting the supply chain process of the dairy industry in Pakistan. Therefore, it is important to identify those factors, trends and drivers to achieve the desired outcomes.

In light of previous studies, this study focuses on the SFFs of the dairy supply chain. This study may also prove to be useful for dairy industry in Pakistan as regards accomplishing efficient and sustainable supply chain practices, and also might represent an important addition to the supply chain literature. This study used a widely used methodology named interpretive structural modeling (ISM) to analyze the SFFs of the dairy supply chain, which could help prioritize important factors that might need to be addressed for the improvement of the dairy supply chain. The ISM approach has been used to identify the complex relationship between different elements, and a supporting methodology, MICMAC, is adopted to illustrate the driving and dependence powers of each element. Further information about ISM and MICMAC are given in the methodology section.

This study contributes both theoretically and practically. Theoretically, this study contributes to the safety literature in the dairy supply chain context of developing countries, especially in Pakistan. This is the first study to identify SFFs in the dairy supply chain. The use of ISM and MICMAC for analyzing SFFs also makes the study novel. Further, the dairy supply chain has also been affected by safety issues in the Pakistani context due to diseases emerging in the COVID-19 scenario. Therefore, the dairy supply chain requires more attention from researchers.

The rest of the paper is structured as follows: Section 2 describes the Literature Review, Section 3 consists of methodology, Section 4 included results and discussion and Section 5 presents the conclusion.

2. Literature Review

2.1. SFFs in Dairy Supply Chain

In recent years, technological innovation, inventory management and supplier management have been the main determinants of quality in the dairy industry [14]. Different developed countries, such as the UK, the US and Australia, have been able to improve their dairy supply chain practices and overcame critical barriers given their plentiful resources [15–17]. However, developing countries are yet to achieve these critical factors. For success in the food supply chain, and to improve the performance of an organization, it is necessary to identify and highlight the critical success factors (CSFs) [18].

Many studies have identified SFFs in the supply chain of the dairy sector. Perron [19] examined four categories of barriers that impede the adoption of supply chain safety measures in SMEs, including attitudinal and perception barriers, information barriers, resource barriers and technical barriers. Ref. [20] described the market challenges and potential losses related to the cold chain in the frozen food industry in the Indian retail sector. The author states that, when considering the end consumers' knowledge, behavior and preferences in food, some key challenges arise regarding unpacking, knowledge optimization strategies and lack of effectiveness in food supply chain. Chandrasekaran and Raghuram [21] analyzed different enablers of risk management in the dairy sector and found that there are a lot of risks in both the dairy farming industry and the dairy processing industry.

There are different uncertainties at different levels of supply chain management in the dairy industry [22]; therefore, a sustainable framework is needed to promote green practices in the manufacturing industry [23]. Kumar and Staal [24] outlined that farmers are not being educated appropriately as advanced methods are being used in modern milk supply chains. The relocation of dairies from regions rich in water resources to regions with limited resources is likely to be shortsighted. Pant and Prakash [25] found that the quality control system in the dairy production process in developing countries is one of the main SFFs in dairy supply chains. Berem and Obare [26] found that the illegal and improper distribution of milk is one of the main causes of lower productivity. Buzby and Hyman [27] identified that food wastage must be stopped as the world is facing a serious issue of food shortage. Lemma and Kitaw [28] proposed the modeling and optimization approaches used in the perishable food supply chain literature. Park and Kim [13] used coding to systematically analyze food safety incidents, and concluded at which point the breakdown in food safety is likely to occur.

Due to the high focus of the supply chain on productivity, on-time delivery and better order filling rates, competition is higher in the food processing sector [6]. External barriers are more impactful than internal barriers, as external barriers include poor regulations, poor supplier commitment and industry-specific barriers, whereas internal barriers include cost and legitimacy [29]. Hemme and Otte [30] described a lack of supervision from relevant authorities in the dairy sector. Many SMEs see the adoption of environmental testing as a cause of high financial cost to the business, which could not be passed on to the end user. Finally, the study found that the government can play an important role in this scenario by improving awareness.

Worldwide, many studies have suggested different techniques and methods to reduce the SFFs in the supply chain of the dairy sector. Chalupkova [31] suggested that appropriate decision-making, following environmental testing standards and regulations, can improve the safety in supply chains. It is necessary and possible to improve productivity and develop the dairy chain using proper indicators [32]. Lemma and Kitaw [28] proposed modeling and optimization approaches that focused on the perishable food supply chain literature, as well as waste and loss assessment in food supply chains. For assessing the adequacy of various innovations in dairy supply chain practices, Ali and Lynch [33] applied the Q-methodology to identify the patterns of low-input and organic dairy supply chain members in four European countries. Kumar [34] described a conceptual model of dairy supply chains, and determined the importance of a novel conceptual model for supply chain performance measurement. Prakash and Pant [35] stated that the balance score card (BSC) approach can be used to measure the safety performance of the dairy supply chain.

Developing countries, including India, that are facing the same problems in the dairy sector have launched dairy processing cooperatives programs in order to improve smallholder dairy products (FAO 2010). Kumar and Kumar [36] suggested that milk procurement potential may be determined using Geographical Information Systems (GISs), and stated that the analytical application of GIS, such as in proximity analysis, is very useful in various business decisions, such as identifying new villages to be used as procurement centers. However, a productive technique for dairy quality control requires systematic risk analysis, which should be based on comprehensive studies from "farm to fork". A questionnaire-based survey was conducted to rank the identified SFFs in the Indian context [37,38].

Based on the literature mentioned in Section 2.1, it is also necessary to summarize the dairy industry of Pakistan and outline the aims of the current study in the Pakistani dairy supply chain context.

2.2. Dairy Industry of Pakistan

Agriculture is the lifeline of Pakistan's economy, and livestock plays a vital role by providing items that are essential to human diets [39]. It contributes 18.9% to the GDP, and consumes 42.3% of the labor force [40]. Pakistan is one of the largest milk producers in the world; nevertheless, only about 3% of it is processed for value addition, while a major sectioned is consumed locally through traditional marketing systems. According to a commission [40], milk production during 2018–2019 was 59,759 tons, which is high compared to the previous year's total of 57,890 tons.

The dairy industry has shaped millions of dairy farmers' lives in Pakistan. The contribution of livestock in the agriculture sector is about 58.92%, and its contribution to GDP growth remains at 0.43% percent, with a share in national GDP of 11.1% (Pakistan Annual Plan 2017–2018). During 2018–2019, the livestock sector grew by 4.00%, and its gross value addition amounted to INR 1430 billion [40]. Pakistan earned USD 528.212 million as foreign exchange through livestock export and allied products during July–March of 2018–2019 [41].

According to Bar [42], in the UNDP's latest survey (September 2018), Pakistan stands at 150th place in the human development index, among 189 countries. Pakistan has the sixth largest population in the world, with approximately 212.242 million occupants, of which 49.08% live in villages (NIPS, Pakistan Demographic and Health Survey 2017–2018). Pakistan's economy is the second biggest economy in the South Asian region, valued at USD 305,000 (The U.S. Central Intelligence Agency, 2017), and Punjab is one of the biggest provinces in terms of dairy milk production in Pakistan, as it produces three-fourths of its total milk. Punjab is also one of the largest milk districts in Asia, with 15 private companies competing to collect farmers' milk for processing, including global giants Nestle, Haleeb foods and Halla. Its per capita production is improving in terms of the number of dairy cows, rather than any increases in milk production.

Based on these above two sub-sections, it has been found that literature on SFFs in the dairy supply chain is scarce, and most of the previous studies have ignored this serious issue. The literature also indicates that earlier studies analyzing SFFs were not carried out via a sound and systematic methodology, such as ISM or MICMAC. Therefore, this study is novel, adding value to the safety literature by evaluating SFFs via the ISM and MICMAC methodologies.

2.3. Study Objective

After a comprehensive literature review, it was found that studies on the dairy industry are limited, and have not concentrated on SFFs in the dairy supply chain, especially in the Pakistani context. Moreover, existing studies focus on supporting the farmers and linking them up with urban markets. However, no study has yet identified major SFFs in the Pakistani dairy supply chain. Therefore, the aim of this study is:

➢ To address the SFFs in the supply chain of the dairy industry in Pakistan;
➢ To establish the interaction among SFFs in the dairy industry using the ISM technique, and classify the barriers through MICMAC analysis;
➢ To propose policy recommendations based on the severity of factors.

3. Research Methodology

The methodology of this research has two main components. In the first part, a detailed literature review derived the key factors of safety failures. In the second, the relevant SFFs were selected for further analysis. After that, the ISM and MICMAC approaches were used to examine the expert opinions through brainstorming sessions. ISM is a methodical and interactive technique that depends on a group of independent professionals and that helps in understanding the interrelationships among variables. ISM and MICMAC analyses also help in addressing the binary relationships among the described factors. However, the relationships among these factors vary; some relations are strong, some of them are normal, and some may be weak [43,44]. In this sense, ISM analysis works as a communicative tool to understand and explain the complex interrelated relationships among factors [45]. Moreover, the experts selected in our study are highly skilled in decision-making and applying ISM techniques. Additionally, the combined use of ISM and MICMAC analyses make this study simpler for readers/managers to understand, and thus use to manage sustainability initiatives in supply chains in the Pakistani context, as well as in other developing countries (with marginal modifications).

All steps of the methodology, along with its goals and output, are explained in Figure 1.

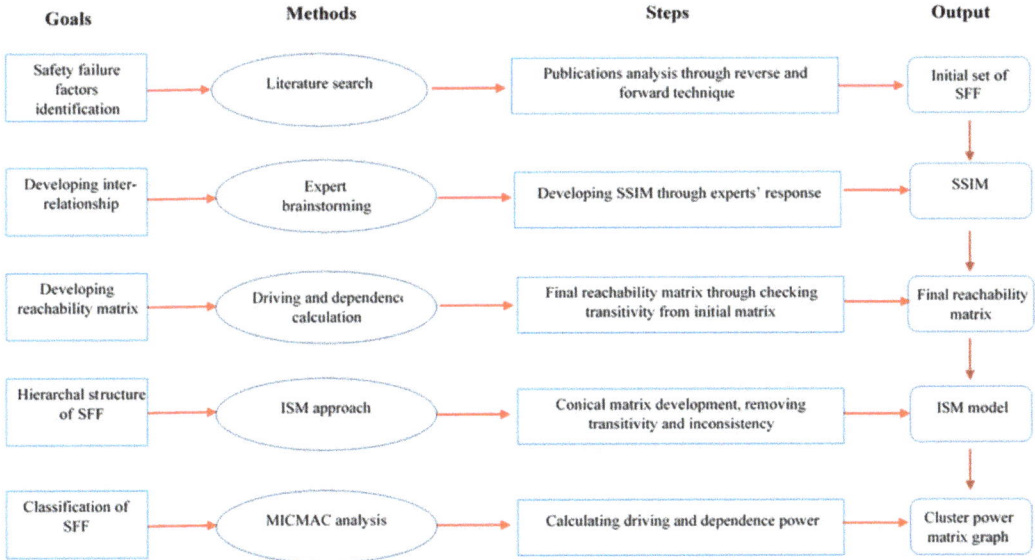

Figure 1. Detailed methodological approach to identifying SFFs in the dairy supply chain.

3.1. SFFs Identification through Extensive Literature Review

To identify the factors in the dairy industry in Pakistan, an extensive literature review was performed with the help of many research articles. Those research articles were found through different databases, including Science Direct, Springer, Emerald, Taylor and Francis, JSTOR, PubMed and Google Scholar. The keywords used to find the related articles are shown in Table 1. Significant keywords were identified in the literature, but their sub-keywords, such as milk delivery, transport and storage, were not considered in this novel study.

Table 1. Literature search criteria.

Keywords	"Dairy industry" OR "Critical issues" OR "Supply chain safety issues" OR "Safety barriers" OR "Dairy Industry issues" OR "Disaster of Risk" OR "Dairy production" OR "Dairy Farming" OR "Dairy product safety failures" OR "Milk production" OR "Dairy Policies" OR "Dairy industry downfall" OR "Dairy industry barriers"
Exclusion criteria	Articles that have only title, author name, keywords, and abstract. A paper that does not feature a review, surveys, different sound methodologies, strong discussion, or dairy issues criteria

Initially, 150 articles were analyzed. Later on, using the evaluation criteria shown in Table 1, 70 articles from 30 journals were identified. After examining the contents and abstracts of these articles, irrelevant articles and journals were removed. Finally, 30 articles from 21 journals remained, with the addition of four conferences. Some of the more popular journals are *Nature, World Applied Sciences Journal, Journal of Social, Behavioral and Health Sciences, Journal of Cleaner Production, Journal of Management Sciences and Technology, International Journal of Scientific and Engineering Research, Journal of Business Management, Journal of Dairy Sciences, Journal of Applied Economic Sciences, Indian Journal of Agricultural Economics, Journal of Clinical Oncology, Journal of Industrial Engineering and Management, Journal of Advanced Operations Management, Journal of Basic and Applied Scientific Research, Journal of Productivity and Performance Management, International Journal of Services and Operations Management, British Food Journal, International Journal of Environmental Studies* and *Journal of Building Engineering*.

The identified factors are explained in Table 2.

Table 2. Significant SFFs in the supply chain of the dairy industry.

No.	Safety Failure Factors
A1	Poor quality control in production process
A2	Employees are the carriers of some diseases and chances of transfer to dairy
A3	Illness of employees
A4	No clinical examination of employees before being officially employed
A5	Inadequate cold storage facility during mobility of dairy food
A6	Unhygienic and unsafe transportation of dairy food
A7	Inappropriate company location
A8	Lack of qualified storehouse
A9	Unsafe milk from the dairy station
A10	Bad health conditions of farmers
A11	Unqualified animals' food and veterinary drugs
A12	Companies purchase unsafe dairy food
A13	Invalid sampling
A14	Non-standardized packaging
A15	Companies sell unsafe dairy products
A16	Improper management
A17	No compliance with the rules and regulations
A18	Farmers are not equipped with the latest farming technology
A19	Lack of feedback mechanism
A20	Illegal supply of raw milk
A21	Wholesalers and retailers promote unsafe dairy food
A22	Unqualified system of milk collection and delayed delivery
A23	Unhealthy cows
A24	Lack of environmental testing by EPA
A25	Lack of supervision by relevant authorities

3.2. Interpretive Structural Modeling (ISM)

The interpretive structural modeling (ISM) technique was established and presented by Warfield in 1973, and its roots are in graph theory. The ISM technique is mainly proposed as an interactive learning process, which collects a set of different but directly related variables into an inclusive systematic model. ISM is a systematic approach, and it gives a structure to the complex relationship among variables. Refs. [46–49] stated that ISM transforms erroneous and unclear models into precise and visible models. Different studies employing ISM and MICMAC techniques are listed in Table 3.

Table 3. Earlier related studies using ISM.

References	Objective	Country	Methodology
[14]	To bring out the barriers in the dairy supply chain and establish the interaction among barriers in the dairy industry.	India	ISM and MICMAC methodology
[50]	To investigate the effects of the barriers and benefits on the e-procurement adoption decisions.	Turkey	ISM and SEM approaches
[51]	To analyze the barriers in green supply chain management.	India	ISM and MICMAC techniques
	To examine the determinants that influences the growth of Indian SMEs in the food industry and to identify the most important variables affecting growth.	India	ISM and SEM approaches
[52]	To identify the factors influencing consumers' decisions when buying beef products and consumers' information from twitter in the form of big data.	India	ISM and Fuzzy MICMAC techniques
[53]	To investigate the technical barriers in the dairy industry in context of Saudi Arabia.	Saudi Arabia	ISM methodology

The ISM–MICMAC approach has been employed in this study to identify the safety failure factors that impact the dairy supply chains of developing countries, especially Pakistan. This approach is used to draw a contextual relationship among different SSFs. It helps to demonstrate the relationships of different elements in the hierarchical structure [54]. However, various MCDM methods can perform the same analysis, e.g., data mining, TOPSIS, game theory, analytical hierarchy processing (AHP) and Bayesian theory. The comparison of these approaches with [48,55–58] is given in Table 4. Raj and Shankar [59] defined some attractive features of the ISM technique, which are given below:

- The ISM interprets the expert's judgement regarding various factors' relationships;
- ISM is a hierarchical structure-based model that justifies the connection of various complex factors;
- This approach helps to show the hierarchical structure of different factors in a diagraph model;
- ISM works on the philosophy of group decision-making (expert opinion), but it is also useful for individual responses.

Table 4. Comparison of ISM–MICMAC with other methodologies.

ISM-MICMAC	Data Mining	TOPSIS	Game Theory	AHP	Bayesian Theory
This technique assists in identifying the interrelations between variables on the bases of their driving and dependence powers.	In this approach, firms try to convert their raw data into useful information through software.	This technique is used to compare alternatives through the identification of their weight criteria for the best possible solution.	In this mathematical approach, different strategies are employed in competitive situations in which respondents' actions are related to the actions of other respondents.	This mathematical approach is applied in the pairwise comparison between variables.	Bayesian theory is used to examine conditional probability through the interpretation of mathematical formulas.

Raj and Rifkin [60] described the characteristics of ISM as follows:

i. This methodology is interpretive, as the opinions of the experts describe why and how dissimilar variables are related;
ii. It is structural, as on the basis of the relationship, a structure is extracted from a complex set of variables;
iii. It is a modeling approach, as the specific relationships and overall structure are illustrated in a diagraph;
iv. It is mainly proposed as a group learning process, but individuals can also use it;
v. It helps to impose the directions and orders on the complex contextual relations among elements of the system.

Despite the advantages of ISM, it has some limitations. The relationships of different variables rely on experts' experience. Hence, the experts' bias during the observation of variables could affect the final model. Moreover, ISM does not apply any weight to the variables either [61]. Karamat and Shurong [62] described the different steps of ISM as follows:

1. Variables affecting the system are listed at first;
2. Secondly, relationships are established among the listed variables to classify which pairs should be examined;
3. The next step is to establish a structural self-interaction matrix (SSIM), which identifies pair-wise relationships among those variables;
4. In this step, the initial reachability matrix is developed to check the transitivity of variables in the binary form;
5. The partition of the initial reachability matrix over different levels is done in this step, and the final reachability matrix is obtained as a result;
6. A diagraph is drawn using the contextual relationships given in the final reachability matrix;
7. The transitive links are mitigated in this step by replacing the variable nodes with problematic elements;
8. The ISM model is to be reviewed in the last step to check the inconsistency, and then necessary modifications are made for improvement.

The above-mentioned steps of the ISM methodology are illustrated in Figure 2.

3.2.1. Application of Interpretive Structural Modeling

Structural Self-Interaction Matrix (SSIM)

A structural self-interaction matrix (SSIM) is obtained from the interactions among the described factors. The existence of a relation between any two factors (i, j) and the associated direction of said relation is questioned. After finding the SFFs in the dairy industry of Pakistan, the contextual relationships among these factors are determined via a discussion amongst experts (developers, academicians, dairy companies and farms).

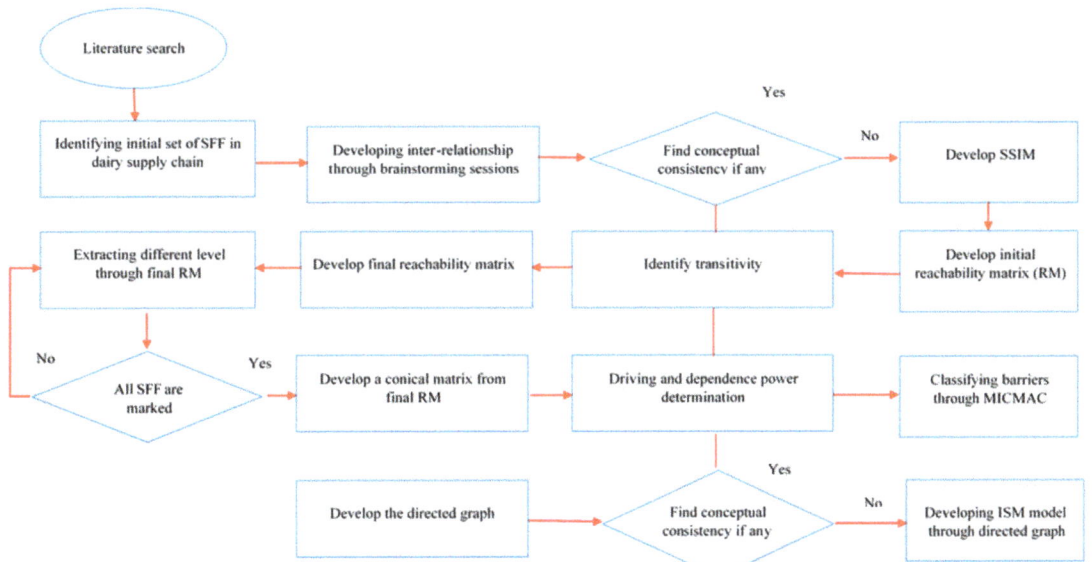

Figure 2. Methodology to develop an ISM model for SSF in the dairy supply chain.

Four letters have been used to denote the direction of the relationship between barriers i and j—V,A,X,O—similar to the previous studies, e.g., [63–65]. The description of each variable is shown below:

V: Factor "I" is related to factor "j";
A: Factor "j" is related to factor "I";
X: Factors "I" and "j" are related to each other;
O: Factors "I" and "j" are not related to each other.

This study was conducted using experts' opinions derived through brainstorming sessions. A group of experts, including one director, one dairy operation manager, three academicians (in the field of operations and supply chain management) and two dairy companies' managers (having direct links with dairy supply chain practices), were invited to rate the contextual relationship among the factors.

The rest of the experts were from Pakistan. The data were collected through a brainstorming session with different experts. All the experts were professionals, with sound knowledge in their fields. Initially, they were approached through sending emails and making phone calls. In total, 25 experts were approached, but due to their busy schedules, 7 experts agreed to participate in the brainstorming session. The sample size of seven experts is enough to meet the criteria of the ISM approach. Tan and Chen [66] used five experts as their sample to determine the barriers to building information modeling from the perspective of the Chinese construction industry. Malek and Desai [67] employed seven professionals to investigate the strategies of sustainable manufacturing, while Ravi and Shankar [63] discovered that a minimum of two experts is enough to meet the criteria of ISM.

Data were collected through a self-structured interaction matrix (SSIM)-based questionnaire, and this helped us to prioritize the identified SFFs in the Pakistani dairy industry SSIM of the SSFs is given in Table 5.

Table 5. SSIM of SFFs.

Critical Factors	A1	A2	A3	A4	A5	A6	A7	A8	A9	A10	A11	A12	A13	A14	A15	A16	A17	A18	A19	A20	A21	A22	A23	A24	A25
A1		O	O	O	A	O	A	O	O	O	O	A	X	V	O	A	A	O	O	O	O	O	O	A	A
A2			X	A	O	V	A	O	O	O	O	O	O	O	O	A	A	O	A	O	O	O	O	O	A
A3				A	O	O	A	O	O	O	A	V	O	O	X	A	O	O	O	O	O	O	O	O	O
A4					O	O	O	O	O	O	O	V	O	O	A	A	O	O	O	O	O	O	O	O	A
A5						V	A	V	O	O	O	O	O	O	A	A	O	O	O	O	O	O	O	A	A
A6							O	A	X	O	O	O	A	O	A	A	A	O	O	O	X	O	O	O	O
A7								V	O	O	O	O	O	O	O	A	A	O	O	O	O	O	O	O	A
A8									O	O	O	O	O	O	O	A	A	O	A	O	O	O	O	A	A
A9										A	A	V	A	O	V	A	A	A	A	A	V	A	A	A	A
A10											V	O	O	O	O	O	A	X	O	V	O	V	X	O	A
A11												O	V	O	A	A	A	O	O	O	O	O	V	A	A
A12													O	O	V	A	A	O	O	A	A	O	O	A	A
A13														V	O	A	A	O	O	O	O	O	O	A	A
A14															O	A	A	O	O	O	O	O	O	A	A
A15																A	A	O	O	A	X	O	O	O	A
A16																	X	O	V	V	O	V	O	A	A
A17																		V	V	V	V	V	O	X	A
A18																			O	V	O	V	V	V	V
A19																				O	O	O	O	A	A
A20																					V	X	O	A	A
A21																						O	O	A	A
A22																							O	O	A
A23																								A	A
A24																									A
A25																									

The details of all the experts are shown in Table 6.

Table 6. Experts' demographics.

Expert	Occupation	Gender	Age	Organization	Qualification	Work Experience	Firm Size
E1	Director	Male	60	Olpers dairy farm	PhD	15 years	300
E2	Diary operation manager	Male	63	Punjab dairy industry	Master	12	900
E3	professor	Male	48	Research institute	PhD	18	3500
E4	Associate professor	Male	37	Research institute	PhD	10	3500
E5	Associate professor	Female	38	Research institute	PhD	5	2200
E6	manager	Male	53	Dairy farm	Bachelor	17	35
E7	Dairy supply manager	Male	57	Dairy farm	Bachelor	13	27

3.2.2. Initial Reachability Matrix (IRM)

Once the SSIM has been developed, it is transformed into binary digits 0 and 1, known as the initial reachability matrix. The directions for transforming the SSIM into the IRM are given below:

- Suppose factors i and j are listed in SSIM as "V", then in IRM, (i,j) will be listed as 1 and (j,i) as 0;
- Suppose factors i and j are listed in SSIM as "A", then in IRM, (i,j) will be listed as 0 and (j,i) as 1;

- Suppose factors i and j are listed in SSIM as "X", then in IRM, (i,j) will be listed as 1 and (j,i) as 1;
- Suppose factors i and j are listed in SSIM as "O", then in IRM, (i,j) will be listed as 0 and (j,i) as 0.

The transformation of SSIM into IRM is shown in Table 7.

Table 7. Initial reachability matrix.

VAR	1	2	3	4	5	6	7	8	9	10	11	12	13	14	15	16	17	18	19	20	21	22	23	24	25
1	1	0	0	0	0	0	0	0	0	0	0	0	1	1	0	0	0	0	0	0	0	0	0	0	0
2	0	1	1	0	0	1	0	0	0	0	0	0	0	0	0	0	0	0	0	0	0	0	0	0	0
3	0	1	1	0	0	0	0	0	0	0	0	1	0	0	1	0	0	0	0	0	0	0	0	0	0
4	0	1	1	1	0	0	0	0	0	0	0	1	0	0	0	0	0	0	0	0	0	0	0	0	0
5	1	0	0	0	1	1	0	1	0	0	0	0	0	0	0	0	0	0	0	0	0	0	0	0	0
6	0	0	0	0	0	1	0	0	1	0	0	0	0	0	0	0	0	0	0	0	1	0	0	0	0
7	1	1	1	0	1	0	1	1	0	0	0	0	0	0	0	0	0	0	0	0	0	0	0	0	0
8	0	0	0	0	0	1	0	1	0	0	0	0	0	0	0	0	0	0	0	0	0	0	0	0	0
9	0	0	0	0	0	1	0	0	1	0	0	1	0	0	1	0	0	0	0	1	0	0	0	0	0
10	0	0	0	0	0	0	0	0	1	1	1	0	0	0	0	0	0	1	0	1	0	1	1	0	0
11	0	0	0	0	0	0	0	0	1	0	1	0	1	0	0	0	0	0	0	0	0	1	0	0	0
12	1	0	1	0	0	0	0	0	0	0	0	1	0	0	1	0	0	0	0	0	0	0	0	0	0
13	1	0	0	0	0	0	0	0	1	0	0	0	1	1	0	0	0	0	0	0	0	0	0	0	0
14	0	0	0	0	0	1	0	0	0	0	0	0	1	0	0	0	0	0	0	0	0	0	0	0	0
15	0	0	0	0	0	0	0	0	0	0	0	0	0	1	0	0	0	0	1	0	0	0	0	0	0
16	1	1	1	1	1	1	0	1	1	0	1	1	1	1	1	1	0	1	1	0	1	0	0	0	0
17	1	1	0	1	1	1	1	1	1	0	1	1	1	1	1	1	1	1	1	1	1	0	1	0	0
18	0	0	0	0	0	1	0	0	1	1	0	0	0	0	0	0	1	0	1	0	1	1	1	1	1
19	0	1	0	0	0	0	0	1	1	0	0	0	0	0	0	0	0	1	0	0	0	0	0	0	0
20	0	0	0	0	0	0	0	0	1	0	0	1	0	1	0	0	0	0	1	1	1	0	0	0	0
21	0	0	0	0	0	0	0	0	0	0	1	0	0	1	0	0	0	0	0	1	0	0	0	0	0
22	0	0	0	0	0	1	0	0	1	0	0	0	0	0	0	0	0	0	1	0	1	0	0	0	0
23	0	0	0	0	0	0	0	0	1	1	0	0	0	0	0	0	0	0	0	0	0	1	0	0	0
24	1	0	0	0	1	0	0	1	1	0	1	1	1	1	0	1	1	0	1	1	1	0	1	1	0
25	1	1	0	1	1	0	1	1	1	0	1	1	1	1	1	1	0	1	1	1	1	1	1	1	1

3.2.3. Final Reachability Matrix (FRM)

After the construction of the initial reachability matrix, the final reachability matrix is obtained by adding 1* entries into the initial reachability matrix, to incorporate transitivity. Transitivity is a basic assumption at this stage, which states that if variable "X" is associated with variable "Y" and "Y" is associated with "Z", then "X" must be associated with "Z". The conversion of IRM into FRM is shown in Table 8.

3.2.4. Level Partition

After developing the final reachability matrix, the reachability sets and antecedent sets of each factor were obtained from the final reachability matrix. The reachability set includes the factor itself and other factors that it may help to determine, while the antecedent set includes the factor itself and other factors that may help it to be achieved. The similar values of the reachability set and the antecedent set were added into another set called the intersection set. Thereafter, level partition was performed. After the allocation of each factor, their levels were also eliminated accordingly. This process was continued until all factors were allocated levels. Through this process, the SFFs were divided into twelve levels. The twelve levels of iteration are shown in Table 9.

Table 8. Final reachability matrix.

VAR	1	2	3	4	5	6	7	8	9	10	11	12	13	14	15	16	17	18	19	20	21	22	23	24	25	Driving Power
1	1	0	0	0	0	1*	0	0	1*	0	0	0	1*	1	0	0	0	0	0	0	0	0	0	0	0	5
2	0	1	1	0	0	1	0	0	1*	0	0	0	1*	0	0	1*	0	0	0	0	0	1*	0	0	0	7
3	1*	1	1	0	0	1*	0	0	1*	0	0	0	1	1*	0	1*	0	0	0	0	0	1*	0	0	0	9
4	1*	1	1	1	0	1	0	0	1*	0	0	0	1	1*	0	1*	0	0	0	0	1*	1*	0	0	0	10
5	1	0	0	0	1	1	0	1	1*	0	0	1*	0	1*	0	0	0	0	0	1*	1*	1*	0	0	0	8
6	0	0	0	0	0	1	0	0	1	0	0	1*	1	0	0	0	0	0	1*	1*	1*	1*	0	0	0	7
7	1	1	1	1*	1	1	1	1	1	1	1*	1*	1	1*	1*	1*	1*	1	0	1*	1*	1	0	0	0	20
8	0	0	0	0	0	1	0	1	1	1	0	1	0	0	1*	0	0	0	0	1*	0	1	0	0	0	8
9	1*	0	1*	0	0	1*	0	1	1	1	1	1	1*	0	1*	0	0	0	0	1*	0	1*	0	0	0	9
10	1*	1	1*	0	0	1*	0	1	1	1	1	1*	1*	1*	1	1*	0	1	0	1	1*	1	1	1	1*	16
11	1*	0	0	0	0	1*	0	0	1	0	1	1*	1*	1	1	1*	1*	0	0	1*	1*	1	1	1	0	13
12	1	1*	1	0	0	1*	0	1	1	0	1	1	1*	0	1*	1	0	0	0	1*	0	1*	0	0	0	11
13	1	0	1*	0	0	1	0	1	1	0	1	1*	1	1	1*	0	0	0	0	1*	1*	1*	0	0	0	11
14	0	0	0	0	0	1	0	0	0	0	0	1	0	1	1	0	0	0	0	1*	1*	1*	0	0	0	8
15	0	0	0	0	0	1	1*	0	1*	0	0	1*	0	0	1	0	0	0	0	0	0	0	0	0	0	3
16	1	1	1*	1	1*	1	1*	1*	1	1	1*	1	1*	1	1	1	1	1*	1	1	1	1	0	0	0	23
17	1*	1	1*	1	0	1	1*	1	1	1	1*	1*	1*	1*	1*	1*	1	1	1*	1*	1*	1*	1*	1*	1*	25
18	1*	1	1*	1	0	1*	0	1	1	1	1*	1*	1*	1*	1*	1*	1*	1	1	1*	1*	1*	1	1	1	25
19	1*	1	1*	0	0	1*	0	0	1	0	0	1	1*	0	1	1*	0	0	1	1	1	1	1	1	0	14
20	1*	1*	1*	0	0	1*	0	0	1*	0	0	1*	1*	1*	1	1*	0	0	0	1	1*	1	0	0	0	13
21	1*	1*	1*	0	0	1*	0	0	1	0	0	1	1*	1*	1*	1*	0	0	0	0	1	1	0	0	0	12
22	1*	1*	1*	0	0	1*	1*	0	1	0	1*	1*	1*	1*	1*	1*	0	0	0	1*	1*	1	0	0	0	13
23	1*	0	1*	0	0	1*	0	1	1	1	1	1	1	0	1*	0	1	1*	1	1	1*	1	1	1	1*	16
24	1	1	1*	1	1	1*	1	1	1	1*	1	1	1	1	1*	1	1	1*	1	1	1	1*	1	1	1*	24
25	1	1	1*	1	1	1*	1	1	1	1*	1	1	1	1	1*	1	1	1*	1	1	1	1	1	1	1	25
Dependence Power	20	14	19	6	7	24	6	9	24	6	9	20	20	17	20	14	6	7	7	17	19	23	8	7	6	335

* transitivity links.

Table 9. Deriving levels from FRM.

Sr.#	Reachability Set	Antecedent Set	Intersection Set	Level
1	1,6,9,13,14	1,3,4,5,7,9,10,11,12,13,16,17,18,19,20,21,22,23,24,25	1,9,13	IV
2	2,3,6,9,13,16,22	2,3,4,7,12,16,17,18,19,20,21,22,24,25	2,3,16,22	V
3	1,2,3,6,9,13,14,16,22	2,3,4,7,9,10,11,12,13,16,17,18,19,20,21,22,23,24,25	2,3,9,13,16,22	V
4	1,2,3,4,6,9,13,14,16,22	4,7,16,17,18,25	4,16	VI
5	1,5,6,8,9,13,14,22	5,7,16,17,18,24,25	5	X
6	6,9,12,15,20,21,22	1,2,3,4,5,6,7,8,9,10,11,12,13,14,16,17,18,19,20,21,22,23,24,25	6,9,12,20,21,22	II
7	1,2,3,4,5,6,7,8,9,11,12,13,14,15,16,17,19,20,21,22	7,16,17,18,24,25	7,16,17	XI
8	6,8,9,12,15,20,21,22	5,7,8,16,17,18,19,24,25	8	IX
9	1,3,6,9,12,15,20,21,22	1,2,3,4,5,6,7,8,9,10,11,12,13,14,16,17,18,19,20,21,22,23,24,25	1,3,6,9,12,20,21,22	III
10	1,3,6,9,10,11,12,13,15,18,20,21,22,23,24,25	10,17,18,23,24,25	10,18,23,24,25	X
11	1,3,6,9,11,12,13,14,15,20,21,22,23	7,10,11,16,17,18,23,24,25	11,23	IX
12	1,2,3,6,9,12,13,14,15,16,22	6,7,8,9,10,11,12,13,14,15,16,17,18,19,20,21,22,23,24,25	6,9,12,13,14,15,16,22	VI
13	1,3,6,9,12,13,14,15,20,21,22	1,2,3,4,5,7,10,11,12,13,16,17,18,19,20,21,22,23,24,25	1,3,12,13,20,21,22	IV
14	6,9,12,14,15,20,21,22	1,3,4,5,7,11,12,13,14,16,17,18,20,21,22,24,25	12,14,20,21,22	III
15	12,15,21	6,7,8,9,11,12,13,14,15,16,17,18,19,20,21,22,23,24,25	12,15,21	I
16	1,2,3,4,5,6,7,8,9,11,12,13, 14,15,16,17,18,19,20,21,22,23,24	2,3,4,7,12,16,17,18,19,20,21,22,24,25	2,3,4,7,12,16,17,18,19,20,21, 22,24	XI
17	1,2,3,4,5,6,7,8,9,10,11,12, 13,14,15,16,17,18,19,20,21,22,23, 24,25	7,16,17,18,24,25	7,16,17,18,24,25	XI
18	1,2,3,4,5,6,7,8,9,10,11,12,13,14,15,16,17,18,19,20,21,22,23, 24,25	10,16,17,18,23,24,25	10,16,17,18,23,24,25	XII
19	1,2,3,6,8,9,12,13,15,16,19,20,21,22	7,16,17,18,19,24,25	16,19	X
20	1,2,3,6,9,12,13,14,15,16,20,21,22	6,7,8,9,10,11,13,14,16,17,18,19,20,21,22,23,24,25	6,9,13,14,16,20,22	VIII
21	1,2,3,6,9,12,13,14,15,16,21,22	6,7,8,9,10,11,13,14,15,16,17,18,19,20,21,22,23,24,25	6,9,13,14,15,16,21,22	VII
22	1,2,3,6,9,12,13,14,15,16,20,21,22	2,3,4,5,6,7,8,9,10,11,12,13,14,16,17,18,19,20,21,22,23,24	2,3,6,9,12,13,14,16,20,21,22	
23	1,3,6,9,10,11,12,13,15,18,20,21,22,23,24,25	10,11,16,17,18,23,24,25	10,11,18,23,24,25	IX
24	1,2,3,5,6,7,8,9,10,11,12,13,14,15,16,17,18,19,20,21,22,23,24,25	10,16,17,18,23,24,25	10,16,17,18,23,24,25	XII
25	1,2,3,4,5,6,7,8,9,10,11,12,13,14,15,16,17,18,19,20,21,22,23,24,25	10,17,18,23,24,25	10,17,18,23,24,25	XII

3.2.5. ISM-Based Hierarchal Model

After level partitioning, the hierarchal structure of the SFFs in the supply chain of the dairy industry in Pakistan were developed, resulting in a diagraph. Thereafter, the transitive links were removed based on the relationships given in the final reachability matrix. The final ISM-based model is shown in Figure 3.

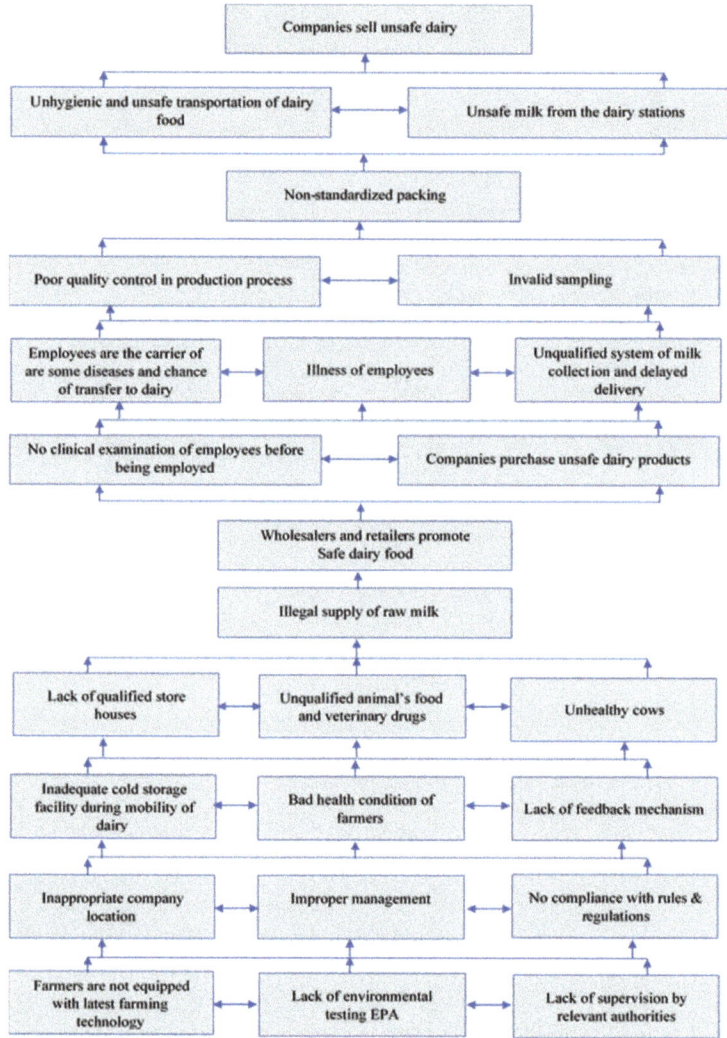

Figure 3. ISM-based model of safety failure factors of the dairy supply chain.

3.3. MICMAC Analysis

MICMAC was introduced by [68]. It is based on the multiplication properties of matrices. In this study, we used MICMAC with ISM in the problem evaluation. In addition, the combined use of these two approaches can assist in understanding the level of importance of each of the considered variables through well-described diagrams (ISM-based hierarchical diagram and classification-based MICMAC analysis). MICMAC analysis was carried out for the validation of the hierarchically structural model of the described implementation factors. In the ISM technique, we considered four probable relations to examine the interactions among the decision variables; however, we could not classify the strength of the contextual relations among these variables. MICAC analysis, on the other hand, can effectively classify the contextual relations among the decision variables and describe the directions and levels of considered variables. Finally, the purpose of this approach is to examine the power of the driving and dependence forces of the described barriers. This is used to formulate dairy supply chains by dividing the barriers into four clusters. These four clusters are as follows.

3.3.1. Autonomous Factors

This includes factors whose driving power and dependence power are both weak. These barriers are relatively disconnected from each other, but have some links that might be strong. The factors in the autonomous clusters are A4, A5, A8 and A11.

3.3.2. Dependent Factors

The cluster of dependent factors has insufficient driving power to drive other barriers, but their dependence power is strong. Dependent factors include A1, A2, A3, A6, A9, A12, A13, A14, A15, A20, A21, and A22.

3.3.3. Linkage Factors

The driving power and dependence power of the linkage factors are both strong, and therefore, these factors are considered to be unstable. In this sense, any impact on these factors also influences other factors. The A16 factor is a linkage factor.

3.3.4. Independent Factors

Independent factors represent the factors with strong driving power but weak dependence power. These include A7, A10, A17, A18, A19, A23, A24 and A25.

The results of the MICMAC analysis are shown in Figure 4.

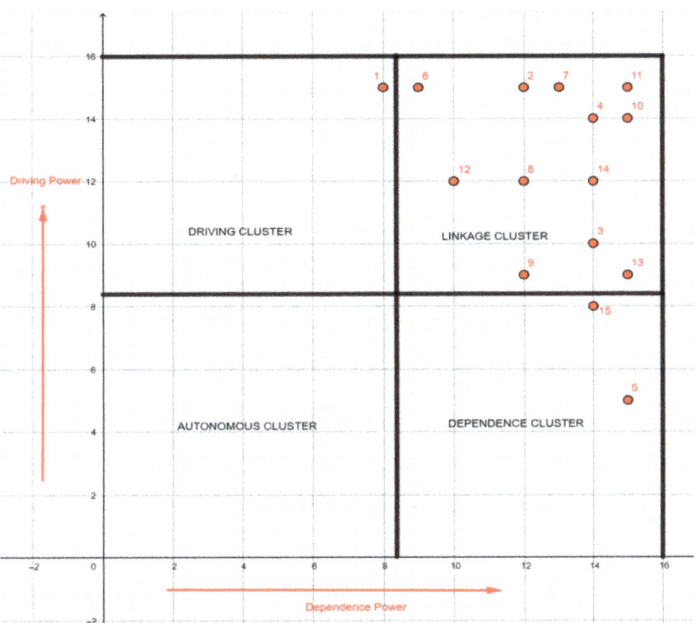

Figure 4. MICMAC analysis of safety failure factors of the dairy supply chain.

4. Results and Discussion

The SFFs (A25) Lack of supervision by relevant authorities, (A24) Lack of environmental testing by the environmental protection agency (EPA) and (A18) Farmers are not equipped with the latest farming technology emerged as the most critical according to the ISM hierarchical framework. These factors are included in level twelve. There is a lack of sufficient supervision by relevant authorities (A25) in the dairy sector of Pakistan. The restricted flow of information across the hierarchy of organizations affects the milk supply chain system. The collaboration of research and support by relevant authorities is required in this context. Organizations require coordination with between personnel and stakeholders to show them how the company's goals are aspired to in their day-to-day functions. (A24) Lack of environmental testing by environmental protection agency (EPA) is another critical safety factor in the dairy industry, because environmental rules and regulations are not implemented properly in this sector, despite how significant they are. There is lack of environmental literacy. The implementation of environmental testing is important in maintaining effectiveness in supply chain systems [69]. The Pakistani government should take action to regulate the EPA's polices appropriately in the dairy industry. Another critical factor is (A18) Farmers are not equipped with the latest farming technology. The adoption of advanced technology and processes is very slow in Pakistan's dairy industry. Farmers are unfamiliar with the latest technology and keep using old methods. Moreover, farmers are hesitant to adopt the latest technology and processes because of their superstitious beliefs and traditional organizational structure. Technologies including milk meters, weight scales, mastitis detection and activity meters can be used to assist on-farm decision-making, and also improve the safety performance of the dairy supply chain. Dairy labor efficiency has been improved recently using automatic cluster removers (ACRs) on farms [70]. The government should educate farmers about the potential benefits of various technologies, along with process automation, and ensure appropriate investment decisions are made.

Level eleven includes three factors, namely, (A17) No compliance with the rules and regulations, (A16) Improper management and (A7) Inappropriate company location, which are also significant SFFs in the dairy supply chain of Pakistan. There is severe negligence of the rules and regulations in the dairy sector of Pakistan, which is one of the big issues. The government should develop effective courses and regulations, and ensure the dairy sector follows them properly in order to stabilize the supply chain system. Improper management (A16) is another issue that makes milk production low. There is resistance to introducing change into the existing dairy supply chain system, including investments, information and production systems, etc. The top management should strengthen its leadership skills by ensuring the right person is employed in the right place. To enhance the productivity of the organization, the top management should mitigate managerial-level conflicts [71]. Inappropriate company location (A7) is one of the big factors influencing safety performance and milk production. Most dairy companies are located out of cities in Pakistan, and this huge distance between company and end customer creates a big gap between demand and supply. Poorly constructed roads, weather challenges and traffic jams are the most common factors affecting the milk supply chain system. The relocation of dairies from regions rich in water resources to regions with limited resources is likely to be shortsighted [72].

(A19) Lack of feedback mechanism, (A10) Bad health condition of farmers and (A5) Inadequate cold storage facility during mobility of dairy food are all included at level ten. These are also important safety factors in the dairy industry of Pakistan. The lack of feedback mechanisms (A19) is also a problem in the dairy industry. There is no trend of giving feedback about the quality of milk, dairy products, production process, supply chain systems, etc., in a formal way. Dairy companies should establish a consumer-oriented feedback mechanism and give immediate responses upon receiving consumer complaints, and they should prevent further deterioration by keeping confidential records. The next factor is (A10) Bad health condition of farmers. Farmers suffer from an increased occurrence of many acute and chronic health conditions, including skin cancer, hearing loss, amputations and respiratory diseases, etc. Other health issues have rarely been studied in the agriculture sector, such as stress and adverse reproductive outcomes [73,74]. The majority of livestock-handling claims were made by males (88%) and by employees on farms employing eleven or more workers (87%) [75]. The government should develop policies of better medical treatment for farmers so they can work effectively. (A5) Inadequate cold storage facilities during mobility of dairy food is also a big issue causing the wastage of milk. Hence, the lack of modern technology, especially decent refrigerator facilities, is the main cause of wastage of milk. As the weather in Pakistan is very hot during summer, and electricity is in short supply, huge amounts of dairy products are wasted during transportation, which leads to high production costs.

Factors including (A23) Unhealthy cows, (A11) Unqualified animals' food and veterinary drugs and (A8) Lack of qualified storehouse fall into level nine. Underdeveloped farms and unbalanced diets are the main reasons for unhealthy cows (A23), which tends to result in the low production of milk. The government should develop easy policies for farmers to access loans so that they can develop their farms. At the same time, farmers should pay attention to animal health by providing proper food and drugs, as well as ensuring some precautions, including noise reduction to reduce animal agitation, nonslip flooring, proper lighting for ease of animal movement, and distraction removal in order to prevent balking. (A11) Unqualified animals' food and veterinary drugs is also a critical issue. Due to the lack of knowledge and training, farmers are unfamiliar with qualified food and veterinary drugs. They are used to traditional treatments. However, the failure to keep to the withdrawal period, including when using potential overdose and long-acting drugs, might be the reason for the presence of unacceptable residues [34]. Government and dairy companies should develop programs to educate farmers regarding vaccination, mastitis, nutrition, and metabolic and reproductive problems. Furthermore, they should hire qualified veterinary doctors on dairy farms. Another factor is (A8) lack of a qualified storehouse, which is one of the big reasons for the wastage of food. The safety of food is a serious problem, even in developed countries, where 15.7 million people are undernourished. There is a need to stop further wastage of food because the world population is growing rapidly [27]. The wastage of food has economic implications in the food supply chain (i.e., farmer, producer and consumer). Food losses have a negative impact on the incomes of both farmers and consumers [76,77]. Further, food loss is also a reason to reduce the financial resources that are applied here and that can be used for investment in other areas. The relevant authorities must be aware of this problem and take adequate measures towards the reduction of dairy food wastage.

Level eight contains a single barrier of (A20) Illegal supply of raw milk. This is a critical barrier in the Pakistan dairy sector. About 97% of dairy farming is not linked with formal dairy channels, which makes economic productivity low (PDDC, 2006). The extra milk leftover after meeting household needs is mostly sold to informal market chains, shopkeepers, directly to consumers, to middlemen and to brokers [26]. The government should take serious action in terms of legalized milk supply in order to avoid monitory losses in the dairy sector.

Level seven includes (A21) Wholesalers and retailers promote unsafe dairy food. The milk products and byproducts in the country include pasteurized milk, powdered milk, ultra-high-temperature (UHT) milk, butter, cream, yogurt, ghee and cheese. Wholesalers and retailers distribute unsafe dairy food by changing the shelf-life or selling at a low price just for the sake of their own short-term interest, which leads to food security issues for consumers [78]. Companies purchase unsafe dairy foods (A12) and carry out no clinical examinations of their employees before being employed (A4). These factors comprise level six. (A12) companies purchase unsafe dairy food because they only focus on their own interests, and they pay much less to dairy farmers. On the other hand, to make a good profit, farmers mix water in milk and use low-quality milk powder in other dairy products to increase production. As a result, consumers pay high, but get poor-quality dairy products, which may lead to diseases. The government and other regulatory departments should carry out inspections of these companies, and devise strict rules related to dairy safety issues. (A4) No clinical examination of employees before being employed is another noticeable issue in the dairy industry of Pakistan. Most dairy farms do not carry out physical examinations when hiring their workers. Some employees with bad health conditions bring different diseases, which may be transferred to other employees and animals as well. Dairy management should require the physical examination of employees before hiring, so the risk of carrying diseases into dairy sector can be reduced.

Level five of the ISM hierarchal model includes (A22) unqualified system of milk collection and delayed delivery, (A3) Illness of employees and (A2) Employees are the carrier of some disease and chance of transfer to dairy. Unqualified systems of milk collection and delayed delivery (A22) is an important SFF. Because of the underdeveloped infrastructure of roads and electricity, and the lack of new technologies in the Pakistani dairy sector, the system of milk collection and milk delivery is poor. Farmers are mostly illiterate, so they do not have a proper record of milk collection, which is one of the big causes of the wastage of milk. Sometime, in milk collection, there is double counting, and sometimes milk collection is not recorded even once, and that is why it is difficult to share the exact quantity of milk production with regulatory bodies. In addition, as most dairy farms in Pakistan live in rural areas, the great distances of farms from urban markets combined with poor transportation systems is a main reason for delayed delivery. This affects the inputs and outputs and the ease of production [79]. The government should make policies regarding milk supply chain systems, and create development programs to increase the ease of milk delivery. (A3) Illness of employees affects the production process. Farm workers are exposed to extreme health risks on a daily basis, related to handling large animals in dairy farms. Employees need to work hard when feeding calves, managing manure and nutrition and using machinery. However, if they are suffering from a bad health condition, they will not be able to perform their duties properly. Occupational safety and health administration (OHSMS) refers to a series of policies, regulations and plans that lay out how an organization can manage occupational health and safety (OHS) issues. The International Labor Organization has also added the guidelines of the OHSM on safety and health into their code of practice (International Labor Organization, 2010). The government of Pakistan should enact health and safety programs for the proper inspection of dairies. (A2) employees are the carrier of some disease and chance of transfer to dairy. This relates to various diseases, including Q fever, rabies, brucellosis, giardiasis, Escherichia coli (E. coli), cryptosporidium, etc., that are transmitted from humans to animals or vice versa. Sometimes, when employees get sick and work with animals, there is a high risk of exposure to various diseases (i.e., infections caused by virus and bacteria). This is a dangerous situation for both dairy employees and animals, causing infections that can easily be transferred from employees to dairy animals. Employee sickness can be prevented by not using unpasteurized dairy products, wearing gloves when handling reproductive tissues, and washing hands after handling animals. The management of the dairy sector should report these cases and undertake safety measures.

(A13) Invalid sampling and (A1) Poor quality control in production process are factors that fall into level four. These are less important SFFs as compared to the above-discussed factors, but they still require the attention of the organization and the government in order to improve the dairy system. (A13) Invalid sampling—the accurate sampling of dairy products is significant, as they have a short lifetime, which causes unstable demand, influenced by dynamic and expressive environmental responses, none of which is addressed in the dairy sector [80,81]. The aim of sampling programs includes quality assurance, regulation, and accurate cost information. The key to deriving accurate results from quality and composition tests include representative sampling and subsequent proper handling. Training and supervision must be supplied to the persons involved in collecting and handling the sampling. The management of the dairy sector should be attentive to invalid sampling, and make strict policies for the valid sampling of milk to improve the dairy industry. (A1) Poor quality control in the production process; there is no provision for quality issues in the milk production process in the dairy sector of Pakistan. The traceability of quality milk inhibits non-value-added (NVA) programs due to the sampling and testing of milk, this issue is a significant factor in the dairy sector [17]. In addition, national dairy product safety test standards and detection systems are very poor in the Pakistani dairy industry. The ministry of health and other government regulatory authorities should take measures to improve the quality standards of the dairy production process.

The issue of non-standardized packaging (A14) falls into level three. Non-standardized packaging leads to a reduction in the shelf life of dairy products, which is a common issue in the dairy industry of Pakistan. As milk and other dairy products, such as powdered milk, butter, ice cream, and cheese, are highly perishable, the quality, safety, cost, and marketing of these products rely closely on their packaging material. Recently, interest in smart packaging has developed in the dairy industry, which has affected sustainability and the atmosphere as well. The government of Pakistan should provide the dairy industry with the latest developments in packaging, including modified atmosphere packaging (MAP) and active packaging, to control some of the associated fungal problems and extend the shelf life of dairy products. Level two consists of two factors, including (A9) Unsafe milk from the dairy stations and (A6) Unhygienic and unsafe transportation of dairy food. As most dairy farms are situated in rural areas, and because of the poor transportation system, farmers and local milk sellers use bicycles, motorbikes, and open vehicles for the delivery or collection of dairy products, which is unhygienic and unsafe. Level one includes (A15) companies sell unsafe dairy products; sometimes, even big companies perform unethical acts in selling unsafe dairy products at low prices, or even at market rates, to maintain their profits (earning per share). These are the least important factors in the dairy sector of Pakistan. If the government and dairy sector paid a little attention to these issues, they could improve the supply chain system of the dairy industry.

5. Conclusions

This study investigates the SFFs in the supply chain of the dairy industry in Pakistan. Twenty-five factors have been identified in this study. The ISM technique was used to identify the contextual relationships among different factors. The MICMAC approach assists the researchers in understanding the significance of barriers in a systematic way. SFFs that have a greater influencing capacity are listed in independent and linked quadrants of the MICMAC analysis.

The results of this study show that SFFs (A15) Companies sell unsafe dairy products, (A25) Lack of supervision by relevant authorities, (A24) Lack of environmental testing by the environmental protection agency (EPA) and (A18) Farmers are not equipped with the latest farming technology are the main SFFs in the dairy supply chain of Pakistan. These factors need more attention from the government and relevant authorities. The manual approach to milk handling is one of the most critical reasons for the wastage of a high quantity of milk. Dairy industries must improve their management systems, especially as regards supervision, information, technical, operational, wastage and transportation. Moreover, a remuneration system could also be fruitful in improving the overall productivity of the dairy industry.

Earlier studies have examined the productivity barriers and critical factors in dairy supply chains, but they have not proposed any rankings among those SFFs. Thus, the current study has generalized the application of ISM for assessing the interaction among SFFs in the supply chain of the dairy industry. Finally, this study is significant for both the dairy industry and academics, because no study has yet related the factors/barriers and their rankings in a real-time industrial scenario for Pakistan. The proposed study can help decision-makers in the dairy industry of Pakistan eradicate the SFFs in the supply chain, plan their supply chain activities more effectively, and gain an advantage over competitors.

5.1. Theoretical and Practical Implications

In conducting this study, many other studies were analyzed. Many address the critical factors in the dairy supply chain in different areas, but none so far have identified SFFs in the Pakistani context. As such, the purpose of this study is to address the critical factors in the supply chain of the dairy industry in Pakistan by applying ISM and MICMAC techniques.

The theoretical implication of the ISM approach is that it shows the interrelationships among variables in a hierarchical model with multiple, complex levels. Theoretically, this study is novel in the context of the dairy supply chain. The lack of literature in this area makes this research essential to academicians and researchers. Many studies relate to the critical barriers in the supply chain of the dairy industry, but none have recognized the SFFs and their causes in detail in the Pakistani context. Additionally, previous studies have not highlighted such issues, meaning these SSFs are novel in the literature on developing countries.

The practical steps that should be taken to overcome the safety failures in the dairy supply chain include farmers not being equipped with the latest technology. If the top management of the dairy industry would consider such factors when designing long-term strategic policies, and equip their workers with advanced technology, then the production of milk could be increased. To manage the environmental concerns, such as the testing of dairy foods according to environmental concerns, dairy professionals need to focus on local, national and global policies. Technical issues should be mitigated through the adoption of innovative technology in the dairy sector; therefore, the top leadership should recruit technically skilled employees who can help to redevelop the production process of dairy products. Additionally, the lack of supervision by relevant authorities is a major issue that could be removed through the development of proper policies. Professionals and other concerned authorities should formulate policies and enforce them to the benefit of the dairy industry.

5.2. Limitations of Study and Directions for Future Research

There are some limitations to this study; for example, this study has been conducted from the perspective of Pakistan, and critical factors have been identified through the opinions of experts. Thus, a new framework can be established based on the data collected from stakeholders, which can provide a different view. In addition, the ISM technique has been used in this study, which assists in formulating the initial model, and shows that there is no authenticity in the statistical relationships among different barriers. Further, any model that assigns weightage to the identified barriers with statistical tools, such as structural equation modeling (SEM), can be used in future studies. Moreover, the barriers were selected with reference to experts' opinions. Finally, a different view can be derived by collecting data from stakeholders.

In Pakistan, there is a dire need to improve production per animal per head. In this regard, different application programs can encourage through the efficient use of local feed resources, the application of improved feed management, and the development of alternatives. Research institutes should communicate with private sectors to deliver mechanisms and technologies at the grass roots level. The methodology of this study may be generalized for other perishable food processing industries, such as meat, poultry, fishery, etc. However, the government and other concerned authorities should cooperate with each other to mitigate barriers. Future studies can be conducted for different interfaces of the dairy supply chain.

Author Contributions: Writing—original draft preparation, A.H.; conceptualization, M.I. (Muhammad Ishtiaq); visualization, N.A.; formal analysis, A.H.; methodology, M.I. (Muhammad Ishtiaq); validation, M.I. (Muzaffar Iqbal) and M.A.; writing—review and editing, M.I. (Muzaffar Iqbal) and M.A; investigation, M.I. (Muzaffar Iqbal) and M.A; supervision, L.C.-X.; project administration, L.C.-X.; reviewing, preparing final draft, N.A.; formatting, N.A.; writing original draft and review, K.H. All authors have read and agreed to the published version of the manuscript.

Funding: All authors are grateful for the supported research funding of National Natural Science Foundation of China (Project No.71673042); Philosophy and Social Science Research Program of Heilongjiang Province (18JYC257).

Institutional Review Board Statement: Not applicable.

Informed Consent Statement: Not applicable.

Data Availability Statement: It will be provided upon request.

Conflicts of Interest: The authors declare no conflict of interest.

References

1. Haghighat, F. The impact of information technology on coordination mechanisms of supply chain. *World Appl. Sci. J.* **2008**, *3*, 74–81.
2. Sartorius, K.; Kirsten, J. A framework to facilitate institutional arrangements for smallholder supply in developing countries: An agribusiness perspective. *Food Policy* **2007**, *32*, 640–655. [CrossRef]
3. Handford, C.E.; Campbell, K.; Elliott, C.T. Impacts of milk fraud on food safety and nutrition with special emphasis on developing countries. *Compr. Rev. Food Sci. Food Saf.* **2016**, *15*, 130–142. [CrossRef]
4. Mor, N.; Dardeck, K.L. Mitigation of posttraumatic stress symptoms from chronic terror attacks on southern Israel. *J. Soc. Behav. Health Sci.* **2017**, *11*, 2.
5. Mor, R.; Singh, S.; Bhardwaj, A.; Singh, L. Technological implications of supply chain practices in agri-food sector: A review. *Int. J. Supply Oper. Manag.* **2015**, *2*, 720–747.
6. Mor, R.S.; Bhardwaj, A.; Singh, S. A structured literature review of the Supply Chain practices in Food Processing Industry. In Proceedings of the 2018 International Conference on Industrial Engineering and Operations Management, Bandung, Indonesia, 6–8 March 2018.
7. Mathiyazhagan, K.; Agarwal, V.; Appolloni, A.; Saikouk, T.; Gnanavelbabu, A. Integrating lean and agile practices for achieving global sustainability goals in Indian manufacturing industries. *Technol. Forecast. Soc. Chang.* **2021**, *171*, 120982. [CrossRef]
8. Pan, Y.; Wu, D.; Luo, C.; Dolgui, A. User activity measurement in rating-based online-to-offline (O2O) service recommendation. *Inf. Sci.* **2019**, *479*, 180–196. [CrossRef]
9. Mor, R.; Singh, S.; Bhardwaj, A. Learning on lean production: A review of opinion and research within environmental constraints. *Oper. Supply Chain Manag. Int. J.* **2015**, *9*, 61–72. [CrossRef]
10. Bhardwaj, A.; Mor, R.S.; Singh, S.; Dev, M. An investigation into the dynamics of supply chain practices in Dairy industry: A pilot study. In Proceedings of the 2016 International Conference on Industrial Engineering and Operations Management, Detroit, MI, USA, 23–25 September 2016.
11. Yorifuji, T.; Kashima, S.; Higa Diez, M.; Kado, Y.; Sanada, S. Prenatal exposure to traffic-related air pollution and child behavioral development milestone delays in Japan. *Epidemiology* **2016**, *27*, 57–65. [CrossRef]
12. Kumar, N.; Kumar, H.; Mann, B.; Seth, R. Colorimetric determination of melamine in milk using unmodified silver nanoparticles. *Spectrochim. Acta Part A Mol. Biomol. Spectrosc.* **2016**, *156*, 89–97. [CrossRef]
13. Park, M.S.; Kim, H.N.; Bahk, G.J. The analysis of food safety incidents in South Korea, 1998–2016. *Food Control* **2017**, *81*, 196–199. [CrossRef]
14. Mor, R.S.; Bhardwaj, A.; Singh, S. Benchmarking the interactions among performance indicators in dairy supply chain. *Benchmarking Int. J.* **2018**, *25*, 3858–3881. [CrossRef]
15. Nicholas, P.K.; Mandolesi, S.; Naspetti, S.; Zanoli, R. Innovations in low input and organic dairy supply chains—What is acceptable in Europe? *J. Dairy Sci.* **2014**, *97*, 1157–1167. [CrossRef]
16. Nicholson, C.F.; Stephenson, M.W. Milk price cycles in the US dairy supply chain and their management implications. *Agribusiness* **2015**, *31*, 507–520. [CrossRef]
17. Issar, G.S.; Cowan, R.T.; Woods, E.J.; Wegener, M. Dynamics of Australian dairy-food supply chain: Strategic options for participants in a deregulated environment. In Proceedings of the Sixth International Conference on Chain and Network Management in Agribusiness and the Food Industry, Ede, The Netherlands, 27–28 May 2004; pp. 458–464.
18. Mangla, S.K.; Govindan, K.; Luthra, S. Critical success factors for reverse logistics in Indian industries: A structural model. *J. Clean. Prod.* **2016**, *129*, 608–621. [CrossRef]
19. Perron, G.M. *Barriers to Environmental Performance Improvements in Canadian SMEs*; Dalhousie University: Halifax, NS, Canada, 2005.
20. Bharti, M.A. Examining market challenges pertaining to cold chain in the frozen food industry in Indian retail sector. *J. Manag. Sci. Technol.* **2014**, *2*, 33–40.
21. Chandrasekaran, N.; Raghuram, G. *Agribusiness Supply Chain Management*; CRC Press: New York, NY, USA, 2014.
22. Goswami, M.; De, A.; Habibi, M.K.; Daultani, Y. Examining freight performance of third-party logistics providers within the automotive industry in India: An environmental sustainability perspective. *Int. J. Prod. Res.* **2020**, *58*, 7565–7592. [CrossRef]
23. Choudhary, A.; De, A.; Ahmed, K.; Shankar, R. An integrated fuzzy intuitionistic sustainability assessment framework for manufacturing supply chain: A study of UK based firms. *Ann. Oper. Res.* **2021**, 1–44. [CrossRef]

24. Kumar, A.; Staal, S.J.; Singh, D.K. Smallholder dairy farmers' access to modern milk marketing chains in India. *Agric. Econ. Res. Rev.* **2011**, *24*, 243–254.
25. Pant, R.; Prakash, G.; Farooquie, J.A. A framework for traceability and transparency in the dairy supply chain networks. *Procedia Soc. Behav. Sci.* **2015**, *189*, 385–394. [CrossRef]
26. Berem, R.M.; Obare, G.; Bett, H. Analysis of factors influencing choice of milk marketing channels among dairy value chain actors in Peri-urban Areas of Nakuru County, Kenya. *Kenya Eur. J. Bus. Manag.* **2015**, *7*, 174–179.
27. Buzby, J.C.; Hyman, J. Total and per capita value of food loss in the United States. *Food Policy* **2012**, *37*, 561–570. [CrossRef]
28. Lemma, Y.; Kitaw, D.; Gatew, G. Loss in perishable food supply chain: An optimization approach literature review. *Int. J. Sci. Eng. Res.* **2014**, *5*, 302–311.
29. Walker, S.L.; Smith, R.F.; Routly, J.E.; Jones, D.N.; Morris, M.J.; Dobson, H. Lameness, activity time-budgets, and estrus expression in dairy cattle. *J. Dairy Sci.* **2008**, *91*, 4552–4559. [CrossRef]
30. Hemme, T.; Otte, J. *Status and Prospects for Smallholder Milk Production: A Global Perspective*; Food and Agriculture Organization of the United Nations (FAO): Rome, Italy, 2010.
31. Chalupkova, E. Application of Multiple Criteria Method of Analytic Hierarchy Process and Sensitivity Analysis in Financial Services in The Czech Republic. *J. Appl. Econ. Sci.* **2014**, *9*, 221–230.
32. Kawaguchi, T.; Ando, M.; Asami, K.; Okano, Y.; Fukuda, M.; Nakagawa, H.; Ibata, H.; Kozuki, T.; Endo, T.; Tamura, A.; et al. Randomized phase III trial of erlotinib versus docetaxel as second-or third-line therapy in patients with advanced non–small-cell lung cancer: Docetaxel and Erlotinib Lung Cancer Trial (DELTA). *J. Clin. Oncol.* **2014**, *32*, 1902–1908. [CrossRef]
33. Ali, Y.M.; Lynch, N.J.; Haleem, K.S.; Fujita, T.; Endo, Y.; Hansen, S.; Holmskov, U.; Takahashi, K.; Stahl, G.L.; Dudler, T.; et al. The lectin pathway of complement activation is a critical component of the innate immune response to pneumococcal infection. *PLoS Pathog.* **2012**, *8*, e1002793. [CrossRef] [PubMed]
34. Kumar, R. Performance measurement in dairy Supply chain management. *Indian J. Res.* **2014**, *3*, 100–101. [CrossRef]
35. Prakash, G.; Pant, R. Performance measurement of a dairy supply chain: A balance scorecard perspective. In Proceedings of the 2013 IEEE International Conference on Industrial Engineering and Engineering Management, Bangkok, Thailand, 10–13 December 2013; IEEE: Tianjin, China, 2013.
36. Kumar, A.; Kumar, R.; Rao, K. Enabling efficient supply chain in dairying using GIS: A case of private dairy industry in Andhra Pradesh state. *Indian J. Agric. Econ.* **2012**, *67*. [CrossRef]
37. Mudgal, R.K.; Shankar, R.; Talib, P.; Raj, T. Greening the supply chain practices: An Indian perspective of enablers' relationships. *Int. J. Adv. Oper. Manag.* **2009**, *1*, 151–176. [CrossRef]
38. Luthra, S.; Kumar, V.; Kumar, S.; Haleem, A. Barriers to implement green supply chain management in automobile industry using interpretive structural modeling technique: An Indian perspective. *J. Ind. Eng. Manag.* **2011**, *4*, 231–257. [CrossRef]
39. Ali, S.; Ahmad, N. Livestock Development and Poverty in Pakistan: Evidence from the Punjab Province. *J. Basic Appl. Sci. Res.* **2014**, *4*, 269–276.
40. Planning Commission. Annual Plan 2019–2020. 2019. Available online: https://www.pc.gov.pk/uploads/annualplan/AnnualPlan2019-20.pdf (accessed on 11 February 2021).
41. The State Bank of Pakistan. Economic Data of Pakistan. 2020. Available online: https://www.sbp.org.pk/ecodata/index2.asp (accessed on 26 March 2021).
42. Bar, F. Pakistan Human Development Report 2017 Team. 2018. Available online: https://www.undp.org/content/dam/pakistan/docs/HDR/NHDR_Summary%202017%20Final.pdf (accessed on 26 July 2020).
43. Iqbal, M.; Ma, J.; Ahmad, N.; Hussain, K.; Usmani, M.S.; Ahmad, M. Sustainable construction through energy management practices in developing economies: An analysis of barriers in the construction sector. *Environ. Sci. Pollut. Res.* **2021**, *28*, 34793–34823. [CrossRef]
44. Iqbal, M.; Ma, J.; Ahmad, N.; Hussain, K.; Usmani, M.S. Promoting sustainable construction through energy-efficient technologies: An analysis of promotional strategies using interpretive structural modeling. *Int. J. Environ. Sci. Technol.* **2021**, 1–24. [CrossRef]
45. Hussain, K.; He, Z.; Ahmad, N.; Iqbal, M. Green, lean, six sigma barriers at a glance: A case from the construction sector of Pakistan. *Build. Environ.* **2019**, *161*, 106225. [CrossRef]
46. Jharkharia, S.; Shankar, R. IT enablement of supply chains: Modeling the enablers. *Int. J. Product. Perform. Manag.* **2004**, *53*, 700–712. [CrossRef]
47. Sage, A.P. Methodology for Large-Scale Systems. 1977. Available online: https://books.google.com.pk/books/about/Methodology_for_Large_scale_Systems.html?id=Om5RAAAAMAAJ&redir_esc=y (accessed on 27 June 2021).
48. Iqbal, M.; Ma, J.; Ahmad, N.; Ullah, Z.; Ahmed, R.I. Uptake and Adoption of Sustainable Energy Technologies: Prioritizing Strategies to Overcome Barriers in the Construction Industry by Using an Integrated AHP-TOPSIS Approach. *Adv. Sustain. Syst.* **2021**, *5*, 2100026. [CrossRef]
49. Iqbal, M.; Ahmad, N.; Waqas, M.; Abrar, M. COVID-19 pandemic and construction industry: Impacts, emerging construction safety practices, and proposed crisis management. *Braz. J. Oper. Prod. Manag.* **2021**, *18*, 1–17. [CrossRef]
50. Toktaş-Palut, P.; Baylav, E.; Teoman, S.; Altunbey, M. The impact of barriers and benefits of e-procurement on its adoption decision: An empirical analysis. *Int. J. Prod. Econ.* **2014**, *158*, 77–90. [CrossRef]
51. Ali, S.S.; Kaur, R.; Jaramillo, A.B. An assessment of green supply chain framework in Indian automobile industry using interpretive structural modelling and its validation using MICMAC analysis. *Int. J. Serv. Oper. Manag.* **2018**, *30*, 318–356. [CrossRef]

52. Mishra, N.; Singh, A.; Rana, N.P.; Dwivedi, Y.K. Interpretive structural modelling and fuzzy MICMAC approaches for customer centric beef supply chain: Application of a big data technique. *Prod. Plan. Control* **2017**, *28*, 945–963. [CrossRef]
53. Khan, W.; Akhtar, A.; Ansari, S.A.; Dhamija, A. Enablers of halal food purchase among Muslim consumers in an emerging economy: An interpretive structural modeling approach. *Br. Food J.* **2020**, *122*, 2273–2287. [CrossRef]
54. Sonar, H.; Khanzode, V.; Akarte, M. Investigating additive manufacturing implementation factors using integrated ISM-MICMAC approach. *Rapid Prototyp. J.* **2020**, *26*, 1837–1851. [CrossRef]
55. Alkahtani, M.; Choudhary, A.; De, A.; Harding, J.A. A decision support system based on ontology and data mining to improve design using warranty data. *Comput. Ind. Eng.* **2019**, *128*, 1027–1039. [CrossRef]
56. Ray, A.; De, A.; Mondal, S.; Wang, J. Selection of best buyback strategy for original equipment manufacturer and independent remanufacturer–game theoretic approach. *Int. J. Prod. Res.* **2020**, 1–30. [CrossRef]
57. Goswami, M.; Daultani, Y.; De, A. Decision modeling and analysis in new product development considering supply chain uncertainties: A multi-functional expert based approach. *Expert Syst. Appl.* **2021**, *166*, 114016. [CrossRef]
58. Mukeshimana, M.C.; Zhao, Z.Y.; Ahmad, M.; Irfan, M. Analysis on barriers to biogas dissemination in Rwanda: AHP approach. *Renew. Energy* **2021**, *163*, 1127–1137. [CrossRef]
59. Raj, T.; Shankar, R.; Suhaib, M. An ISM approach for modelling the enablers of flexible manufacturing system: The case for India. *Int. J. Prod. Res.* **2008**, *46*, 6883–6912. [CrossRef]
60. Raj, A.; Rifkin, S.A.; Andersen, E.; Van Oudenaarden, A. Variability in gene expression underlies incomplete penetrance. *Nature* **2010**, *463*, 913–918. [CrossRef]
61. Kannan, G.; Pokharel, S.; Kumar, P.S. A hybrid approach using ISM and fuzzy TOPSIS for the selection of reverse logistics provider. *Resour. Conserv. Recycl.* **2009**, *54*, 28–36. [CrossRef]
62. Karamat, J.; Shurong, T.; Ahmad, N.; Afridi, S.; Khan, S.; Khan, N. Developing Sustainable Healthcare Systems in Developing Countries: Examining the Role of Barriers, Enablers and Drivers on Knowledge Management Adoption. *Sustainability* **2019**, *11*, 954. [CrossRef]
63. Ravi, V.; Shankar, R. Analysis of interactions among the barriers of reverse logistics. *Technol. Forecast. Soc. Chang.* **2005**, *72*, 1011–1029. [CrossRef]
64. Govindan, K.; Kannan, D.; Mathiyazhagan, K.; Jabbour, A.B.; Jabbour, C.J. Analysing green supply chain management practices in Brazil's electrical/electronics industry using interpretive structural modelling. *Int. J. Environ. Stud.* **2013**, *70*, 477–493. [CrossRef]
65. Ahmad, N.; Zhu, Y.; Hongli, L.; Karamat, J.; Waqas, M.; Mumtaz, S.M. Mapping the obstacles to brownfield redevelopment adoption in developing economies: Pakistani Perspective. *Land Use Policy* **2020**, *91*, 104374. [CrossRef]
66. Tan, T.; Chen, K.; Xue, F.; Lu, W. Barriers to Building Information Modeling (BIM) implementation in China's prefabricated construction: An interpretive structural modeling (ISM) approach. *J. Clean. Prod.* **2019**, *219*, 949–959. [CrossRef]
67. Malek, J.; Desai, T.N. Interpretive structural modelling based analysis of sustainable manufacturing enablers. *J. Clean. Prod.* **2019**, *238*, 117996. [CrossRef]
68. Godet, M. Introduction to la prospective: Seven key ideas and one scenario method. *Futures* **1986**, *18*, 134–157. [CrossRef]
69. Rao, P.; Holt, D. Do green supply chains lead to competitiveness and economic performance? *Int. J. Oper. Prod. Manag.* **2005**, *25*, 898–916. [CrossRef]
70. Totty, V.K.; Greenwood, S.L.; Bryant, R.H.; Edwards, G.R. Nitrogen partitioning and milk production of dairy cows grazing simple and diverse pastures. *J. Dairy Sci.* **2013**, *96*, 141–149. [CrossRef]
71. Olawumi, T.O.; Chan, D.W.; Wong, J.K.; Chan, A.P. Barriers to the integration of BIM and sustainability practices in construction projects: A Delphi survey of international experts. *J. Build. Eng.* **2018**, *20*, 60–71. [CrossRef]
72. Zhou, Q.; Han, R.; Li, T. A two-step dynamic inventory forecasting model for large manufacturing. In Proceedings of the 2015 IEEE 14th International Conference on Machine Learning and Applications (ICMLA), Miami, FL, USA, 9–11 December 2015; IEEE: Miami, FL, USA, 2015.
73. Brackbill, R.M.; Cameron, L.L.; Behrens, V. Prevalence of chronic diseases and impairments among US farmers, 1986–1990. *Am. J. Epidemiol.* **1994**, *139*, 1055–1065. [CrossRef] [PubMed]
74. Ahmad, M.; Iqbal, M.; Drissi, J.; Hassan, A. Modeling of Supply Chain Sustainability Enablers by Considering the Impact of COVID-19 on Developing Countries. *N. Am. Acad. Res. J.* **2021**, *4*, 264–279.
75. Marsalis, M.A.; Hagevoort, G.R.; Lauriault, L.M. *Survey of Silage Crop Nutritive Value in New Mexico and West Texas*; NM State University Cooperative Extension Service, College of Agricultural: Tucumcari, Mexico, 2012.
76. Gustavsson, J.; Cederberg, C.; Sonesson, U.; Van Otterdijk, R.; Meybeck, A. *Global Food Losses and Food Waste*; FAO: Rome, Italy, 2011.
77. Waqas, M.; Honggang, X.; Khan, S.A.; Ahmad, N.; Ullah, Z.; Iqbal, M. Impact of reverse logistics barriers on sustainable firm performance via reverse logistics practices. *LogForum* **2021**, *17*, 213–230.
78. Velázquez-Ordoñez, V.; Valladares-Carranza, B.; Tenorio-Borroto, E.; Talavera-Rojas, M.; Varela-Guerrero, J.A.; Acosta-Dibarrat, J.; Puigvert, F.; Grille, L.; Revello, Á.G.; Pareja, L. Microbial contamination in milk quality and health risk of the consumers of raw milk and dairy products. In *Nutrition in Health and Disease-Our Challenges Now and Forthcoming Time*; IntechOpen: London, UK, 2019.
79. Chamberlin, J.; Jayne, T.S. Unpacking the meaning of 'market access': Evidence from rural Kenya. *World Dev.* **2013**, *41*, 245–264. [CrossRef]

80. Muatip, K.; Sugiarto, M. Farmer Children's Willingness for Dairy Farming Succession in Banyumas Regency. *Anim. Prod.* **2016**, *18*, 118–124. [CrossRef]
81. Usmani, M.S.; Wang, J.; Ahmad, N.; Iqbal, M.; Ahmed, R.I. Mapping green technologies literature published between 1995 and 2019: A scientometric review from the perspective of the manufacturing industry. *Environ. Sci. Pollut. Res.* **2021**, *28*, 28848–28864. [CrossRef]

Article

Developing Country-Specific Methane Emission Factors and Carbon Fluxes from Enteric Fermentation in South Korean Dairy Cattle Production

Ridha Ibidhi [1], Tae-Hoon Kim [2], Rajaraman Bharanidharan [3], Hyun-June Lee [1], Yoo-Kyung Lee [4], Na-Yeon Kim [5] and Kyoung-Hoon Kim [1,2,*]

1. Department of Eco-Friendly Livestock Science, Institutes of Green Bio Science and Technology, Seoul National University, Pyeongchang, Gangwon-do 25354, Korea; ridha@snu.ac.kr (R.I.); Dadim922@snu.ac.kr (H.-J.L.)
2. Department of International Agricultural Technology, Graduate School of International Agricultural Technology, Seoul National University, Pyeongchang, Gangwon-do 25354, Korea; whyhoon1@snu.ac.kr
3. Department of Agricultural Biotechnology, College of Agriculture and Life Sciences, Seoul National University, Seoul 08826, Korea; bharanidharan7@snu.ac.kr
4. National Institute of Animal Sciences, Rural Development Administration, Jeonju-si, Jeollabuk-do 54875, Korea; yoo3930@korea.kr
5. Asia Pacific Ruminant Institute, Icheon, Gyeonggi-do 17385, Korea; narziss924@hanmail.net
* Correspondence: khhkim@snu.ac.kr

Abstract: Dairy cattle farming contributes significantly to greenhouse gas (GHG) emissions through methane (CH_4) from enteric fermentation. To complement global efforts to mitigate climate change, there is a need for accurate estimations of GHG emissions using country-specific emission factors (EFs). The objective of this study was to develop national EFs for the estimation of CH_4 emissions from enteric fermentation in South Korean dairy cattle. Information on dairy cattle herd characteristics, diet, and management practices specific to South Korean dairy cattle farming was obtained. Enteric CH_4 EFs were estimated according to the 2019 refinement of the 2006 Intergovernmental Panel on Climate Change (IPCC) using the Tier 2 approach. Three animal subcategories were considered according to age: milking cows >2 years, 650 kg body weight (BW); heifers 1–2 years, 473 kg BW; and growing animals <1 year, 167 kg BW. The estimated enteric CH_4 EFs for milking cows, heifers, and growing animals, were 139, 83 and 33 kg/head/year, respectively. Currently, the Republic of Korea adopts the Tier 1 default enteric CH_4 EFs from the North America region for GHG inventory reporting. Compared with the generic Tier 1 default EF of 138 (kg CH_4/head/year) proposed by the 2019 refinement to the 2006 IPCC guidelines for high-milking cows, our suggested value for milking cows was very similar (139 kg CH_4/head/year) and different to heifers and growing animals EFs. In addition, enteric CH_4 EFs were strongly correlated with the feed digestibility, level of milk production, and CH_4 conversion rate. The adoption of the newly developed EFs for dairy cattle in the next national GHG inventory would lead to a potential total GHG reduction from the South Korean dairy sector of 97,000 tons of carbon dioxide-equivalent per year (8%). The outcome of this study underscores the importance of obtaining country-specific EFs to estimate national enteric CH_4 emissions, which can further support the assessment of mitigation actions.

Keywords: emission factor; enteric fermentation; greenhouse gas; gross energy; milking cows

Citation: Ibidhi, R.; Kim, T.-H.; Bharanidharan, R.; Lee, H.-J.; Lee, Y.-K.; Kim, N.-Y.; Kim, K.-H. Developing Country-Specific Methane Emission Factors and Carbon Fluxes from Enteric Fermentation in South Korean Dairy Cattle Production. *Sustainability* **2021**, *13*, 9133. https://doi.org/10.3390/su13169133

Academic Editor: Giuseppe Todde

Received: 6 July 2021
Accepted: 10 August 2021
Published: 15 August 2021

Publisher's Note: MDPI stays neutral with regard to jurisdictional claims in published maps and institutional affiliations.

Copyright: © 2021 by the authors. Licensee MDPI, Basel, Switzerland. This article is an open access article distributed under the terms and conditions of the Creative Commons Attribution (CC BY) license (https://creativecommons.org/licenses/by/4.0/).

1. Introduction

Concerns about climate change have increased during the last decade, and the issue has become one of the world's most serious challenges, threatening the sustainability of agricultural production [1,2]. Global climate change is caused by the accumulation of greenhouse gas (GHG) emissions in the atmosphere [3]. The livestock sector is one of the most significant sources of GHG emissions, contributing 14.5% of global GHG emissions [4]. In addition, methane (CH_4) is recognized as the second most important

GHG emitted by anthropogenic sources and is a major driver of climate change [4]. Enteric fermentation (CH_4) emissions from ruminants account for approximately 17% of total global anthropogenic CH_4 emissions [5].

Accurate estimations of GHG emissions are of primary importance when reporting inventories, calculating carbon footprints, identifying GHG sources and sinks, and developing mitigation strategies for GHG emissions at the national scale. Improved estimations using country-specific emission factors (EFs) would support more accurate GHG inventories. The Intergovernmental Panel on Climate Change (IPCC) has developed comprehensive guidelines for national GHG inventories for livestock enteric CH_4 emissions and has proposed different levels (Tier 1, Tier 2, and Tier 3) of estimations according to the quantity of information required and the level of complexity [6]. Tier 1 is an empirical method to calculate enteric CH_4 emissions that use default EFs per head of livestock. Tier 2 is an improved method that requires information about animal categories, feeding, and production systems. The Tier 3 approach is used when a country-specific methodology for enteric CH_4 emission estimation has been developed by the IPCC [7]. Methods for estimating enteric CH_4 EFs that were included in the 2006 IPCC guidelines were partly refined in 2019 to provide an updated scientific basis for supporting the preparation and continuous improvement of national GHG inventories. The refinement mainly comprised the update of the CH_4 conversion factor (Ym, %) and default EFs [6]. The Republic of Korea (ROK) ranks 10th in the world in terms of GHG emissions and plans to reduce its GHG emissions by more than 200 million tons of carbon dioxide-equivalent (CO_2-eq) (37%) by 2030. This plan affects eight different economic sectors, including agriculture [8]. Despite the low contribution of the agriculture sector (3%) to total GHG emissions, the ROK is working towards low-cost strategies to reduce GHG emissions, especially from the livestock sector, which contributes 42% of agriculture emissions (9.4 million tons CO_2-eq per year), mainly from the CH_4 produced from enteric fermentation. Hence, the ROK has set a target to reduce emissions from livestock production by 0.6 million tons per year by 2030 and to become carbon neutral by 2050 [9]. On the other hand, dairy cattle production is recognized as a strategic sector in the ROK, accounting for 23% of the total enteric fermentation emissions [9,10]. The South Korean dairy sector has grown and undergone significant changes since 1990, with the annual milk production increasing considerably due to the increase in animal productivity. Milk productivity per cow increased from 5500 kg/cow/year in 1990 to more than 9000 kg/cow/year in 2019, and the ROK is considered to have one of the highest milk productivities per cow in the world [11]. This increase in milk production can be explained by the adoption of milking cows with a high milk production potential and the use of a total mixed ration (TMR) based on high feed quality imported from the international market [10]. However, the ROK communicates its enteric CH_4 emissions from livestock using default EFs based on the IPCC Tier 1 method from North America [9,12].

To the best of our knowledge, there have been no studies conducted on the development of enteric CH_4 EFs in dairy cattle production using the most recently updated IPCC Tier 2 approach in the ROK. In addition, most industrialized countries such as Canada, Japan, Denmark, and Ireland adopted the Tier 2 approach to calculate enteric CH_4 EFs instead of the Tier 1 method for reporting their GHG inventory from dairy cattle [13,14]. The synthesis of appropriate data would provide a much-needed improvement over the current IPCC Tier 1 approach, leading to an inventory (Tier 2) that reflects the country-specific feeding management practices as well as animal productivity and animal categories. Therefore, the objective of the current study was to develop CH_4 EFs from enteric fermentation in the main dairy cattle subcategories in the ROK using the 2019 refinement to the 2006 IPCC Tier 2 approach and to assess the impact on the national GHG inventory.

2. Materials and Methods

2.1. Development of Tier 2 Enteric Methane Emission Factors for Dairy Cattle

The EFs from enteric CH_4 fermentation for each dairy cattle subcategory were developed using the IPCC [6] Tier 2 approach based on gross energy intake (GEI) and Ym (%) as follows:

$$EF = \left[\frac{GE \times (Y_m/100) \times 365 \text{ days/year}}{55.65 \text{ MJ/kg } CH_4}\right] \quad (1)$$

where, EF is the CH_4 emission factor (kg CH_4/head/year) and Ym represents the CH_4 conversion rate (%), which is the fraction of gross energy in feed converted to CH_4 (CH_4 yield).

The calculation procedures involve the following four steps: calculation of net energy (NE) requirements for different functions (maintenance, lactation, pregnancy and growth), conversion of NE requirement to gross energy (GE) requirement, estimation of CH_4 energy output using Ym as the proportion of GE intake, and conversion of CH_4 energy output to CH_4 emissions [6,15] (Figure 1). Activity data are required to calculate the GE based on IPCC guidelines [6] and the effect of input data on CH_4 EFs was determined using the random forest procedure [16], which was operated using SAS software (version 9.4). The equation used to calculate daily GEI for dairy cattle subcategories is as follows:

$$GE = \left[\frac{\left(\frac{NE_m+NE_a+NE_l+NE_p}{REM}\right) + \left(\frac{NE_g}{REG}\right)}{\frac{DE\%}{100}}\right] \quad (2)$$

where, GE, Gross energy (MJ/head/day); NE_m, Net energy for maintenance (MJ/day); NE_a, Net energy for activity (MJ/day); NE_l, Net energy for lactation (MJ/day); NE_p, Net energy for pregnancy (MJ/day); NE_g, Net energy for growth (MJ/day); DE%, Digestible energy expressed as a percentage of gross energy; REM, Ratio of net energy available in diet for maintenance to digestible energy consumed; REG, Ratio of net energy available for growth in a diet to digestible energy consumed.

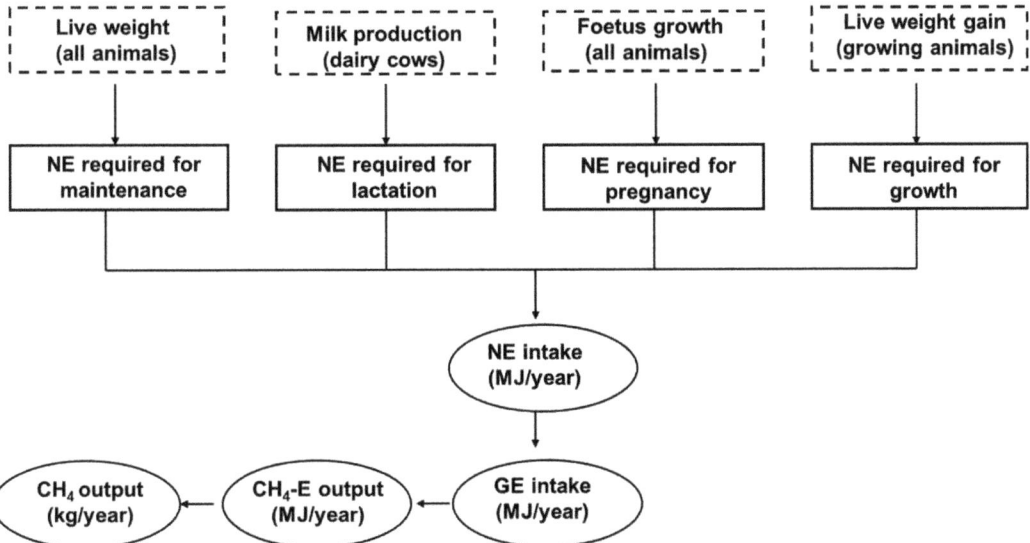

Figure 1. Flow chart presenting the calculation of enteric methane emission factors for dairy cattle using the IPCC [6] Tier 2 approach. NE, net energy; GE, gross energy; CH_4, methane; CH_4-E, energy emitted by methane.

Equations used to estimate NE requirements (NE_m, NE_l, NE_p, and NE_g), REM and REG are listed as follows:

$$NE_m = Cf_i \times (\text{weight})^{0.75} \quad (3)$$

where, NE_m, Net energy for maintenance (MJ/day); Cf_i, maintenance coefficient.

$$NE_l = \text{Milk} \times (1.47 + 0.40 \times \text{Fat}) \quad (4)$$

where, NE_l, Net energy for lactation (MJ/day).

$$NE_p = C_{pregnancy} \times NE_m \quad (5)$$

where, NE_p, Net energy for pregnancy (MJ/day); $C_{pregnancy}$, pregnancy coefficient; NE_m, Net energy for maintenance (MJ/day).

$$NE_g = 22.02 \times \left(\frac{BW}{C \times MBW}\right)^{0.75} \times WG^{1.097} \quad (6)$$

where, NE_g, Net energy for growth (MJ/day); BW, body weight (kg); C, coefficient with a value of 0.8 for females, 1.0 for castrates, and 1.2 for bulls; MBW, mature body weight (kg); WG, average daily weight gain (kg/day).

$$REM = \left[1.123 - \left(4.092 \times 10^{-3} \times DE\right) + \left(1.126 \times 10^{-5} \times (DE)^2\right) - \left(\frac{25.4}{DE}\right)\right] \quad (7)$$

where, REM, Ratio of net energy available in diet for maintenance to digestible energy consumed; DE, Digestible energy expressed as a percentage of gross energy.

$$REG = \left[1.164 - \left(5.16 \times 10^{-3} \times DE\right) + \left(1.308 \times 10^{-5} \times (DE)^2\right) - \left(\frac{37.4}{DE}\right)\right] \quad (8)$$

where, REG, Ratio of net energy available for growth in a diet to digestible energy consumed; DE, Digestible energy expressed as a percentage of gross energy.

2.2. Identification of Animal Subcategories, Breeds, and Body Weights

The IPCC guidelines [6] recommended that livestock subcategories should be defined to generate relatively homogeneous subcategories of animals that reflect country-specific variations in animal characteristics, performance, and feeding systems [6]. The dairy cattle population is reported in South Korean national statistics as an aggregation of animals based on their age (months) in three main subcategories: milking cows, corresponding to animals aged >2 years; heifers, representing animals 1–2 years of age; and growing animals, corresponding to animals aged <1 year [17]. Although many countries raise several dairy cattle breeds, the ROK raises only a single breed (Holstein), which is characterized by a high potential milk production. Due to the lack of data on the average body weight (BW) and average daily gain (ADG) of dairy cattle subcategories in the ROK, data were gathered from various literature sources. Average BW, mature BW, and ADG were calculated from the South Korean feeding standards for dairy cattle [18] and an updated version of the growing chart of dairy cattle in the USA [19].

2.3. Milk Production and Fat Content

The average milk production from milking cows was obtained from a milk recording dataset of the NongHyup Agribusiness Group Inc. [20]. The NongHyup Agribusiness Database is an online recording system where dairy farmers report their milk production and milk fat content (%) monthly. The national average milk production in 2019 was defined based on milk yield data from 1300 dairy farms. The average milk production and milk fat content of milking cows in the ROK were 10,517 (kg/305 days/head) and 3.9%, respectively. The annual milk production was refined to be equal to the quantity of milk produced per 365 days.

2.4. Feed Intake and Digestibility Assumption

The TMR is the typical feeding system for dairy cattle production in the ROK, with animals fed in stalls [18]. The dry matter intake (DMI) of different dairy cattle subcategories was estimated from daily nutrient requirements reported in the South Korean feeding standard [18]. For animals >2 years, the DMI was determined according to the average milk production and milk fat levels. However, for animals aged 1–2 years and animals <1 year, the DMI was defined according to the average BW and ADG of each category [18]. Data on DE (%) of TMR were not available. The DE was assumed based on the total digestible nutrient content (TDN, %) of TMR feed and the DMI (kg) of each subcategory. The DE (%) assumption process is shown in Figure 2 [6,21].

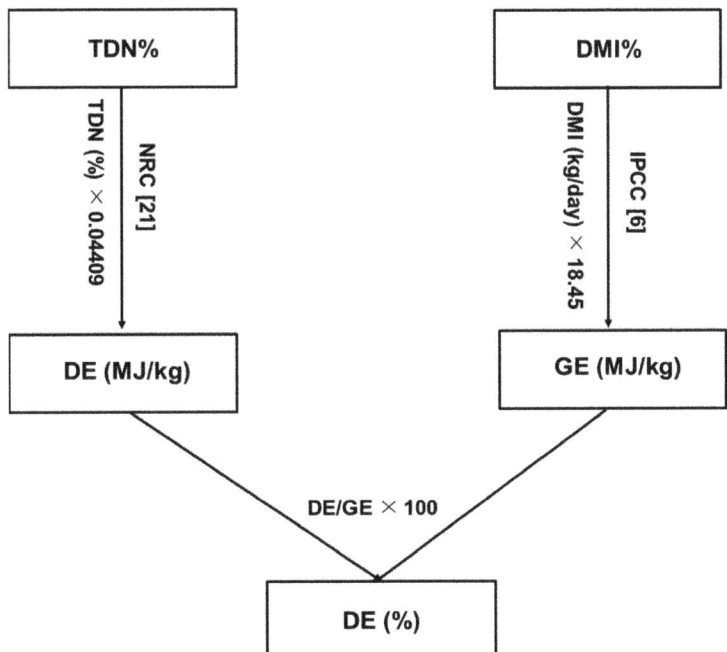

Figure 2. Flowchart of the feed digestibility assumption. The TDN (%) and DMI (kg) were obtained from RDA [18]. TDN, total digestible nutrients; DMI, dry matter intake; DE, digestibility; GE, gross energy.

2.5. Methane Conversion Rate Selection Process

The Ym (%) was defined as the percentage of GEI converted to CH_4. The magnitude of the feed energy converted by each subcategory to CH_4 depends on several interacting factors, such as feed and animal performance. The selection of the Ym (%) was based on the IPCC recommendation considering the milk yield, the neutral detergent fiber (NDF) content, and the DE (%) of feed [6]. For milking cows receiving a high-quality feed (DE ≥ 70 and NDF ≤ 35), the Ym (%) was a weighted annual value (5.8), using a high-productivity value of 5.7 for the lactating period of 305 days and a value of 6.3 for the dry period (60 days). For non-dairy animals (aged 1–2 years and animals > 1 year old) fed a TMR diet with a DE (%) lower than 72%, an average Ym (%) of 6.3 was selected for both categories. Activity data used to calculate the enteric CH_4 EFs are presented in Table 1.

Table 1. Input parameters used to estimate CH_4 emission factors from enteric fermentation in dairy cattle subcategories using the 2019 refined 2006 IPCC Tier 2 methodology and their reference sources.

Parameter	Symbol	Unit	Subcategories			Reference
			<1 Year	1–2 Years	>2 Years	
Net energy for maintenance						
Body weight	BW	kg	167	473	650	RDA [18]
Maintenance coefficient	Cfi	MJ day/kg	0.077	0.077	0.093	Jones and Heinrichs [19] IPCC [6]
Net energy for lactation						
Milk yield	-	kg/305 days	-	-	10517	NH [20]
Milk yield	-	kg/head/day	-	-	28.8	NH [20]
Milk fat content	-	%	-	-	3.9	NH [20]
Net energy for growth						
Average daily gain	ADG	kg/day	0.79	0.66	0	RDA [18]
Mature body weight	MBW	kg	680	680	680	RDA [18]
Net energy for pregnancy						
Pregnancy coefficient	Cp	%	-	-	0.10	IPCC [6]
Feeding system						
Total digestible nutrients	TDN	%	64.7	57.8	73	RDA [18]
Dry matter intake	DMI	kg/day	4.1	11.0	22.0	RDA [18]
Methane conversion rate	Ym	%	6.3	6.3	5.8	IPCC [6]

3. Results and Discussion

3.1. National Enteric Methane Emission Factors of Dairy Cattle

The estimated GEI and annual enteric CH_4 EFs from dairy cattle subcategories using the 2019 refined IPCC Tier 2 methodology are reported in Table 2. The national values of the EFs of milking cows (>2 years), heifers (1–2 years), and growing animals (<1 year) were 139, 83 and 33 (kg CH_4/head/year), respectively. Compared with the default enteric CH_4 EF of dairy cattle used for the South Korean GHG inventory (118 kg CH_4/head/year) [22], our estimated enteric CH_4 EF for milking cows was 15% higher (Table 3). These differences between our predicted CH_4 EF and the default value proposed by the IPCC [22] could be explained by the fact that the input data used for Tier 1 are defined to be representative at the continental level. The ROK uses the North America default CH_4 EF to report their enteric fermentation emissions from dairy cattle [9] due to the similarity of farm management and animal productivity [22]. The Tier 2 estimations in the current study are specific to the dairy cattle production system in the ROK. Similarly, the enteric CH_4 EFs of milking cows (animals >2 years) generated in this study using the IPCC [6] Tier 2 approach were 8% higher than the previous EF default value reported by the IPCC [7]. The default EF (138 kg/head/year) using the IPCC [6] Tier 1 approach was quite similar to the milking cow EF and different to heifers and growing animals EFs generated in this study. Despite the relative similarity between EFs for milking cows generated using the Tier 1 and 2 approaches, the IPCC Tier 2 methodology is preferred to IPCC Tier 1 in estimating the CH_4 EF for enteric fermentation of dairy cattle because it provides specific EFs for each dairy cattle subcategory (milking cows, heifers, and growing animals), in addition to the fact that Tier 1 default EF is based on regional data while Tier 2 EFs are based on national data.

Table 2. Estimated net energy requirements, feed digestibility, gross energy intake, and enteric CH_4 emission factors by dairy cattle subcategory.

Parameter	Animal Subcategories		
	<1 Year	1–2 Years	>2 Years
NEm (MJ/day)	15	32.7	49.7
NEg (MJ/day)	7	12.6	-
NEl (MJ/day)	-	-	87.3
NEp (MJ/day)	-	-	5
REM (%)	0.51	0.48	0.54
REG (%)	0.31	0.26	-
DE (%)	64.4	57.6	72.7
GEI (MJ/day)	81	201	365
CH_4 EF (kg/head per year)	33	83	139

NEm, net energy for maintenance; NEg, net energy for growth; NEl, net energy for lactation; NEp, net energy required for pregnancy; REM, ratio of net energy available in a diet for maintenance to digestible energy consumed; REG, ratio of net energy available for growth in a diet to digestible energy consumed; GEI, gross energy intake; DE, digestible energy; CH_4, methane; EF, emission factor.

Table 3. Comparison of enteric CH_4 emission factors generated from this study with emission factors from previous IPCC emission factor estimation approaches.

Approach	Enteric CH_4 EF (kg CH_4/Head/Year)		
	<1 Year	1–2 Years	>2 Years
IPCC [22], Tier 1	-	-	118
IPCC [7], Tier 1	-	-	128
IPCC [6], Tier 1	-	-	138
IPCC [6], Tier 2	33	83	139

EF, emission factor; CH_4, methane; IPCC, Intergovernmental Panel on Climate Change.

The enteric CH_4 EF for the milking cows subcategory calculated in this study was lower than those of countries with heavier and higher-yielding cows, such as the USA, Israel, Denmark, and Sweden, where the average enteric CH_4 EFs were 146, 150, 161, and 141.6, respectively [23–25]. The differences in dairy cow enteric CH_4 EFs were related mainly to the differences in input data, such as milk productivity, feed quality, and CH_4 conversion rate. By contrast, the enteric CH_4 EF for milking cows generated in this study was higher than other reported values for highly productive milking cows in countries such as Japan and Saudi Arabia, where the enteric CH_4 EFs were 131.5 and 124.8 kg CH_4/head/year, respectively [26,27].

3.2. Effect of Input Data on Enteric Methane Emission Factors

To investigate the importance of using accurate input parameter values in the Tier 2 enteric EFs, we used the random forest procedure [16] available in SAS v9.4. The results (Figure 3) showed that DE (%), milk production, and Ym (%) were the most important parameters that affected and determined the CH_4 EFs from enteric fermentation in dairy cattle. For example, applying a higher DE (%) and milk production resulted in a decrease of enteric CH_4 EFs.

Several studies have reported that activity data used for the calculation of the enteric CH_4 EFs such as DE (%), Ym (%), and milk production could vary across animals, breeds, and feeding and production systems [5,26–28]. Parra and Mora-Delgado [29] reported that the enteric CH_4 EF from dairy cattle in Colombia is largely influenced by animal performance and management system characteristics, especially dietary composition. In the same context, Ibidhi and Calsamiglia [28] reported that increasing the productivity of milking cows is an effective strategy to mitigate GHG emissions, especially enteric CH_4 emissions, which may allow for a reduction in animal numbers while providing the same edible product output at a reduced level of GHG emissions. Parra and Mora-Delgado [29] reported that an increase of ~1500 L/cow/year would reduce the enteric CH_4 emissions of milk by 6.4%. In addition, improving the DE (%) by providing high-quality feed to dairy cattle to raise milk productivity per cow will dilute the CH_4 costs associated with maintenance energy requirements [5]. Feed supply provides substrates for microbial fermentation, and differences in feed digestibility and chemical composition affect the amount of energy, patterns of volatile fatty acids, and CH_4 generated [5].

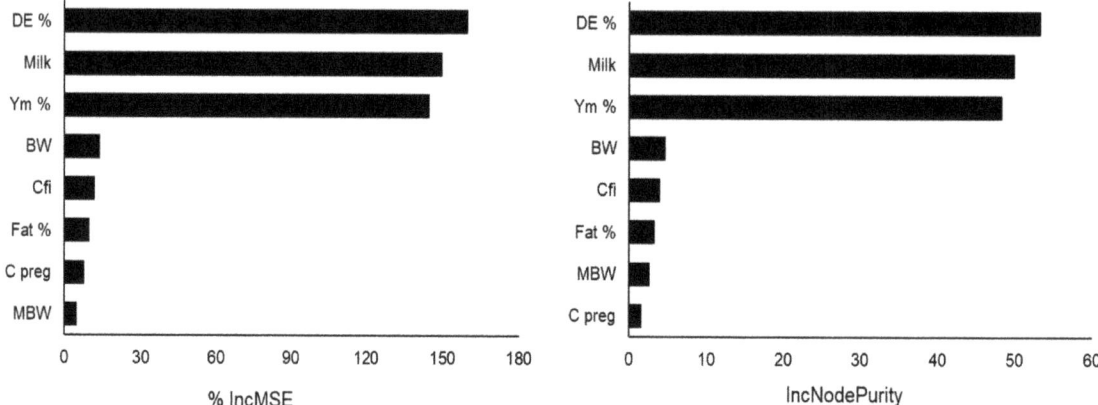

Figure 3. A decision forest tree indicating the importance of input parameters used in the Tier 2 approach to estimate the enteric methane emission factor in milk cows (animals > 2 years). % IncMSE, increase in the mean square error in terms of percentage; IncNodePurity, increase in the mean square error in terms of quality.

3.3. Mapping the Methane Flux from Enteric Fermentation

The spatial allocation of enteric fermentation CH_4 fluxes using the IPCC Tier 2 methodology and dairy cattle population of 2019/2020 across South Korean districts is presented in Figure 4. The contribution of dairy cattle at the district level to the total national CH_4 emissions varied considerably. More than 45% of cattle were concentrated in the Gyeonggi-do and Chungcheongnam-do districts, and therefore the maximum CH_4 emissions occurred in those districts. Other districts had lower CH_4 emissions due to their lower cattle population. Therefore, strategies to reduce CH_4 from enteric fermentation should be adopted, especially in districts with a high emission intensity.

3.4. Implications of the Development of Enteric Methane Emission Factors

The ROK ratified the United Nations Framework Convention on Climate Change (UNFCCC) in September 2002 and made a commitment to become carbon neutral by 2050. The UNFCCC requires involves parties to the convention to develop, publish, and regularly update their national GHG emissions inventories from different sectors including agricultural and livestock production. In the fourth biennial update report of GHG emissions submitted to the UNFCCC in 2019, the ROK reported their enteric CH_4 emissions from dairy cattle using the default EF from the IPCC [22] (118 kg CH_4/head), which was applied to all dairy cattle subcategories. The annual total enteric CH_4 emissions from dairy cattle was 1,204,000 tons CO_2-eq. Using our suggested values of 139, 88, and 33 kg CH_4 for milking cows (>2 years), heifers (1–2 years), and growing animals (<1 year), respectively, which are based on specific data from the South Korean dairy cattle production system, the enteric CH_4 emissions were assumed to average 1,107,000 tons CO_2-eq. Moreover, the use of the Tier 1 approach overestimated the enteric CH_4 emissions from dairy cattle by 8%. Therefore, it is strongly recommended that the Tier 2 EF values generated in the current study are adopted in the next national GHG inventory communication to the UNFCCC. Finally, the findings of this study strengthen the accuracy of the estimation of GHG emissions from the South Korean dairy industry and provide policy-makers with a global overview of the contribution of this sector to global warming, which could help to develop mitigation strategies at the national level. Specific research-based CH_4 EFs are required to define a specific Ym (%) value for each feeding system in the dairy production system. In addition, the national livestock statistics and feed quality data should be improved, which would enhance the quality and the availability of activity data used as inputs in the Tier 2 approach. These improvements could include the use of more subcategories in national statistics, such as dry cows, pregnant heifers, and female calves, and build a DE (%) database specific to each feed ingredient.

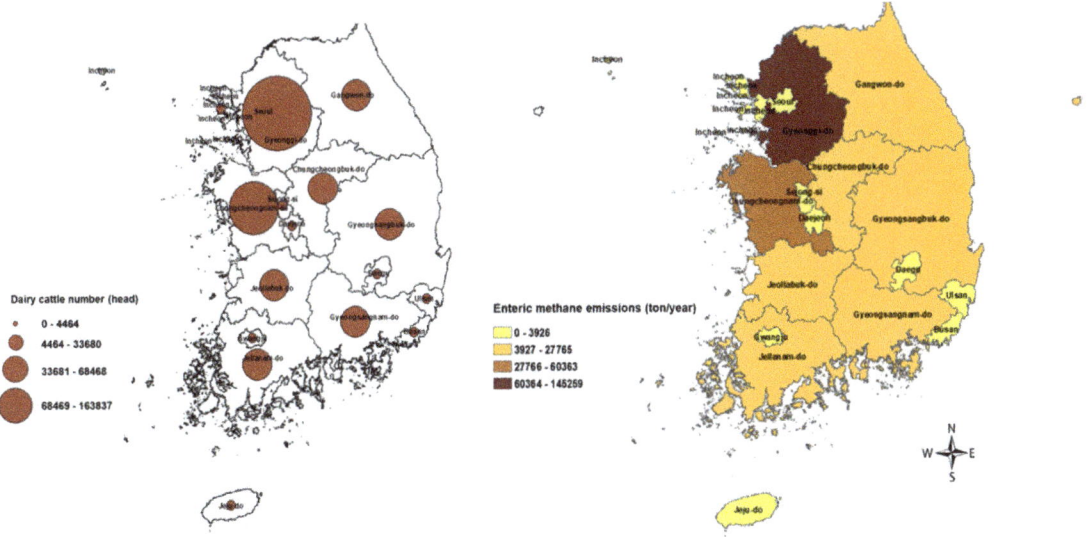

Figure 4. The dairy cattle numbers and enteric methane emission fluxes from dairy cattle (ton/year) at the provincial level.

3.5. Emission Factor Uncertainty

Generating EFs using the Tier 2 methodology may be limited by the availability and quality of data and carries a certain level of uncertainty [30]. The IPCC recommends that an uncertainty analysis should be conducted at the 95% confidence interval for establishing GHG inventories and the EFs of livestock production. There are many uncertainties in the enteric CH_4 generated from the Tier 1 approach compared to values generated from the Tier 2 approach because the results are based on default values [6]. Penman et al. [31] reported that EFs calculated using the Tier 1 approach might be as uncertain as ±50%. The uncertainty using the Tier 2 approach was dependent on the accuracy of input data required to calculate enteric CH_4 EFs such as livestock characterization, milk production, and the feeding system. The present study was subject to limitations such as gaps in the BW and ADG data in each subcategory, in addition to the determination of the DE (%) of feed. These data gaps contribute to the uncertainty in estimated EFs. Therefore, an uncertainty analysis of the CH_4 EFs from enteric fermentation is difficult to be implemented because of the limited data availability, with much of the input data represented by single values, and a lack of available expert judgment. Studies to evaluate the enteric CH_4 EF uncertainties are needed in the future.

4. Conclusions

Country-specific CH_4 EFs for the enteric fermentation of dairy cattle were developed using the 2019 IPCC Tier 2 approach, and the carbon flux from CH_4 emission variations across districts was determined for the ROK. The EFs of the three main subcategories, animals aged > 2 years, animals aged 1–2 years, and growing animals aged < 1 year, were 139, 83, and 33 kg CH_4/head/year, respectively. The IPCC Tier 2 methodology is recommended instead of IPCC Tier 1 in estimating the CH_4 EF for enteric fermentation of dairy cattle because it can cover specific EFs for different dairy cattle subcategories (milking cows, heifers, and growing animals), which can increase the accuracy of the GHG inventory. In this essence, improving activity data compilation is a key factor for producing more accurate country-specific emission factors. The outputs of this study will be useful for the preparation of future national GHG inventories. Specific CH_4 EFs by region and feeding system considering different Ym (%) values are needed to improve the accuracy of the EFs.

Author Contributions: Conceptualization, R.I. and K.-H.K.; methodology, R.I., T.-H.K. and K.-H.K.; validation, R.I. and T.-H.K.; formal analysis, R.B., H.-J.L., Y.-K.L. and N.-Y.K.; investigation, R.I. and T.-H.K.; data curation, R.I. and T.-H.K.; writing—original draft preparation, R.I.; writing—review and editing, R.I. and K.-H.K.; supervision, K.-H.K.; project administration, K.-H.K. All authors have read and agreed to the published version of the manuscript.

Funding: This research was funded by the National Institute of Animal Sciences, Rural Development Administration, Republic of Korea, research project number "PJ0149402021".

Institutional Review Board Statement: Not applicable.

Informed Consent Statement: Not applicable.

Data Availability Statement: No data were deposited in an official repository.

Conflicts of Interest: The authors declare no conflict of interest.

References

1. Gornall, J.; Betts, R.; Burke, E.; Clark, R.; Camp, J.; Willett, K.; Wiltshire, A. Implications of climate change for agricultural productivity in the early twenty-first century. *Philos. Trans. R. Soc. B Biol. Sci.* **2010**, *365*, 2973–2989. [CrossRef] [PubMed]
2. Opio, C.; Gerber, P.; Steinfeld, H. Livestock and the environment: Addressing the consequences of livestock sector growth. *Adv. Anim. Biosci.* **2011**, *2*, 601. [CrossRef]
3. Rojas-Downing, M.M.; Nejadhashemi, A.P.; Harrigan, T.; Woznicki, S.A. Climate change and livestock: Impacts, adaptation, and mitigation. *Clim. Risk Manag.* **2017**, *16*, 145–163. [CrossRef]
4. Steinfeld, H.; Henderson, B.; Mottet, A.; Opio, C.; Dijkman, J.; Falcucci, A.; Tempio, G. *Tackling Climate Change Through Livestock: A Global Assessment of Emissions and Mitigation Opportunities*; Food and Agriculture Organization of the United Nations (FAO): Rome, Italy, 2013; ISBN 925107920X.
5. Knapp, J.R.; Laur, G.L.; Vadas, P.A.; Weiss, W.P.; Tricarico, J.M. Invited review: Enteric methane in dairy cattle production: Quantifying the opportunities and impact of reducing emissions. *J. Dairy Sci.* **2014**, *97*, 3231–3261. [CrossRef]
6. Intergovernmental Panel on Climate Change. *2019 Refinement to the 2006 IPCC Guidelines for National Greenhouse Gas Inventories*; Intergovernmental Panel on Climate Change: Kanagawa, Japan, 2019.
7. Intergovernmental Panel on Climate Change. Guidelines for national greenhouse gas inventories. general guidance and reporting. In *IPCC National Greenhouse Gas Inventories Programme*; Eggleston, H.S., Buendia, L., Miwa, K., Ngara, T., Tanabe, K., Eds.; Institute for Global Environmental Strategies: Hayama, Japan, 2006; Volume 1.
8. Ha, S.; Tae, S.; Kim, R. A study on the limitations of South Korea's National Roadmap for Greenhouse Gas Reduction by 2030 and suggestions for improvement. *Sustainability* **2019**, *11*, 3969. [CrossRef]
9. Greenhouse gas inventory report (GIR) of Korea. *Fourth National Communication of the Republic of Korea under the United Nations Framework Convention on Climate Change*; Greenhouse Gas Inventory and Research Center: Seoul, Korea, 2019; Volume 1, pp. 23–24.
10. Corazzin, M.; Schermer, M.; Park, S.-Y. Tools to retain added value in dairy farms: The South Korea case. *J. Asian Rural. Stud.* **2017**, *1*, 81–96. [CrossRef]
11. The International Committee for Animal Recording Biennial Statistics of Cow Milk, Sheep MILK and Beef Recording: Yearly Survey on the Situation of Milk Recording Systems in ICAR Member Countries for Cow, Sheep and Goats 2021. Available online: www.icar.org (accessed on 3 March 2021).
12. Greenhouse gas inventory report (GIR) of Korea. Korea's third national communication under the United Nations Framework Convention on climate change: Low carbon, green growth. *Greenh. Gas Invent. Res. Cent.* **2012**, *1*, 1–200. Available online: https://unfccc.int/resource/docs/natc/kornc3.pdf (accessed on 3 March 2021).
13. Ominski, K.H.; Boadi, D.A.; Wittenberg, K.M.; Fulawka, D.L.; Basarab, J.A. Estimates of enteric methane emissions from cattle in Canada using the IPCC Tier-2 methodology. *Can. J. Anim. Sci.* **2007**, *87*, 459–467. [CrossRef]

14. Wilkes, A.; Dijk, S. *Tier 2 Inventory Approaches in the Livestock Sector: A Collection of Agricultural Greenhouse Gas Inventory Practices*; CGIAR Research Program on Climate Change, Agriculture and Food Security (CCAFS): Wageningen, The Netherlands, 2018.
15. Tongwane, M.I.; Moeletsi, M.E. Emission factors and carbon emissions of methane from enteric fermentation of cattle produced under different management systems in South Africa. *J. Clean. Prod.* **2020**, *265*, 121931. [CrossRef]
16. Breiman, L. Random forests. *Mach. Learn.* **2001**, *45*, 5–32. [CrossRef]
17. KOSTAT. Livestock Statistics Survey. 2019. Available online: http://kostat.go.kr/portal/eng/surveyOutline/1/10/index.static (accessed on 12 October 2020).
18. Rural Development Administration. *Korean Feeding Standards for Dairy Cattle*, 3rd ed.; National Institute of Animal Science, Rural Development Administration (RDA): Suwon, Korea, 2017; Available online: http://lib.rda.go.kr/search/searchDetail.do?ctrl=000000217076&marcType=m&lang=kor&loca=rda00001&sysDiv=TOT&siteCode=home&sMenu=3010&paramStrlink=siteCode%3dhome%26amp%3bamp%3bsType%3dKWRD%26amp%3bamp%3bq%3d%25EA%25B0%2580%25EC%25B6%2595%25EC%2582%25AC%25EC%2596%2591%25ED%2591%259C%25EC%25A4%2580%26amp%3bamp%3bsysDiv%3dTOT%26amp%3bamp%3bnDt%3d%26amp%3bamp%3bsi%3dTOTAL%26amp%3bamp%3bloca%3drda00001%26amp%3bamp%3bsMenu%3dnull (accessed on 8 January 2021).
19. Jones, C.; Heinrichs, A.J. *Customized Dairy Heifer Growth Chart*; Penn State University: State College, PA, USA, 2016.
20. Nonghyup. Nonghyup (NH) Dairy Farm Association: Inquiry of Verification Score by Farm: NongHyup Agribusiness Group Inc (NH), Korea. 2019. Available online: www.dcic.co.kr (accessed on 15 November 2019).
21. National Research Council. *Nutrient Requirements of Dairy Cattle: 2001*; National Academies Press: Washington, DC, USA, 2001; ISBN 0309069971.
22. Intergovernmental Panel on Climate Change. *Intergovernmental Panel on Climate Change/Organization for Economic Cooperation and Development) Guidelines for National Greenhouse Gas Inventories*; OECD/OCDE: Paris, France, 1997.
23. Koch, J.; Dayan, U.; Mey-Marom, A. Inventory of emissions of greenhouse gases in Israel. *Water. Air. Soil Pollut.* **2000**, *123*, 259–271. [CrossRef]
24. Greenhouse Gas Inventory Report of Sweden. *National Inventory Report Sweden 2019*; Environmental Protection Agency: Stockholm, Sweden, 2019; Volume 5, pp. 314–346. Available online: https://www.naturvardsverket.se/upload/miljoarbete-i-samhallet/internationellt-miljoarbete/miljokonventioner/FN/nir-sub-15-april.pdf (accessed on 4 April 2021).
25. O'Brien, D.; Shalloo, L. *A Review of Livestock Methane Emission Factors*; EPA Research Report; Environmental Protection Agency: Johnstown Castle, Ireland, 2019; pp. 1–62.
26. Aljaloud, A.A.; Yan, T.; Abdukader, A.M. Development of a national methane emission inventory for domestic livestock in Saudi Arabia. *Anim. Feed Sci. Technol.* **2011**, *166*, 619–627. [CrossRef]
27. Greenhouse gas inventory report of Japan. National greenhouse gas inventory report of Japan. *Cent. Glob. Environ. Res.* **2019**, *5*, 1–63. Available online: https://www.env.go.jp/earth/ondanka/ghg-mrv/unfccc/NIR-JPN-2019-v3.0.pdf (accessed on 4 April 2021).
28. Ibidhi, R.; Calsamiglia, S. Carbon footprint assessment of Spanish dairy dattle farms: Effectiveness of dietary and farm management practices as a mitigation strategy. *Animals* **2020**, *10*, 2083. [CrossRef] [PubMed]
29. Parra, A.S.; Mora-Delgado, J. Emission factors estimated from enteric methane of dairy cattle in Andean zone using the IPCC Tier-2 methodology. *Agrofor. Syst.* **2019**, *93*, 783–791. [CrossRef]
30. Kouazounde, J.B.; Gbenou, J.D.; Babatounde, S.; Srivastava, N.; Eggleston, S.H.; Antwi, C.; Baah, J.; McAllister, T.A. Development of methane emission factors for enteric fermentation in cattle from Benin using IPCC Tier 2 methodology. *Animal* **2015**, *9*, 526–533. [CrossRef] [PubMed]
31. Penman, J.; Kruger, D.; Galbally, I.E.; Hiraishi, T.; Nyenzi, B.; Emmanuel, S.; Buendia, L.; Hoppaus, R.; Martinsen, T.; Meijer, J. *Good Practice Guidance and Uncertainty Management in National Greenhouse Gas Inventories*; Intergovernmental Panel on Climate Change: Kanagawa, Japan, 2000.

Article

The Whole and the Parts—A New Perspective on Production Diseases and Economic Sustainability in Dairy Farming

Susanne Hoischen-Taubner [1,*], Jonas Habel [1], Verena Uhlig [2], Eva-Marie Schwabenbauer [1], Theresa Rumphorst [1], Lara Ebert [1], Detlev Möller [2] and Albert Sundrum [1]

[1] Department of Animal Nutrition and Animal Health, University of Kassel, Nordbahnhofstraße 1a, 37213 Witzenhausen, Germany; Jonas.habel@uni-kassel.de (J.H.); e.m.schwabenbauer@uni-kassel.de (E.-M.S.); theresa.rumphorst@uni-kassel.de (T.R.); laraebert@outlook.de (L.E.); sundrum@uni-kassel.de (A.S.)

[2] Department of Farm Management, University of Kassel, Steinstraße 19, 37213 Witzenhausen, Germany; v.uhlig@uni-kassel.de (V.U.); d.moeller@uni-kassel.de (D.M.)

* Correspondence: susanne.hoischen@uni-kassel.de; Tel.: +49-5542-98-1652

Abstract: The levels of production diseases (PD) and the cow replacement rate are high in dairy farming. They indicate excessive production demands on the cow and a poor state of animal welfare. This is the subject of increasing public debate. The purpose of this study was to assess the effect of production diseases on the economic sustainability of dairy farms. The contributions of individual culled cows to the farm's economic performance were calculated, based on milk recording and accounting data from 32 farms in Germany. Cows were identified as 'profit cows' when they reached their individual 'break-even point'. Data from milk recordings (yield and indicators for PD) were used to cluster farms by means of a principal component and a cluster analysis. The analysis revealed five clusters of farms. The average proportion of profit cows was 57.5%, 55.6%, 44.1%, 29.4% and 19.5%. Clusters characterized by a high proportion of cows with metabolic problems and high culling and mortality rates had lower proportions of profit cows, somewhat irrespective of the average milk-yield per cow. Changing the perception of PD from considering it as collateral damage to a threat to the farms' economic viability might foster change processes to reduce production diseases.

Keywords: profit cows; economic sustainability; knowledge transfer; production disease; production disease economics

1. Introduction

1.1. Production Diseases Affect Animal Welfare and Economic Viability

Animal health and welfare is the subject of increasing importance in social discourse in western European societies and an important feature in the consideration of the external social sustainability of dairy farming [1–3]. It is closely linked to the concept of "one-health" and ultimately to the question of acceptance of intensive livestock production by citizens and consumers [4,5]. The short lifespan of dairy cows and the level of production diseases are a starting point for consumers' and scientists' criticism of modern dairy systems [6–8].

Production diseases reflect production induced stressors and indicate an overstressing of the ability of animals to adapt and cope with suboptimal living conditions. They cause pain, suffering and injury and indicate poor animal welfare [9,10]. Lameness, metabolic disorders, or mammary and uterine infection are not related to a single cause but are affected by multiple factors. In this context, farm management plays a pivotal role in guarding against production diseases [11,12]. At the same time, production diseases have a substantial impact on the economic performance of the farm due to a reduced milk yield and an increase in involuntary culling [2,13,14]. Additionally, antimicrobials and other pharmaceutical substances are used to mitigate economic losses reinforcing concerns about antimicrobial resistance and residues [15]. An important and often underestimated cost due to diseases is the cost of culling. The term culling is used in different ways in the

literature [16]. Referring to the definition used in Germany, the term culling here includes on-farm death of cows and all sales for slaughter. Sales for breeding purposes are excluded. Voluntary and involuntary culling needs to be regarded as distinctively different. Voluntary culling occurs when the farmer decides to replace a cow for reasons other than disease or injury. This may be economically desirable when a cow has exceeded the peak of her milk production which is usually after her fifth parity [17]. However, few cows reach this age. According to Hare et al. [18], the average number of completed lactations for cows leaving the herd in the USA is three with a trend towards a shorter productive life. Vries and Marcondes [17] refer to an average productive life of 2.5 to 4 years in developed countries and of less than 3 years in the USA. Dairy cows have an average productive life of 3 lactations and are on average 5.4 years old when culled in Germany [19]. About 35% of the dairy herd is replaced each year in Germany [20]. Overton and Dhuyvetter [21] reported an average herd turnover due to mortality and culling for 50 US dairy herds of 39%, ranging from 25 to 51%. If 50% of calves are female, even raising every single female calf as a replacement heifer for a culled cow would not be enough to keep the herd size stable where the replacement rate exceeds one third [21]. Only the use of sex-sorted semen resulting in more female calves or purchasing heifers can close the gap. Raising heifers requires substantial resources in feed, labor, and housing. Increased culling rates require substantially greater numbers of heifers to be raised and in consequence a higher consumption of resources in order to provide enough replacements [22,23].

1.2. Insufficient Knowledge Transfer Regarding Animal Health

From the perspective of animal science, huge efforts have been made in order to gain knowledge on factors affecting the health of dairy cows. Several research projects investigated production diseases from different perspectives such as animal nutrition and metabolism [9,24,25], economics [15,26], veterinary science [27–29] and breeding and genetics [30–32] in addition to fundamental approaches [33]. Further research in the field of social science [34–36], veterinary advice [37–39] and agricultural extension and knowledge exchange [40–42] focused on barriers to the implementation of new knowledge aiming to reduce the prevalence of production diseases and discussed ethical perspectives [7,43,44]. Some studies explicitly addressed the systemic nature of animal health in the farming context [45] and followed a transdisciplinary and participatory research design involving expertise from different disciplines as well as stakeholders' knowledge [46,47].

Despite the vast amount of knowledge on hygiene, nutrition, milking technology etc. that is accessible in the literature and has been disseminated [33], the levels of (subclinical) production diseases remain high in modern dairy farms [48–51]. The cause(s) for the perceived "know-do" gap, i.e., the gap between what is known and what is done [52] are numerous. The linear model of knowledge transfer is often criticised for not being able to adequately deal with complex real-world situations [53–55]. A lack of success in transferring knowledge was disclosed in other domains where the impact of new knowledge was intended to have an effect in practice [56,57]. The complexity of the interactions between biological (animal) and social systems (farm) hampers the implementation of knowledge to reduce the level of production diseases in livestock farming [47,58]. In agriculture the complex nature of animal health, the significance of the (farm) context and the socio-economic environment is seldom accounted for [59].

1.3. Complexity in Dairy Farming

Although the concept of different systemic levels in dairy farming such as biological (cows) and socio-economic (farm) systems is commonly accepted (Figure 1), it is rarely accounted for in research and extension, where increases in performance are preferred to improvements in sustainability [59].

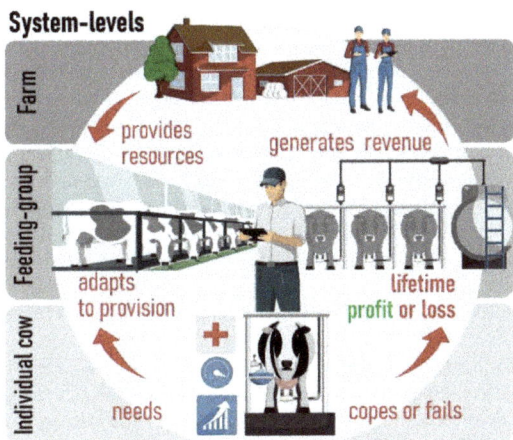

Figure 1. Concept of system levels in a farm relevant for biological and economic sustainability.

Accounting for the systemic nature of dairy farming is at the core of this study. At the level of the farm as well as at the level of the individual animal, the aim is self-preservation [9,60]. At the same time, the sub- and superordinate levels are mutually supporting and dependent. The enterprise revenue from marketable products depends largely on the amount of milk produced as the aggregated output of the individual cows. Each cow represents a single biological system with several functional regulatory circuits (e.g., metabolism and behaviour) that aim to sustain the system [61–63], thereby relying on sub-systems such as organs or the immune system and at the same time on superordinate levels of the group and the farm (social system, resources). Lactating cows are challenged by an increasing energy demand due to the onset of milk production. Cows adjust to this demand through massive changes in their metabolism, prioritising energy (glucose) flow to the udder and mobilizing body reserves from fat and tissue [64,65]. Consequently, many cows suffer from production diseases such as ketosis, uterine and udder infections in early lactation [64,66,67]. However, cows vary considerably in their metabolism and related factors such as milk yield, feed intake and loss of body weight [68,69]. For practical reasons cows are treated as a group of animals when it comes to housing and feeding [70]. Consequently, feeding strategies are generally targeted to the average energy need of cows in a group or herd, rather than the specific needs of single individuals which vary from the groups average [71].

The detection of production diseases depends on signals given by the biological systems of the cows and on a receiver that is sensitive to these signals and able to decide and act on these information (which depends on subjective personal knowledge, attitudes and the availability of resources) [72,73]. Some signals can be found in the milk which is tested regularly for specific constituents. Data on yield and content of fat, protein and from somatic cell counts amongst others provides information on the cows' health [74,75].

Consideration of the dairy herd usually relates to all the dairy cows currently living in the herd (e.g., for calculating energy requirements to adjust feed). However, the reduced lifespan of dairy cows with about 35% of a dairy herd replaced each year in Germany [20] means that information on productivity for a specific time frame (e.g., a year) includes a significant proportion of data from cows that have died. These cows form the "dead herd" of a farm. From an animal welfare perspective these mainly involuntarily culled cows failed to cope with their environmental conditions [10]. From an economic perspective, the "dead herd" must be paid for by rearing heifers. Although genetic opportunity costs must be considered, the costs associated with high rates of herd turnover are a major barrier to economic success [23].

The farm is an important unit and level when it comes to changing environmental conditions for the dairy herd to allow more animals to sustain. At the same time, efforts related to such changes have to be justified in terms of the farms' economic sustainability. Financial analysis for taxation is based on the farm level, whereas the efficiency of each enterprise within a farm may be calculated for management and controlling reasons by allocating the financial surplus and costs to each enterprise. At the individual enterprise level, the accounting of costs and revenues is based on accumulated and averaged figures such as the total milk yield in a year and the average milk produced per cow and year. The regulatory environment and the market for products are shaped by society and politics and affected by social perceptions on the sociocultural sustainability of dairy farming [76].

1.4. Challenges in Evaluating the Contribution of Individual Cows to the Economic Sustainability of a Farm

Management efforts quite often focus on ways to improve or ensure a high milk average yield per cow per lactation or year. It is often used as an indicator or even an objective in farm management, extension and for ranking/comparing farms. However, the costs and revenues of a dairy enterprise on a farm with other enterprises cannot easily be allocated to individual animals by simple division. Subject to individual factors such as the age at first calving, milk yield, days in the herd, and diseases, the contribution of an individual cow to the dairy enterprise's total result show substantial variation. The costs of diseases include the expenditures for diagnosis and treatment which show up in the financial data at herd level. In addition, production diseases are associated with reduced milk yield (e.g., due to changes in the milk glandular tissue subsequent to a mastitis), milk discarded several days after treatments to avoid drug residues and reduced life expectancy of cows due to on-farm mortality and involuntary culling and the subsequent greater need for replacement heifers. These costs are summarized in the term 'failure costs' [15,77]. These costs vary between the cows.

From the economic point of view, individual cows can only contribute positively to the financial performance of the farm system when they are able to reach beyond their individual "break-even point". That is, when the revenues for milk and slaughter value exceed the costs for raising the heifer, the full costs for feed and keep and the proportional share of the fix costs of the farm. Those cows are referred to as "profit cows" from here on in this article. A recent study developed approaches the farm level and the individual cow simultaneously by addressing the need to sustain performance at both levels. The core of the concept is the assignment of revenues and costs to the individual animal [78,79].

A survey of 32 milk producing farms in Germany revealed that a considerable share (about 60%) of culled dairy cows had not reached their break-even point [78]. The proportion of profit cows in the herd (i.e., culled cows which generated more revenue than costs during lifetime) varied between farms (0–74.1%).

Taking the proportion of profit cows as an indicator for economic sustainability, the following assessment examined the effects of production diseases on the farms' economic sustainability. Aiming to account for varying contextual conditions in different farm situations, we aimed to develop a typology of farms, which would generate roughly homogenous groups of farms regarding patterns of milk production and production diseases based on information from milk recording data. This would group farms according to their emerging output regarding yield and production diseases rather than certain input factors such as farm size, production- or milking system. Furthermore, the typology raises the question if the proportion of profit cows was different between groups and which consequences could be drawn for the farm management.

2. Materials and Methods

2.1. Farms and Animals

Milk recording and economic data were collected over the whole financial year from each of 32 dairy farms in Germany between May 2017 and July 2018. Farms were selected as a convenience sample to cover different farm structures and sizes (Table 1).

Table 1. Structure of farms included in the study (n = 32).

Class	n	Farm Size [a] mean	Farm Size [a] min–max	Herd Size [b] mean	Herd Size [b] min–max	Milk Yield [c] mean	Milk Yield [c] min–max	Organic Farms [d] number
<195 cows	11	138	60–290	138	95–183	8030	5557–10,647	4
195–800 cows	10	1103	130–2346	453	239–792	9428	7581–12,739	1
>800 cows	11	2225	750–5000	1327	862–2198	8551	7171–10,793	0

[a] arable land (ha). [b] number of cows after first calving (lactating and dry). [c] kg/305 days in milk. [d] Council Regulation (EC) No 834/2007 [80].

Included were five organic dairy farms. All farms used the herd management program HERDE® (dsp agrosoft, Ketzin, Germany) and were located in different regions in Germany. Holstein Friesian cows were the predominant breed (28 farms). Simmental and Brown Swiss were also kept (2 farms each). All farms kept their cows in loose pens, nine farms offered grazing for lactating cows. Total mixed rations were used in 21 farms and 11 farms followed other feeding regimes such as feeding additional concentrate. Automated milking systems were used in four farms (two farms < 195 cows, two farms 195–800 cows).

2.2. Database and Calculation of Cow Data

Information at the cow level was documented in the herd management software and included: (1) lifecycle data of each cow, including birth date, first calving, age at first calving, last calving, culling date and reasons, lactation number (Lact) and days in milk (DIM). For cows culled during the observation period, DIM_{LL} refers to the day of death in their last lactation. (2) milk yield records of the observation period as well as yield data from previous lactations. Milk recording data from monthly or bimonthly milking records of the period monitored (31 and 1 farm/s respectively) included information on somatic cell count (SCC), fat and protein for each test day. Based on these data, common indices were calculated for each cow: total and daily milk yield during lifetime (MY_L kg resp. kg/day), total and daily milk yield in productive live (MY_{PL} kg resp. kg/day), 305-d milk yield; and average daily milk yield during last/culling lactation (MY_{LL}; kg/day). Milk yield was calculated using the test-day-records based on the German ADR system according to the "Test Interval method" described in the ICAR Guidelines [81]. Overall, data on 20,644 cows were available. Of these 4962 (24%) were cows culled during the observation periods.

Information from cow individual milk recording data were aggregated at farm level to categorise individual cows in relation to certain thresholds reached at least once in one lactation that ended or began in the year of observation. These thresholds related to milk fat and protein (fat > 5%, protein < 3%, fat-protein ratio (FPR) > 1.5, fat-protein ratio at the first test day after calving (FPR_1) >1.5) and SCC (>100,000 cells/mL, three consecutive test days >700,000 cells/mL). The cure rate was calculated from cows with a SCC of more than 100,000 cells/mL before drying off and less than 100,000 cells/mL at the first test day after calving. Variables were calculated separately for culled and retained cows; the latter were those cows that were living in the herd by the end of the observation period.

2.3. Calculation of Costs

Data at the farm level included information on the herd size. Based on the sum of the days individual cows were present in the herd during the observation period, cow-years were calculated by dividing the total sum from all cows by 365 days (observation period). Days present was based on test-day milk record information and calving dates.

Data relevant to farm economic performance were collected from financial accounting data. Farm-specific enterprise accounts were developed, following a widely used farm business budget approach of enterprises [82] that summarizes revenues from milk and slaughter as well as feed and other production costs. Factor costs for labour and capital were included. Full costs and revenues of the observation period were used to calculate the average farm-specific milk price (cent/kg), farm-specific slaughter value (EUR per cow), farm-specific average rearing costs (EUR per heifer), average farm-specific production costs without costs of rearing dairy heifers (EUR per day). Production costs were divided into feed costs (EUR per day) and other farm-specific production costs to calculate cow individual production costs, accounting for varying feed costs according to differences in milk yield. Individual cow profit (EUR) was calculated as the difference between individual revenues (from milk and slaughter) and individual costs (rearing costs and production costs per day). For more details see Habel et al. [78].

2.4. Typification of Farming Systems

Farms were selected to cover a variety of farming systems in Germany. The farm level is relevant for providing the specific conditions under which cows show certain indicators of production diseases. To establish a typology of farms we applied a factor and cluster analysis following the procedure of six stages described by Köbrich et al. [83]: (1) determination of the specific theoretical framework for typification, (2) selection of variables, (3) collection of data, (4) factor analysis, (5) cluster analyses and (6) validation.

(1) In the first step we decided to use information from milk recording data on yield and specific constituents, since these data are not subjectively biased and are available for individual cows. (2) Daily milk yield during lifetime (MY_L), daily milk yield in productive live (MY_{PL}) and the average daily milk yield during last lactation (MY_{LL}) were selected to cover aspects of yield and include effects from age of first calving and differences in the lengths of dry periods. Variables with information on a fat to protein ratio of more than 1.5 in the first test day after calving and in the first 100 days of lactation (FPR_1 > 1.5%; FPR > 1.5%) and a fat content of more than 5% in the first 100 days of lactation were selected to represent indications of metabolic problems [84,85]. Data from test days with a somatic cell count exceeding 100,000 cells/mL milk (SCC > 100,000), on SCC exceeding 700,000 cells/mL on more than three test days (SCC 3 x > 700,000), and information on the cure rate were selected to represent information on udder ill-health [20,86]. With the exception of cure rate, variables included in the analysis referred to the culled cows. Cure rate refers to all cows in the herd. This was due to the small number of cows in some farms which prevented the calculation of cure rate for the culled cows separately. (3) data were collected as described in Section 2.2. (4) We applied a principal component analysis (PCA) with varimax rotation to extract the most important independent factors from test-day milk records. Kaiser-Meyer-Olkin (KMO) measure of sampling adequacy and Bartlett's test of sphericity were used to assess the suitability of data. Only factors with eigenvalues ≥ 1 were considered [87–89]. Since the sample size of 32 farms was small, the communalities of variables, the number of factors, and the simplicity of the structure were taken into account to justify the application of multivariate statistics [90]. (5) Factors were used in a hierarchical cluster analysis using Ward's method [91] aiming for groups of farms with high internal homogeneity and maximum heterogeneity between groups [83]. Starting from individual cases with this method in each step those clusters are merged, which result in the smallest increase of total variance in the new cluster. The method aims to reduce the loss of homogeneity when combining clusters and leads in consequence to homogenous groups. (6) A comparison of means with ANOVA was used to further examine the identified clusters and their interrelation with the proportion of profit cows and other characteristics of the whole herd (culled and living). p-values below 0.05 were set as an indication of statistical significance. Statistical calculations were performed using IBM® SPSS® Statistics.

3. Results

The Kaiser–Meyer–Olkin measure of sampling adequacy was 0.616, representing a medium sampling adequacy, and Bartlett's test of Sphericity was significant ($p < 0.001$), indicating that correlations between variables were sufficiently large for performing a PCA [87,88]. The principal component analysis yielded three factors with eigenvalues, exceeding 1 which accounted for 78.4% of the total variance. (Table 2).

Table 2. Communalities and Factor loadings, resulting from a principal component analysis on milk recording variables, KMO = 0.616.

	Communalities (Extraction)	Rotated Component Matrix		
		Component		
		"Milk Yield"	"Metabolism"	"Udder Ill-Health"
MY_{PL} ^{CC} (kg/day)	0.880	**0.936**	−0.049	0.029
MY_L ^{CC} (kg/day)	0.879	**0.920**	0.006	0.183
MY_{LL} ^{CC} (kg/day)	0.835	**0.902**	−0.003	−0.144
FPR > 1.5% ^{CC} (%) [1]	0.906	0.003	**0.952**	0.032
FPR_1 > 1.5% ^{CC} (%) [1]	0.884	0.069	**0.933**	0.092
Fat > 5% ^{CC} (%) [1]	0.733	−0.120	**0.846**	−0.045
SCC > 100,000 ^{CC} (%) [1]	0.680	0.079	−0.151	**0.807**
SCC 3 x > 700,000 ^{CC} (%) [1]	0.682	0.187	0.102	**0.798**
Cure rate ^{AC} (%) [2]	0.579	0.330	−0.150	**−0.669**
% of the total variance explained		30.8	28.0	19.6

Extraction method: Principal component analysis. Rotation method: Varimax with Kaiser Normalization. Rotation converged in 4 iterations. ^{CC} = culled cows; ^{AC} = all cows, culled and persistent; MY_L = daily milk yield during lifetime; MY_{PL} = daily milk yield in productive live; MY_{LL} = average daily milk yield during last lactation; FPR = fat to protein ratio in the first 100 days of lactation; FPR_1 = fat to protein ratio in the first test day after calving; SCC = somatic cell count. [1] Proportion of cows that exceeded the threshold in the first 100 days of the lactation on at least one test day. [2] Proportion of cows with SCC < 100,000 cells/mL at the first test day from cows with SCC > 100,000 cells/mL at the last test day before dry-off. Highest loadings for each variable are in bold.

The first component of aggregated variables describing the average daily performance level based on (1) the milk yield per day of living (MY_L), (2) the milk yield per day of milking (MY_{PL}) and (3) their average daily milk yield during last/culling lactation (MY_{LL}), calculated as the milk yield in their last lactation (kg). The second component covered aspects of metabolism, combining information on the percentage of cows showing (1) a FPR above the threshold of 1.5 during the monitoring period (FPR) and (2) in the first test day after calving (FPR_1)) and (3) milk fat above 5% (Fat > 5%). The third component aggregated information on udder ill-health, represented by variables on (1) SCC > 700,000/mL milk in three consecutive test days, (2) the cure rate and (3) the percentage of cows with a SCC above 100,000 at one test day in their last lactation.

Based on the three components, the hierarchical cluster analysis identified five clusters represented in the dendrogram in Figure 2.

Table 3 shows the final cluster centres representing the average value of components in each cluster, based on variable values estimated in the PCA for each case (farm). The factor milk-yield was dominant in defining Cluster 3 (highest absolute value within the cluster); the factor metabolism had a major influence on defining Cluster 1, and Cluster 5, while the factor udder ill-health was most important to define Cluster 2 and Cluster 4.

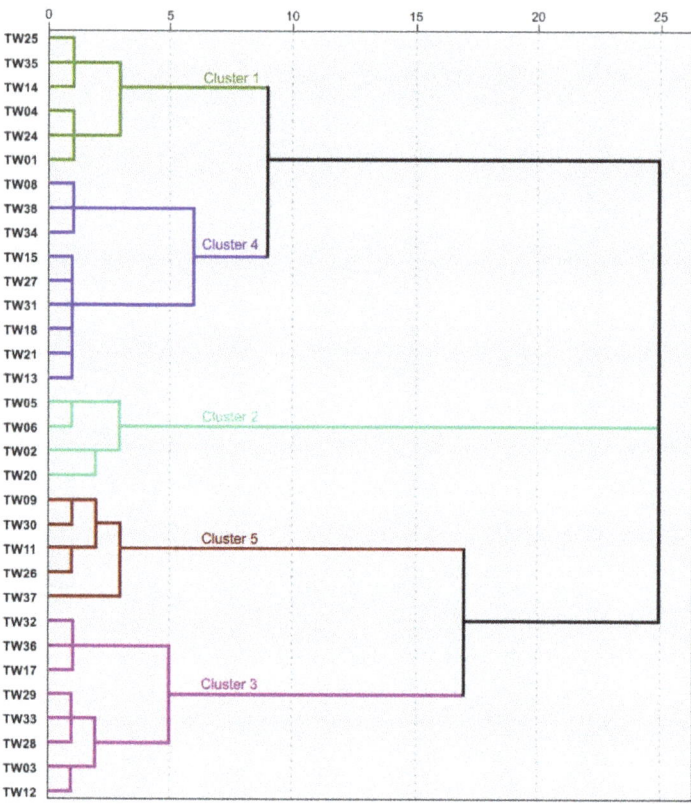

Figure 2. Dendrogram (using Ward linkage) for 32 dairy farms (named TW01–TW38; 6 farms missing) with five clusters.

Table 3. The contribution of the three classification factors to the five cluster centres.

	Cluster				
	1	2	3	4	5
Number of farms	6	4	8	9	5
Milk-yield	0.444	−0.951	1.206	−0.513	−0.777
Metabolism	−0.973	−1.140	0.412	0.043	1.343
Udder ill-health	−0.622	1.632	0.541	−0.919	0.230

Clusters were compared with respect to the proportion of profit cows and other attributes (Table 4). Profit cows accounted for 57.5% of culled cows on six farms in Cluster 1. These farms were characterized by the second-highest value for the factors milk yield and metabolism (referring to a small proportion of cows with an indication of metabolic diseases in their test-day results) and the second lowest values for the factor udder ill-health (i.e., high cure rates and few cows with high cell counts) compared to the other clusters (Table 3). Accordingly, the cluster can be described as *high performer with good health status*. The six farms in this cluster realized high 305-day milk yield in the whole herd (culled and living, 9248 kg) and had a low percentage of cows with an FEQ higher than 1.5 in the first test day in a lactation or test day results showing a fat content > 5%, indicating few cows with metabolic problems. Furthermore, the cure rate in this cluster was the highest

(68.7%). Farms in this cluster had the second highest production costs per day (EUR 9.23) but on average realized a quite high milk price compared to the other clusters (n.s.) and the highest slaughter value (n.s.). One farm in this cluster was an organic farm with a higher milk price than non-organic farms. In this cluster the average culling rate was 29% in this cluster which was slightly lower than the average culling rate of all farms (29.5%).

The second highest proportion of profit cows (55.6%) was detected in farms of Cluster 2. The four farms of Cluster 2 were characterized by the lowest values for the factors milk yield and metabolism and the highest value for the factor udder ill-health (Table 3). The cluster can be described as *low performer with impaired udder health*. This low performing cluster had the largest proportion of cows with increased SCC and the lowest cure rate. On average these farms had the highest average cow age of all clusters (n.s.). The age at first calving (28.5 month) and the last calving interval (426 days) was highest in this cluster. The highest milk price of all clusters (45 c/kg) and combined with the lowest production costs per day (EUR 8.00) contributed to a positive effect on the economic performance. Two farms in this cluster were organic farms with a higher milk price, but also both non-organic farms realized relatively high milk prices (39.2c, 42.0c) above the average milk price for non-organic milk (36.9c). This cluster had the lowest culling rate (27.2%). The cows left the herd later in lactation (DIM_{LL} 258) than in other clusters. Farms in this cluster tended to be smaller.

Cluster 3 ranked at the third position with 44% of profit cows on average, however showing a quite large variation. Cluster 4 aggregated eight farms with the highest value for milk yield and at the same time the second highest values for the factors of udder ill-health and metabolism, indicating higher proportions of cows with impaired udder health and metabolism (Table 3). Accordingly, the cluster is described as *high performer with impaired health status*. This cluster includes the highest performing farms (average 305-d milk yield 10,138 kg/cow/year) and with the lowest age at first calving (25.7 month). The highest production costs per day per cow (EUR 10.20) and the lowest milk price (no organic farm in this cluster) were unfavourable conditions regarding the economic success as well as the effect of the second highest on-farm mortality (6.3%, n.s.). Farms in this cluster managed a low rate of culling of primiparous cows (4.4%).

In the biggest cluster (Cluster 4) with nine farms, less than a third of all culled cows were profit cows (29.4%). This cluster was characterized by the lowest values for the factor udder ill-health (indicating few cows with an indication of impaired udder health in their test-day results) in all clusters. The value for the factor milk yield was below average, while the factor metabolism was at a medium level compared to the other clusters. This cluster is described as *average performer with good udder health*. At a medium level of milking performance these farms showed the best results on SCC indicators, and the second highest cure rate (62.3%). With 3.05 lactations culled cows were the youngest in all clusters (n.s.) and they left the herd early in the lactation (DIM_{LL} 178). This corresponded with the highest culling rate (32%, n.s.), especially for primiparous cows (8.4%). In this cluster, the milk price was at the lowest level for the eight non-organic farms (36.1c). In addition, this group realized the lowest slaughter value (n.s.). With quite some variation, the biggest farms were in this cluster.

In the five farms of Cluster 5 only 19.5% of culled cows were profit cows. The farms were characterized by the highest values for the factor metabolism and second lowest values for milk yield (Table 3). This cluster is described as *poor performer with metabolic problems*. At a low performance level farms in this cluster had the highest proportion (35.4%) of cows showing a fat content of more than 5% in at least one test day result. At the same time, the calving interval (414 days) was above average (404 days) and the second highest in the five clusters. Farms in this cluster had the second highest culling rate (30.9%) and the highest on farm mortality (7.3%), and a high rate of culling of primiparous cows (7.5%).

In summary, farms with a high proportion of profit cows were found in Cluster 1 and Cluster 2 with either high (Cluster 1) or low (Cluster 2) performance levels. The above average milk price plays an important role for farms to realize a larger proportion of profit cows, especially at a low performance level. This differentiated Cluster 2 from Cluster 5. Metabolic problems were low in Cluster 1 and Cluster 2 in contrast to 4 and 5 which were at a high- and low performance level, respectively.

Smaller proportions of profit cows were found in the clusters with the highest culling rate (Cluster 4 and Cluster 5) and high rates of on-farm mortality (Cluster 3 and Cluster 5). Very high milk performance levels were not associated with the highest proportions of profit cows where production costs were high and the milk price low (Cluster 3). Furthermore, death or culling early in the lactation and culling of quite young cows countered higher proportions of profit cows, even with lower production costs (Cluster 4). A very low share of profit cows was associated with high proportions of cows with metabolic disorders as observed from in test day milk results, especially when this was accompanied by a low performance level, even with production costs at a medium level (Cluster 5).

A large share of profit cows was incompatible with high incidences of metabolic problems, a low milk price and/or high culling and mortality rates.

Table 4. The number of farms and the averages ± standards deviation of a range of characteristics of dairy farming systems identified by cluster analysis.

	Eta^2	p	Cluster 1		Cluster 2		Cluster 3		Cluster 4		Cluster 5		Total	
Number of farms			6		4		8		9		5		32	
Profit cows [CC] (%)	0.442	0.003	57.5	±11.9	55.6	±10.6	44.1	±20.9	29.4	±17.5	19.5	±16.7	40.1	±21.0
Cow years [AC]	0.193	0.198	434	±394	111	±55	486	±276	745	±658	398	±275	488	±451
305-d milk yield [AC] (kg)	0.435	0.003	9248	±1128	7580	±786	10,138	±1696	7822	±1126	7876	±1130	8646	±1578
Milk price (ct)	0.243	0.099	0.40 [1]	±0.07	0.45 [2]	±0.05	0.37	±0.02	0.37 [3]	±0.04	0.39 [4]	±0.06	0.39	±0.05
slaughter value (€)	0.123	0.452	732	±121	766	±107	680	±166	636	±129	703	±34	692	±127
production costs/day €	0.367	0.012	9.23	±0.83	8.00	±1.22	10.20	±1.37	8.74	±0.76	9.06	±0.57	9.15	±1.17
FPR > 1.5, 1st test day [AC] (%)	0.400	0.007	9.8	±3.0	8.5	±3.4	17.6	±5.8	15.4	±4.8	16.5	±4.5	14.2	±5.6
FPR > 1.5 [AC] (%)	0.236	0.110	15.8	±6.0	22.7	±12.9	29.1	±11.9	27.1	±7.5	29.8	±11.6	25.4	±10.6
Fat > 5% [AC] (%)	0.429	0.003	18.4	±6.8	24.4	±4.3	30.9	±9.9	32.3	±4.1	35.4	±9.2	28.8	±9.0
SCC > 100,000 [AC] (%)	0.538	0.000	61.1	±8.2	80.1	±3.4	69.2	±8.5	58.2	±5.3	67.2	±7.8	65.6	±9.7
SCC 3 x > 700,000 [AC] (%)	0.553	0.000	3.7	±1.6	9.4	±4.0	6.7	±1.9	3.1	±1.8	5.8	±1.0	5.3	±2.9
Cure rate [AC] (%)	0.462	0.002	68.7	±4.6	46.7	±10.4	58.8	±9.8	62.3	±4.6	57.5	±6.8	59.9	±9.3
Age at first calving [AC]	0.369	0.012	26.2	±1.2	28.5	±1.8	25.7	±0.7	26.0	±1.6	27.0	±1.0	26.4	±1.5
Calving interval [AC]	0.423	0.004	397	±8	426	±23	399	±10	396	±14	414	±11	404	±16
Culling rate (%)	0.076	0.697	29.0	±6.6	27.2	±4.3	27.8	±6.6	31.8	±7.3	30.9	±7.7	29.5	±6.6
Culling rate primiparous (%)	0.366	0.013	7.7	±3.2	4.1	±1.0	4.4	±2.1	8.4	±2.8	7.8	±3.1	6.7	±3.1
On farm mortality	0.099	0.573	5.1	±2.6	5.2	±3.6	6.3	±3.5	4.8	±2.9	7.3	±1.4	5.7	±2.9
Lact [CC]	0.200	0.181	3.3	±0.5	3.9	±0.6	3.4	±0.3	3.1	±0.5	3.5	±0.9	3.4	±0.6
DIM_{LL} [CC]	0.309	0.035	196	±35	258	±54	213	±38	178	±37	204	±38	204	±44

CC = culled cows; AC = all cows, culled and persistent; FPR = fat to protein ratio; SCC = somatic cell count; Lact = lactation number; DIM_{LL} = days in milk in the last/culling lactation [1] one organic farm (53.4c,), five non-organic farms (Ø37.7c). [2] two organic farms (Ø48.8c), two non-organic farms (Ø40.6c). [3] one organic farm (48.5c), eight non-organic farms (Ø36.1c). [4] one organic farm (48.5c), four non-organic farms (Ø36.1c).

4. Discussion

4.1. Test Day Results as Systemic Farm Output

Farms included in this study covered a wide range of farm types for example regarding their herd size, structure of the farm business (family farm/farm cooperative) and production systems (conventional/organic). They provided their dairy cows with conditions resulting in five different patterns of milk yield and proportions of cows showing an indication on metabolic diseases and impaired udder health evident from milk recording data.

The advantage of using milk records lies in the fact that they are recorded routinely monthly for many dairy farms. They provide information on health traits [74,84,92] while not being affected by subjective judgements as is the case for the documentation of diagnoses and treatments. Test-day milk somatic cell count (SCC) is an established indicator for udder infections [93–95]. Various thresholds to classify a cow with udder infection are used worldwide. However, at a level of more than 100,000 cells/mL (SCC100) an inflammation is the likely cause [96] and this level is established to distinguish cows with healthy udder in Germany [20,86]. Even though more accurate blood tests are required for diagnosis when assessing the metabolic health status of an individual cow, high milk fat and low milk protein percent as well as a fat-to-protein ratio above 1.5 are associated with increasing risks of subclinical ketosis [84,85,97] and provide an information on the metabolic status of the herd.

The proportion of cows showing at least one test day with an SCC100 in this investigation was 65.6% and 25.4% of the cows had a fat-to-protein ratio above 1.5 in the early lactation (first 100 days) (Table 4). According to the assessment scheme applied, a cow was counted as a cow with SCC100 if at least one test-day result per lactation reached this threshold. This cow and lactation-based assessment was chosen to capture long time effects of cured inflammations e.g., on reduced milk yield in the ongoing lactation. However, common assessments of udder health indicators focus on the cross-sectional analysis of the herd or cow individual samples at one test day [20,96]. On a test day base, the proportion of milk samples with an increased SCC100 in this study was 36.4% (result not shown). In other studies from Germany, including samples from 2000 to 2008 the proportion was 38% [96] with regional variation between 39.5 and 42.8% in a recent study involving 723 farms. The proportion of cows with a fat-to-protein ratio above 1.5 in early lactation in the same study ranged from 25.0 to 29.7% [20]. On a test day base rather than per cow per lactation the proportion in our data was 23.6% (result not shown). The slightly better results in our data might be affected by the selection of farms, which as a prerequisite had to use the herd management software 'HERDE' by dsp-agrosoft to participate in the study. This could be related to a higher management standard.

While milk recording data can only provide information on some disease complexes and miss others such as lameness, they provide valuable information because they represent an objective measurement of output variables measured on single cows in a herd.

Beside the factor milk yield, the PCA distinguished between factors representing metabolic problems and udder health, which are however related to different management areas at the farm level. Metabolic problems are strongly related to the adequate supply of energy and feeding resources in relation to the milk yield at the cow level [24]. Somatic cell counts (SCC) as a proxy for mastitis are related to various management practices, mostly about milking hygiene, hygienic conditions in the housing, protecting the teats and udder from adverse effects of the milking system and applying control measures as well as effective treatments [75]. However, the effect of various measures was not consistent in different studies, while herd managers attitude on SCC was detected as a meta-factor with an effect on SCC [73,75]. This emphasizes the effect of farm specific conditions and the role of the farmer, steering the system. The proportion of profit cows can function as a starting point for a weak point analysis and the identification of effective measures to improve the situation. It would provide orientation for setting strategic goals in the farm management which serves economic and sociocultural sustainability of dairy farming [76].

4.2. Factor and Cluster Analysis for Farm Typology

A typology of farms aims at homogenous groups of farms out of a diverse range of variables. Multivariate statistics can be used to build a typology of farms [83,98]. The number of farms involved in this study is quite low for the application of multivariate statistics. A sample size of at least 50 cases and/or a specific number of cases per variable is usually recommended [89]. However, according to several authors those propositions were inconsistent and such recommendations on absolute values have gradually been abandoned as misconceived [90,99,100]. Winter et al. [90] revealed that even small sample sizes well below 50 could yield reliable solutions with exploratory factor analysis. For situations with high communalities, few factors, a simple structure, and large number of variables even sample sizes smaller than 10 were sufficient. Small sample sizes should not ban the application of such analysis, since it might reveal valuable latent patterns [90]. In our analysis, the revealed three factors were well defined, with only weak interfactor correlations well below 0.2 for all variables except for the correlation of cure rate with the still low correlation of 0.3. Furthermore, the communality for all variables was above 0.5 with one variable above 0.9, four variables above 0.8 (Table 3). Furthermore, comparing the resulting farm typologies offers explanatory classifications regarding the varying proportion of profit cows for which they were intended. According to Köbrich et al. [83] this confirms the conceptual validity of the farm typology.

Farms in this study were selected to reflect the diversity of dairy farms in Germany, thereby providing the variation needed for segmentation. Furthermore, the data involved were objectively measured. Using these data to create homogenous groups does not alter the information at the farm level. However, the typology classification should not be used to predict results e.g., the proportion of profit cows for individual farms. It shows that for farms of different type (regarding yield and production diseases) different management goals should be implemented to align management strategies to increase or ensure a sufficient proportion of profit cows.

4.3. Profitability at the Cow Level Linked to Production Diseases

As a parameter, the proportion of profit cows brings together economic and welfare criteria at the farm level [79]. Whether a dairy cow reaches her break-even point is affected by cow related factors of production (milk yield) and farm-level economic factors such as the milk price and the farm specific costs for keeping dairy cows [78]. The role of diseases is indirect since diseases affect milk yield, culling decisions, and production costs. At the level of the individual animal, milk yield and early death (lower lactation number and fewer days in milk) determine the economic result of the individual cow [78]. However, from a management perspective the lactation number and especially early culling in the lactation are often the consequence of health problems which are related to metabolic problems, rooted in a negative energy balance in early lactation [24,101]. The results of the cluster analysis indicate that a lower share of cows with indicated metabolic problems in test day results is associated with the bigger share of profit cows in Clusters 1 and Cluster 2. Although milk price was quite high in these clusters (for both conventional and organic production, the low rates of (involuntary) culling and the highest average day of lactation at culling (DIM_{LL}) might explain the higher share of profit cows in those clusters. The smaller proportion of profit cows in Clusters 4 and Cluster 5 seemed to be related to medium milk yield in combination with lower milk prices and an earlier average day of culling (DIM_{LL}). The latter had a significant effect on individual cow profit in some farms [78]. These clusters showed the highest culling rates, especially for primiparous cows. Farms in Cluster 5 also had the highest on-farm mortality of cows. Accordingly, a high proportion of profit cows cannot be achieved with a high rate of culling that reflects the number of cows that ultimately fail to cope within the farming system [79,102]. In the systemic view of dairy farming, the coherence between indicators of metabolic diseases and a lower share of profit cows underlines the nested and interdependent system levels. The lack of energy in early lactation is a major cause for several disease complexes linked to culling decisions. Beside direct consequences of a negative energy balance on metabolic problems, it impacts on reproductive performance and lameness [101,103]. Metabolic problems are a major cause for culling decisions in early lactation [104,105], whereas failure to reproduce and lameness are important culling reasons later in the lactation [106].

While metabolic diseases seemed to have an effect on the proportion of profit cows which could be explained through the effect on early culling in the lactation, an impaired udder health (Cluster 2 and Cluster 4) was not associated with a lower-than-average proportion of profit cows. However, this might be affected by calculating the individual cow profit from the total amount of milk produced (MY_L) rather than the milk sold. Information on discarded milk due to medical treatments (e.g., of udder infections) were not regularly recorded for individual animals and therefore not available for this analysis. At the farm level, however, about 6.4% (0.9–17.0%) of the milk recorded was not sold (data not shown). More data on the discarding of milk from single animals might provide more insight on possible effects of impaired udder health on the proportion of profit cows. The very best results for indicators for udder health were found in Cluster 3 with a low average of 32% of profit cows. It remains an open question whether the lowest proportion of cows with chronical udder diseases (SCC 3 x > 700,000 cells/mL) was a consequence of high culling rates. However, with the 305-day milk yield below average, possible advantages from good udder health were countered by negative effects from the earliest day of culling and the high culling rate.

The cow level is the level where disease parameters are manifested and where action plans must start. Increasing efficiency of dairy farms by intensification was (and still is) a main driver in dairy system policies [107]. Production diseases seem to be regarded as collateral damage: an externality like environmental effects of intensive cropping systems [108]. Negative effects from production diseases on the whole farm system are obscured by aggregated data on costs at the farm level and the lack of information on the costs of invisible failures. Furthermore, the effect of an inefficient dairy enterprise on the whole farm is blurred by EU agricultural payments, which are usually accounted for at the farm level (not the level of the dairy enterprise). Zhu et al. [109] found that a higher degree of public payments in the total farm income reduced the motivation of farmers to improve efficiency. This study points to an approach that production disease are not externalities but should be integrated as an emergent intrinsic factor of production processes and management decisions. The approach provides an option on how to deal with culling and production diseases as an essential intrinsic factor of a farm system, which needs to be addressed appropriately to support self-maintenance of dairy cows and dairy farms.

5. Conclusions

Some realignment is required to overcome the negative side effects established by the productivist approach to dairy farming [108], which culminates (beside the considerable environmental externalities) in a loss of dairy cows. This is due to both the overstressing of the cow's ability to sustain as well as the poor economic results of dairy farms, which is the opposite of sustainability. In the thirty-year period from 1983 to 2013, 6% of the dairy farms in the older EU member states closed down each year amounting to a decline of 81% of farms with dairy cows. Farms of all types decreased by 55% [110] in this period.

Information from milk recording test-day results is a representation of the cows' ability to cope with its environment [9,62] and are in this context the result of complex system interaction. According to Wells and McLean [57] a systemic perception of change requires management that focuses on shaping the environment. From the recognition of the cows failing to cope with their environment as an indication of a lack of animal welfare follows the obligation to design an environment that is better adapted to the needs of the animals. To change the level of production diseases requires the shaping of an environment from which the desired change, a different level of production diseases, may emerge.

The current study highlights the need to shift priorities from milk outputs to a wider range of goals that better sustain the system in the longer term. This takes account of the needs at the different levels in the farming system considering the individual cow level at the same time so that the two levels sustain each other. By taking the single animal into consideration, rather than an average of the herd, the proportion of profit cows is proposed as an indicator of productivity that accounts for the complexity of the dairy farm system. The focus on the economic contribution of cows that left the farm for slaughter or died on the farm addresses the importance of production diseases (the most common reason why cows are culled) for the viability of the farm business. It is a starting point for further analysis (diagnosis) of how cows could be supported, and how the cows' environment could be improved to allow a greater proportion of animals to cope.

We argue that the proportion of profit cows serves as a more sustainable objective for the farm management than other measures of (economic) performance such as 305-day milk yield, milk price or feed costs. It brings more attention to the creation of environmental conditions for the dairy cows that are suited to reduce metabolic stress (Cluster 1 and Cluster 2 with more than 50% of profit cows) and culling (Cluster 4 and Cluster 5 with less than 30% of profit cows). The identification of farm-specific economic benefits, i.e., prevention of losses due to involuntary culling, might foster the awareness for giving attention to single animals and their demands.

The concept of knowledge transfer refers to the paradigm of rational choice, assuming that people will use the information provided (from research findings) to decide on the option with the best utility [56,111,112]. However, social science and psychology have shown that behavioural change does not solely depend on the availability of certain information. Kahneman [113] described the strength of loss aversion as a driver which might support consideration of individual cow profit and the share of profit cows an advantage in supporting changes for improved health and welfare in dairy farms.

The proportion of profit cows was identified for each of the five types of farms characterized by milk yield, an indication of metabolic problems, and impaired udder health. These farm types require different strategic approaches to protect and increase the proportion of profit cows in the herd, thus improving economic performance. Identifying the proportion of profit cows in a farm rather than focusing on average milk production traits, such as 305-day milk yield as a measure for success, uncovers synergies between health and longevity of single animals and economic performance of the farm business. It puts a focus on the context-dependence of output variables and requires and allows for various equifinal individual farm solutions. By this, it qualifies as an approach that deals with the complexity of biological and socio-economical system levels. Future research should assess differences in efficient strategies to increase the proportion of profit cows in various initial and boundary conditions, as reflected by the farm typology. Furthermore, research should improve methods to assess cow individual costs more accurately, e.g., due to discarded milk as well as methods that account for the farm-specificity of both economic and biological conditions.

Raising the proportion of profit cows is a suitable strategic goal to provide orientation for the farm management and validation for implemented measures that consider heterogeneous farming conditions. It addresses the need to shape environmental conditions for the dairy cows to allow for a desired outcome, rather than to strive for measures that generalize farm performance. Joining economic results and animal health and welfare of individual animals is a way to change the perception of production diseases from collateral damage to a cause for losses. This might foster farm individual, iterative change processes, aiming for less production diseases and for a higher farmers' income.

Author Contributions: Conceptualization, A.S.; methodology, J.H., S.H.-T., E.-M.S., D.M.; formal analysis J.H., S.H.-T.; investigation, L.E., V.U., E.-M.S.; resources, T.R.; data curation, J.H., E.-M.S., V.U., L.E. and T.R.; writing—original draft preparation, S.H.-T.; writing—review and editing, A.S., D.M., J.H., E.-M.S., V.U.; supervision, A.S., D.M.; project administration, S.H.-T.; funding acquisition, A.S., D.M., S.H.-T. All authors have read and agreed to the published version of the manuscript.

Funding: The project was supported by funds of the Federal Ministry of Food and Agriculture (BMEL) based on a decision of the Parliament of the Federal Republic of Germany via the Federal Office for Agriculture and Food (BLE) under the Innovation Support Program.

Institutional Review Board Statement: The study was conducted according to the guidelines of the German animal protection act (implementing the Directive 2010/63/EU on the protection of animals used for scientific purposes) and was approved by institutional review. Animals did not undergo any experimental procedure.

Informed Consent Statement: Informed consent was obtained from all subjects involved in the study.

Data Availability Statement: Restrictions apply to the availability of these data. The data are not publicly available due to confidentiality agreements with the providing farmers.

Acknowledgments: We gratefully thank the farms participating in this study for sharing their data, dsp-Agrosoft GmbH (Paretz, Germany) for their support in transferring and preparation of data and Donal Murphy-Bokern for his careful editing of this manuscript.

Conflicts of Interest: The authors declare no conflict of interest. The funders had no role in the design of the study; in the collection, analyses, or interpretation of data; in the writing of the manuscript, or in the decision to publish the results.

References

1. Clark, B.; Stewart, G.B.; Panzone, L.A.; Kyriazakis, I.; Frewer, L.J. A Systematic Review of Public Attitudes, Perceptions and Behaviours Towards Production Diseases Associated with Farm Animal Welfare. *J. Agric. Environ. Ethic.* **2016**, *29*, 455–478. [CrossRef]
2. Van Calker, K.J.; Berentsen, P.; Giesen, G.W.J.; Huirne, R.B.M. Identifying and ranking attributes that determine sustainability in Dutch dairy farming. *Agric. Hum. Values* **2005**, *22*, 53–63. [CrossRef]
3. Segerkvist, K.A.; Hansson, H.; Sonesson, U.; Gunnarsson, S. Research on Environmental, Economic, and Social Sustainability in Dairy Farming: A Systematic Mapping of Current Literature. *Sustainability* **2020**, *12*, 5502. [CrossRef]
4. Sundrum, V.A. Assessment of animal protection services in livestock farming. *Berichte über Landwirtschaft* **2018**, *96*, 1.
5. Hueston, W.D.; van Klink, E.G.; Rwego, I.B. One Health Leadership and Policy. In *Beyond One Health: From Recognition to Results*; Herrmann, J.A., Johnson-Walker, Y.J., Eds.; John Wiley & Sons: Newark, NJ, USA, 2018; pp. 269–278.
6. Krieger, M.; Sjöström, K.; Blanco-Penedo, I.; Madouasse, A.; Duval, J.; Bareille, N.; Fourichon, C.; Sundrum, A.; Emanuelson, U. Prevalence of production disease related indicators in organic dairy herds in four European countries. *Livest. Sci.* **2017**, *198*, 104–108. [CrossRef]
7. Rossi, J.; Garner, S.A. Industrial Farm Animal Production: A Comprehensive Moral Critique. *J. Agric. Environ. Ethic.* **2014**, *27*, 479–522. [CrossRef]
8. Oltenacu, P.A.; Algers, B. Selection for increased production and the welfare of dairy cows: Are new breeding goals needed? *Ambio* **2005**, *34*, 311–315. [CrossRef] [PubMed]
9. Sundrum, A. Metabolic Disorders in the Transition Period Indicate that the Dairy Cows' Ability to Adapt is Overstressed. *Animals* **2015**, *5*, 978–1020. [CrossRef]
10. Broom, D.M. Animal welfare defined in terms of attempts to cope with the environment. *Acta Agric. Scand. Sect. A Anim. Sci. Suppl.* **1997**, *27*, 22–28.
11. Nir, O. The multifactorial approach to fertility problems in dairy herds. In Proceedings of the XXV Jubilee World Buiatrics Congress, WBC 2008, Budapest, Hungary, 6–11 July 2008; pp. 77–81.
12. Payne, J.M. Production disease. *J. R. Agric. Soc. Engl.* **1972**, *133*, 69–86.
13. Bruijnis, M.; Hogeveen, H.; Stassen, E. Assessing economic consequences of foot disorders in dairy cattle using a dynamic stochastic simulation model. *J. Dairy Sci.* **2010**, *93*, 2419–2432. [CrossRef] [PubMed]
14. Hogeveen, H.; van Soest, F.J.; van der Voort, M. Economics for the veterinary practitioner: From burden to blessing. In Proceedings of the 29th World Buiatrics Congress, Dublin, Ireland, 3–8 July 2016; pp. 72–75.
15. Hogeveen, H.; Steeneveld, W.; Wolf, C.A. Production Diseases Reduce the Efficiency of Dairy Production: A Review of the Results, Methods, and Approaches Regarding the Economics of Mastitis. *Annu. Rev. Resour. Econ.* **2019**, *11*, 289–312. [CrossRef]
16. Fetrow, J.; Nordlund, K.; Norman, H. Invited Review: Culling: Nomenclature, Definitions, and Recommendations. *J. Dairy Sci.* **2006**, *89*, 1896–1905. [CrossRef]
17. De Vries, A.; Marcondes, M. Review: Overview of factors affecting productive lifespan of dairy cows. *Animal* **2020**, *14*, s155–s164. [CrossRef]
18. Hare, E.; Norman, H.; Wright, J. Survival Rates and Productive Herd Life of Dairy Cattle in the United States. *J. Dairy Sci.* **2006**, *89*, 3713–3720. [CrossRef]
19. Bundesverband Rind und Schwein e.V. (Ed.) *Rinder- und Schweineproduktion in Deutschland 2019*; Bundesverband Rind und Schwein e.V.: Bonn, Germany, 2020.
20. Hoedemaker, M.; Knubben-Schweizer, G.; Müller, K.E.; Campe, A.; Merle, R. Abschlussbericht: Tiergesundheit, Hygiene und Biosicherheit in Deutschen Milchkuhbetrieben-Eine Prävalenzstudie (PraeRi). Available online: https://www.vetmed.fu-berlin.de/news/_ressourcen/Abschlussbericht_PraeRi.pdf (accessed on 30 June 2020).
21. Overton, M.; Dhuyvetter, K. Symposium review: An abundance of replacement heifers: What is the economic impact of raising more than are needed? *J. Dairy Sci.* **2020**, *103*, 3828–3837. [CrossRef]
22. Nor, N.M.; Steeneveld, W.; Mourits, M.; Hogeveen, H. The optimal number of heifer calves to be reared as dairy replacements. *J. Dairy Sci.* **2015**, *98*, 861–871. [CrossRef]
23. De Vries, A. Economic trade-offs between genetic improvement and longevity in dairy cattle. *J. Dairy Sci.* **2017**, *100*, 4184–4192. [CrossRef]
24. Habel, J.; Sundrum, A. Mismatch of Glucose Allocation between Different Life Functions in the Transition Period of Dairy Cows. *Animals* **2020**, *10*, 1028. [CrossRef]
25. Ingvartsen, K.L.; Moyes, K. Nutrition, immune function and health of dairy cattle. *Animal* **2013**, *7*, 112–122. [CrossRef]
26. van Soest, F.; Santman-Berends, I.M.; Lam, T.J.; Hogeveen, H. Failure and preventive costs of mastitis on Dutch dairy farms. *J. Dairy Sci.* **2016**, *99*, 8365–8374. [CrossRef] [PubMed]
27. Gruber, S.; Mansfeld, R. Herd health monitoring in dairy farms-discover metabolic diseases. An overview. *Tierärztliche Praxis Ausgabe G Großtiere/Nutztiere* **2019**, *47*, 246–255. [CrossRef]

28. Gilbert, R.O. Management of Reproductive Disease in Dairy Cows. *Vet. Clin. N. Am. Food Anim. Pract.* **2016**, *32*, 387–410. [CrossRef] [PubMed]
29. De Vliegher, S.; Fox, L.; Piepers, S.; McDougall, S.; Barkema, H. Invited review: Mastitis in dairy heifers: Nature of the disease, potential impact, prevention, and control. *J. Dairy Sci.* **2012**, *95*, 1025–1040. [CrossRef] [PubMed]
30. König, S.; May, K. Invited review: Phenotyping strategies and quantitative-genetic background of resistance, tolerance and resilience associated traits in dairy cattle. *Animal* **2019**, *13*, 897–908. [CrossRef]
31. Miglior, F.; Fleming, A.; Malchiodi, F.; Brito, L.F.; Martin, P.; Baes, C. A 100-Year Review: Identification and genetic selection of economically important traits in dairy cattle. *J. Dairy Sci.* **2017**, *100*, 10251–10271. [CrossRef]
32. Søndergaard, E.; Sørensen, M.; Mao, I.; Jensen, J. Genetic parameters of production, feed intake, body weight, body composition, and udder health in lactating dairy cows. *Livest. Prod. Sci.* **2002**, *77*, 23–34. [CrossRef]
33. Sundrum, A. Lack of success in improving farm animal health and welfare demands reflections on the role of animal science. *J. Sustain. Organ. Agric. Syst.* **2020**, *70*, 11–15. [CrossRef]
34. Jones, P.; Sok, J.; Tranter, R.; Blanco-Penedo, I.; Fall, N.; Fourichon, C.; Hogeveen, H.; Krieger, M.; Sundrum, A. Assessing, and understanding, European organic dairy farmers' intentions to improve herd health. *Prev. Vet. Med.* **2016**, *133*, 84–96. [CrossRef]
35. Garforth, C. Livestock Keepers' Reasons for Doing and Not Doing Things Which Governments, Vets and Scientists Would Like Them to Do. *Zoonoses Public Health* **2015**, *62*, 29–38. [CrossRef]
36. Jones, P.; Marier, E.; Tranter, R.; Wu, G.; Watson, E.; Teale, C. Factors affecting dairy farmers' attitudes towards antimicrobial medicine usage in cattle in England and Wales. *Prev. Vet. Med.* **2015**, *121*, 30–40. [CrossRef] [PubMed]
37. Lam, T.J.G.M.; Jansen, J.; Borne, B.V.D.; Renes, R.J.; Hogeveen, H. What veterinarians need to know about communication to optimise their role as advisors on udder health in dairy herds. *N. Z. Vet. J.* **2011**, *59*, 8–15. [CrossRef] [PubMed]
38. Janssen, J.; Steuten, C.; Renes, R.-J.; Aarts, N.; Lam, T. Debunking the myth of the hard-to-reach farmer: Effective communication on udder health. *J. Dairy Sci.* **2010**, *93*, 1296–1306. [CrossRef] [PubMed]
39. Hogeveen, H.; Lam, T. *Udder Health and Communication*; Springer: Berlin/Heidelberg, Germany, 2011.
40. Blanco-Penedo, I.; Sjöström, K.; Jones, P.; Krieger, M.; Duval, J.; van Soest, F.; Sundrum, A.; Emanuelson, U. Structural characteristics of organic dairy farms in four European countries and their association with the implementation of animal health plans. *Agric. Syst.* **2019**, *173*, 244–253. [CrossRef]
41. Hoischen-Taubner, S.; Bielecke, A.; Sundrum, A. Different perspectives on animal health and implications for communication between stakeholders. In farming systems facing global challenges: Capacities and strategies. In Proceedings of the 11th European IFSA Symposium, Berlin, Germany, 1–4 April 2014.
42. Brinkmann, J.; March, S.; Winckler, C. 'Stable Schools' to promote animal health in organic dairy farming-First results of a pilot study in Germany. *Agric. For. Res.* **2012**, *360*, 128–131.
43. Alrøe, H.F.; Vaarst, M.; Kristensen, E.S. Does Organic Farming Face Distinctive Livestock Welfare Issues?—A Conceptual Analysis. *J. Agric. Environ. Ethic.* **2001**, *14*, 275–299. [CrossRef]
44. Callicott, J.B. The metaphysical transition in farming: From the newtonian-mechanical to the eltonian ecological. *J. Agric. Environ. Ethic.* **1990**, *3*, 36–49. [CrossRef]
45. Krieger, M.; Hoischen-Taubner, S.; Emanuelson, U.; Blanco-Penedo, I.; De Joybert, M.; Duval, J.E.; Sjöström, K.; Jones, P.J.; Sundrum, A. Capturing systemic interrelationships by an impact analysis to help reduce production diseases in dairy farms. *Agric. Syst.* **2017**, *153*, 43–52. [CrossRef]
46. Krieger, M.; Jones, P.J.; Blanco-Penedo, I.; Duval, J.E.; Emanuelson, U.; Hoischen-Taubner, S.; Sjöström, K.; Sundrum, A. Improving Animal Health on Organic Dairy Farms: Stakeholder Views on Policy Options. *Sustainability* **2020**, *12*, 3001. [CrossRef]
47. Hoischen-Taubner, S.; Bielecke, A.; Sundrum, A. Knowledge transfer regarding the issue of animal health. *Org. Agric.* **2018**, *8*, 105–120. [CrossRef]
48. Brunner, N.; Groeger, S.; Raposo, J.C.; Bruckmaier, R.M.; Gross, J.J. Prevalence of subclinical ketosis and production diseases in dairy cows in Central and South America, Africa, Asia, Australia, New Zealand, and Eastern Europe1. *Transl. Anim. Sci.* **2019**, *3*, 84–92. [CrossRef]
49. Levison, L.; Miller-Cushon, E.; Tucker, A.; Bergeron, R.; Leslie, K.; Barkema, H.; Devries, T. Incidence rate of pathogen-specific clinical mastitis on conventional and organic Canadian dairy farms. *J. Dairy Sci.* **2016**, *99*, 1341–1350. [CrossRef] [PubMed]
50. Compton, C.; Heuer, C.; Thomsen, P.T.; Carpenter, T.; Phyn, C.; McDougall, S. Invited review: A systematic literature review and meta-analysis of mortality and culling in dairy cattle. *J. Dairy Sci.* **2017**, *100*, 1–16. [CrossRef] [PubMed]
51. Knaus, W. Dairy cows trapped between performance demands and adaptability. *J. Sci. Food Agric.* **2009**, *89*, 1107–1114. [CrossRef]
52. Pakenham-Walsh, N. Learning from one another to bridge the "know-do gap". *BMJ* **2004**, *329*, 1189. [CrossRef]
53. Davies, H.; Nutley, S.; Walter, I. Why 'knowledge transfer' is misconceived for applied social research. *J. Health Serv. Res. Policy* **2008**, *13*, 188–190. [CrossRef] [PubMed]
54. Scoones, I. *Beyond Farmer First: Rural People's Knowledge, Agricultural Research and Extension Practice*; Intermediate Technology Publ.: London, UK, 1994.
55. Roux, D.J.; Rogers, K.H.; Biggs, H.; Ashton, P.J.; Sergeant, A. Bridging the science-management divide: Moving from unidirectional knowledge transfer to knowledge interfacing and sharing. *Ecol. Soc.* **2006**, *11*, 1. [CrossRef]
56. Chapman, K. *Complexity and Creative Capacity Rethinking Knowledge Transfer, Adaptive Management and Wicked Environmental Problems*; Routledge: London, UK, 2016.

57. Wells, S.; McLean, J. One Way Forward to Beat the Newtonian Habit with a Complexity Perspective on Organisational Change. *Systems* **2013**, *1*, 66–84. [CrossRef]
58. Hoischen-Taubner, S.; Sundrum, A. Hemmnisse im Wissenstransfer zur Tiergesundheit: Ergebnisse und Implikationen eines Reflexionsprozesses mit Stakeholdern. *Berichte über Landwirtschaft Zeitschrift für Agrarpolitik und Landwirtschaft* **2018**, *96*, 1–38. [CrossRef]
59. Eshuis, J.; Stuiver, M. Learning in context through conflict and alignment: Farmers and scientists in search of sustainable agriculture. *Agric. Hum. Values* **2005**, *22*, 137–148. [CrossRef]
60. Maturana, H.R.; Varela, F.J. Autopoiesis and Cognition: The Realization of the Living. In *Boston Studies in the Philosophy of Science*; Cohen, R.S., Wartofsky, M.W., Eds.; D. Riedel Publishing Company: Boston, MA, USA, 1980.
61. Broom, D.M.; Gillmor, R. *Biology of Behaviour: Mechanisms, Functions and Applications. With Animal Drawings by Robert Gillmor*; Cambridge University Press: London, UK, 1981.
62. Broom, D.M. Behaviour and welfare in relation to pathology. *Appl. Anim. Behav. Sci.* **2006**, *97*, 73–83. [CrossRef]
63. Thompson, T. Relations among Functional Systems in Behavior Analysis. *J. Exp. Anal. Behav.* **2007**, *87*, 423–440. [CrossRef] [PubMed]
64. Leblanc, S. Monitoring Metabolic Health of Dairy Cattle in the Transition Period. *J. Reprod. Dev.* **2010**, *56*, S29–S35. [CrossRef] [PubMed]
65. Friggens, N.; Brun-Lafleur, L.; Faverdin, P.; Sauvant, D.; Martin, O. Advances in predicting nutrient partitioning in the dairy cow: Recognizing the central role of genotype and its expression through time. *Animal* **2013**, *7*, 89–101. [CrossRef]
66. Mallard, B.; Dekkers, J.; Ireland, M.; Leslie, K.; Sharif, S.; VanKampen, C.L.; Wagter, L.; Wilkie, B. Alteration in immune responsiveness during the peripartum period and its ramification on dairy cow and calf health. *J. Dairy Sci.* **1998**, *81*, 585–595. [CrossRef]
67. Burvenich, C.; Bannerman, D.; Lippolis, J.; Peelman, L.; Nonnecke, B.; Kehrli, M.; Paape, M. Cumulative Physiological Events Influence the Inflammatory Response of the Bovine Udder to Escherichia coli Infections During the Transition Period. *J. Dairy Sci.* **2007**, *90*, E39–E54. [CrossRef]
68. Zachut, M.; Moallem, U. Consistent magnitude of postpartum body weight loss within cows across lactations and the relation to reproductive performance. *J. Dairy Sci.* **2017**, *100*, 3143–3154. [CrossRef]
69. Ollion, E.; Ingrand, S.; Delaby, L.; Trommenschlager, J.; Colette-Leurent, S.; Blanc, F. Assessing the diversity of trade-offs between life functions in early lactation dairy cows. *Livest. Sci.* **2016**, *183*, 98–107. [CrossRef]
70. Bewley, J.; Robertson, L.; Eckelkamp, E. A 100-Year Review: Lactating dairy cattle housing management. *J. Dairy Sci.* **2017**, *100*, 10418–10431. [CrossRef]
71. Grant, R.; Albright, J. Effect of Animal Grouping on Feeding Behavior and Intake of Dairy Cattle. *J. Dairy Sci.* **2001**, *84*, E156–E163. [CrossRef]
72. Borne, B.V.D.; Jansen, J.; Lam, T.; Van Schaik, G. Associations between the decrease in bovine clinical mastitis and changes in dairy farmers' attitude, knowledge, and behavior in the Netherlands. *Res. Vet. Sci.* **2014**, *97*, 226–229. [CrossRef]
73. Jansen, J.; Borne, B.V.D.; Renes, R.; van Schaik, G.; Lam, T.; Leeuwis, C. Explaining mastitis incidence in Dutch dairy farming: The influence of farmers' attitudes and behaviour. *Prev. Vet. Med.* **2009**, *92*, 210–223. [CrossRef]
74. Toni, F.; Vincenti, L.; Grigoletto, L.; Ricci, A.; Schukken, Y.H. Early lactation ratio of fat and protein percentage in milk is associated with health, milk production, and survival. *J. Dairy Sci.* **2011**, *94*, 1772–1783. [CrossRef]
75. Dufour, S.; Fréchette, A.; Barkema, H.W.; Mussell, A.; Scholl, D.T. Invited review: Effect of udder health management practices on herd somatic cell count. *J. Dairy Sci.* **2011**, *94*, 563–579. [CrossRef]
76. Boogaard, B.; Oosting, S.; Bock, B.; Wiskerke, J. The sociocultural sustainability of livestock farming: An inquiry into social perceptions of dairy farming. *Animal* **2011**, *5*, 1458–1466. [CrossRef]
77. Langford, F.; Stott, A. Culled early or culled late: Economic decisions and risks to welfare in dairy cows. *Anim. Welf.* **2012**, *21*, 41–55. [CrossRef]
78. Habel, J.; Uhlig, V.; Hoischen-Taubner, S.; Schwabenbauer, E.-M.; Rumphorst, T.; Ebert, L.; Möller, D.; Sundrum, A. In-come over service life cost—Estimation of individual profitability of dairy cows at time of death reveals farm-specific eco-nomic trade-offs. *Livest. Sci.* **2021**, in press.
79. Sundrum, A.; Habel, J.; Hoischen-Taubner, S.; Schwabenbauer, E.-M.; Uhlig, V.; Möller, D. Anteil Milchkühe in der Gewinnphase—Meta-Kriterium zur Identifizierung tierschutzrelevanter und ökonomischer Handlungsnotwendigkeiten. *Berichte über Landwirtschaft Zeitschrift für Agrarpolitik und Landwirtschaft* **2021**. [CrossRef]
80. Council Regulation (EC) No 834/2007 of 28 June 2007 on Organic Production and Labelling of Organic Products and Repealing Regulation (EEC) No 2092/91. Available online: https://www.legislation.gov.uk/eur/2007/834/introduction (accessed on 26 June 2021).
81. International Committee for Animal Recording. Procedure 2 of Section 2 of ICAR Guidelines Computing of Accumulated Lactation Yield, 2017. Available online: https://www.icar.org/Guidelines/02-Procedure-2-Computing-Lactation-Yield.pdf (accessed on 26 June 2021).
82. Deutsche Landwirtschafts-Gesellschaft (DLG) (Ed.) *Die neue Betriebszweigabrechnung: Ein Leitfaden für die Praxis; Vorschlag für bundeseinheitliche Gestaltungen von Betriebszweigabrechnungen auf der Grundlage des BMVEL-Jahresabschlusses*; DLG-Verlag: Frankfurt am Main, Germany, 2011.

83. Köbrich, C.; Rehman, T.; Khan, M. Typification of farming systems for constructing representative farm models: Two illustrations of the application of multi-variate analyses in Chile and Pakistan. *Agric. Syst.* **2003**, *76*, 141–157. [CrossRef]
84. Duffield, T.F.; Kelton, D.F.; Leslie, E.K.; Lissemore, K.D.; Lumsden, J.H. Use of test day milk fat and milk protein to detect subclinical ketosis in dairy cattle in Ontario. *Can. Vet. J.* **1997**, *38*, 713–718.
85. Gross, J.J.; Bruckmaier, R.M. Review: Metabolic challenges in lactating dairy cows and their assessment via established and novel indicators in milk. *Animal* **2019**, *13*, s75–s81. [CrossRef]
86. German Veterinary Association. *Leitlinien: Bekämpfung der Mastitis des Rindes als Bestandsproblem*; Verlag der DVG Service: Gießen, Germany, 2012.
87. Guttman, L. Some necessary conditions for common-factor analysis. *Psychometrika* **1954**, *19*, 149–161. [CrossRef]
88. Kaiser, H.F. The Application of Electronic Computers to Factor Analysis. *Educ. Psychol. Meas.* **1960**, *20*, 141–151. [CrossRef]
89. Hair, J.F. *Multivariate Data Analysis*; Pearson Education Ltd.: Harlow, UK, 2014.
90. de Winter, J.; Dodou, D.; Wieringa, P.A. Exploratory Factor Analysis with Small Sample Sizes. *Multivar. Behav. Res.* **2009**, *44*, 147–181. [CrossRef]
91. Ward, J.H. Hierarchical Grouping to Optimize an Objective Function. *J. Am. Stat. Assoc.* **1963**, *58*, 236–244. [CrossRef]
92. Viguier, C.; Arora, S.; Gilmartin, N.; Welbeck, K.; O'Kennedy, R. Mastitis detection: Current trends and future perspectives. *Trends Biotechnol.* **2009**, *27*, 486–493. [CrossRef] [PubMed]
93. Lipkens, Z.; Piepers, S.; De Visscher, A.; De Vliegher, S. Evaluation of test-day milk somatic cell count information to predict intramammary infection with major pathogens in dairy cattle at drying off. *J. Dairy Sci.* **2019**, *102*, 4309–4321. [CrossRef]
94. Alhussien, M.N.; Dang, A. Milk somatic cells, factors influencing their release, future prospects, and practical utility in dairy animals: An overview. *Vet. World* **2018**, *11*, 562–577. [CrossRef]
95. Ruegg, P.L. A 100-Year Review: Mastitis detection, management, and prevention. *J. Dairy Sci.* **2017**, *100*, 10381–10397. [CrossRef]
96. Schwarz, D.; Diesterbeck, U.; Failing, K.; König, S.; Brügemann, K.; Zschöck, M.; Wolter, W.; Czerny, C.-P. Somatic cell counts and bacteriological status in quarter foremilk samples of cows in Hesse, Germany—A longitudinal study. *J. Dairy Sci.* **2010**, *93*, 5716–5728. [CrossRef]
97. Xu, W.; Saccenti, E.; Vervoort, J.; Kemp, B.; Bruckmaier, R.M.; van Knegsel, A.T. Short communication: Prediction of hyperketonemia in dairy cows in early lactation using on-farm cow data and net energy intake by partial least square discriminant analysis. *J. Dairy Sci.* **2020**, *103*, 6576–6582. [CrossRef]
98. Dos Santos, M.L. Segmenting farms in European Union. *Agric. Econ. Zeměd. Ekon.* **2013**, *59*, 49–57. [CrossRef]
99. Jackson, D.L. Sample Size and Number of Parameter Estimates in Maximum Likelihood Confirmatory Factor Analysis: A Monte Carlo Investigation. *Struct. Equ. Model. Multidiscip. J.* **2001**, *8*, 205–223. [CrossRef]
100. Maccallum, R.C.; Widaman, K.; Zhang, S.; Hong, S. Sample size in factor analysis. *Psychol. Methods* **1999**, *4*, 84–99. [CrossRef]
101. Esposito, G.; Irons, P.C.; Webb, E.C.; Chapwanya, A. Interactions between negative energy balance, metabolic diseases, uterine health and immune response in transition dairy cows. *Anim. Reprod. Sci.* **2014**, *144*, 60–71. [CrossRef]
102. Oltenacu, P.A.; Broom, D.M. The impact of genetic selection for increased milk yield on the welfare of dairy cows. *Anim. Welf.* **2010**, *19*, 39–49.
103. Bicalho, R.; Machado, V.; Caixeta, L. Lameness in dairy cattle: A debilitating disease or a disease of debilitated cattle? A cross-sectional study of lameness prevalence and thickness of the digital cushion. *J. Dairy Sci.* **2009**, *92*, 3175–3184. [CrossRef]
104. Probo, M.; Pascottini, O.B.; LeBlanc, S.; Opsomer, G.; Hostens, M. Association between metabolic diseases and the culling risk of high-yielding dairy cows in a transition management facility using survival and decision tree analysis. *J. Dairy Sci.* **2018**, *101*, 9419–9429. [CrossRef]
105. Seifi, H.A.; LeBlanc, S.J.; Leslie, K.E.; Duffield, T.F. Metabolic predictors of post-partum disease and culling risk in dairy cattle. *Vet. J.* **2011**, *188*, 216–220. [CrossRef]
106. Pinedo, P.J.; De Vries, A.; Webb, D.W. Dynamics of culling risk with disposal codes reported by Dairy Herd Improvement dairy herds. *J. Dairy Sci.* **2010**, *93*, 2250–2261. [CrossRef]
107. Clay, N.; Garnett, T.; Lorimer, J. Dairy intensification: Drivers, impacts and alternatives. *Ambio* **2020**, *49*, 35–48. [CrossRef]
108. Maréchal, K.; Joachain, H.; Ledant, J.-P. The Influence of Economics on Agricultural Systems: An Evolutionary and Ecological Perspective. Working Papers CEB 08-028. RS, 2008. Available online: https://EconPapers.repec.org/RePEc:sol:wpaper:08-028 (accessed on 3 September 2020).
109. Zhu, X.; Demeter, R.M.; Oude Lansink, A.G. Technical efficiency and productivity differentials of dairy farms in three EU countries: The role of CAP subsidies. *Agric. Econ. Rev.* **2012**, *13*, 66–92. [CrossRef]
110. Augère-Granier, M.-L. The EU Dairy Sector Main Features, Challenges and Prospects: Briefing, 2018. Available online: https://www.europarl.europa.eu/RegData/etudes/BRIE/2018/630345/EPRS_BRI(2018)630345_EN.pdf (accessed on 8 September 2020).
111. Greenhalgh, T.; Wieringa, S. Is it time to drop the 'knowledge translation' metaphor? A critical literature review. *J. R. Soc. Med.* **2011**, *104*, 501–509. [CrossRef] [PubMed]
112. van Kerkhoff, L.; Lebel, L. Linking Knowledge and Action for Sustainable Development. *Annu. Rev. Environ. Resour.* **2006**, *31*, 445–477. [CrossRef]
113. Kahneman, D. *Thinking, Fast and Slow*; Penguin Books: London, UK, 2012.

Article

Designing and Distinguishing Meaningful Artisan Food Experiences

Erin Percival Carter *[ID] and Stephanie Welcomer

Maine Business School, University of Maine, Orono, ME 04469, USA; welcomer@maine.edu
* Correspondence: erin.p.carter@maine.edu; Tel.: +1-(207)-581-4944

Abstract: We examine consumer expectations about how specialty versus conventional food products affect well-being and how small, artisan producers can use that information to design better customer experiences. Drawing on recent work examining the costs and benefits of pleasure- and meaning-based consumption, we investigate whether consumer expectations that specialty products are more meaningful lead to increased desire for additional product information. We selectively sampled from the target market of interest: high-involvement consumers who regularly consume a food (cheese) in both more typical and specialty forms. The authors manipulate product type (typical versus special) within participant and measure differences in expected pleasure and meaning as well as a variety of behaviors related to and preference for additional product information. We find that these high-involvement consumers expect special food products to provide both more meaningful (hypothesized) and more pleasurable consumption experiences (not hypothesized) than typical food products. Consistent with our theory, consumer use of, search for, and preference for additional product information was greater for special products. A causal mediation analysis revealed that expectations of meaning mediate the relationship between product type and utility of product information, an effect which persists controlling for the unexpected difference in expected pleasure.

Keywords: food science; customer experience design; food well-being; food psychology; sustainability; hedonia; eudaimonia; meaningful consumption; artisan products; local food

Citation: Percival Carter, E.; Welcomer, S. Designing and Distinguishing Meaningful Artisan Food Experiences. *Sustainability* **2021**, *13*, 8569. https://doi.org/10.3390/su13158569

Academic Editor: Hossein Azadi

Received: 22 June 2021
Accepted: 26 July 2021
Published: 31 July 2021

Publisher's Note: MDPI stays neutral with regard to jurisdictional claims in published maps and institutional affiliations.

Copyright: © 2021 by the authors. Licensee MDPI, Basel, Switzerland. This article is an open access article distributed under the terms and conditions of the Creative Commons Attribution (CC BY) license (https://creativecommons.org/licenses/by/4.0/).

1. Introduction

There is a growing movement among consumers to know more intimately where their food comes from and how it comes to be on their plate [1–3]. Providing consumers with information about how food was produced, efforts made to reduce negative environmental impacts, effects of local food purchases on developing rural economies, can each foster a deeper connection with food [4]. Consumers who value this connection often search outside of traditional outlets and product varieties to satisfy their needs [5]. For some, this can mean shifting their purchasing from grocery stores and chain restaurants to farmers' markets, harvest festivals, and restaurants specializing in local produce [6]. For others, this might mean trying hyper-local fare that does not fit in any traditional classification, or rediscovering old varieties and food traditions [7]. Consumers are motivated to engage in these costly behaviors at least in part due to the belief that consuming these different kinds of food will lead to improved consumer well-being [8].

Consumer researchers are also increasingly interested in better understanding the relationship between consumption and consumer well-being [9]. The term "well-being" can and has been used to refer to many different constructs: positive affect, life satisfaction, a sense of meaning, optimism, physical health, mental health, financial stability, the ability to regulate one's digital media environment and more. In recent years, consumer researchers have found it useful to distinguish between consumption motivated by a desire to maximize momentary, affective, hedonic pleasure-based aspects of well-being, and consumption motivated by a desire to cultivate a sense of purpose, connectivity, and the eudaimonic meaning-based aspects of well-being [10–13]. Many of the factors that drive consumers to consume less conventional food or food from less conventional outlets appear to be more associated with pursuing purpose, connection, and meaning than affective pleasure yet the former has received less attention in the literature [14].

Building on this insight and recent work attempting to better understand meaningful consumption [11–13,15], in this paper, we provide information useful for scholars, as well as small-scale producers attempting to design specialty food products that cater to consumers' diverse needs. These producers and marketers currently face a dilemma. Should they generalize from best practices and research findings developed for more conventional producers competing based on economies of scale and with commodity products or should they seek to design products and consumption experiences better suited to the unique needs they satisfy? We suggest that smaller-scale agricultural producers are uniquely situated to meet consumers' unmet needs, particularly customers' desire to feel that their food provides not only pleasure but meaning. Building on recent work examining meaningful consumption, we analyze data collected from consumers who actively sought out local and artisanal products to examine the role of information in supporting meaningful consumption.

2. Literature Review

2.1. Consumption and Well-Being

What it means to live a good life has been a topic of discussion among philosophers, scientists, religious figures, and everyday people across the world and throughout history. In recent years, marketers and consumer researchers have paid increasing attention to better understanding how consumers think about their own well-being and how best to pursue a fuller, richer life [16]. For several reasons, not least of which were concerns about measurement validity [17], early work on this idea in consumer literature responded to the call for a field of hedonic psychology [18]. According to this tradition, well-being was best defined and measured as an integrated measure of one's affective experience [19–21]. The logic was that a life featuring relatively more positive affective experiences than negative was preferable and individuals' reported affect were more reliable over time than their reported life satisfaction; thus, identifying and communicating strategies for improving the affective quality of one's life would contribute to our understanding and promotion of well-being in a more reliable manner. Correspondingly, a seminal work in consumer behavior examining differences in motivations for consuming different types of products was Hirschman and Holbrook's [22] work distinguishing between hedonic experiences (characterized by the pursuit of pleasure and affective gratification) and utilitarian experiences (characterized by instrumental need fulfillment). The idea that the motivation for consumption can be categorized as either primarily hedonic or utilitarian has remained a bedrock assumption in much of the marketing literature [23–27].

There is, however, a class of experiences for which the hedonic versus utilitarian distinction seems to be inadequate. "Meaning" can be variously defined—we approach the concept of meaning drawing from a long history of scientific and philosophical work seeking to understand eudaimonia, the Greek word for living a full, meaningful, and deeply satisfying life and actualizing one's human potential [28–30]. Meaning is traditionally measured as subjective evaluations of how meaningful one's life is or the extent to which an experience contributes to one's sense of meaning [10,31,32]. Thus, people's lives and experiences can be said to be "meaningful" or "have meaning" to the extent that they contribute to one's sense of meaning [33].

Consumption, too, can be meaningful. Consumers motivated to pursue meaning select books, vacations, movies, concerts, restaurants, and other consumption experiences with an eye to which options they believe are likely to facilitate meaning making [11,12]. Importantly, pleasure- and meaning-based benefits should not be thought of as existing on one continuum but as orthogonal constructs. In other words, the same experience (eating a really delicious piece of local cheese) can provide a mix of utilitarian (eating this makes me less hungry and I can get back to work without being distracted by hunger), pleasure (eating this is enjoyable and makes me feel happier right now) and meaning (I feel connected to the person who made this cheese and like I am learning something new) based benefits but the requirements and outcomes of each differ [12,13].

To illustrate, imagine three people who consume a piece of dark chocolate at the end of every day. Person 1 consumes chocolate every night because she saw a headline stating that doing so would reduce her risk of heart disease. She does not particularly enjoy dark chocolate, she prefers milk or white chocolate, but she understands that higher percentage cacao provides more significant health benefits. Each night she eats a piece of 85% cacao dark chocolate and then immediately brushes her teeth. This consumer is consuming primarily for utilitarian reasons.

Person 2 consumes chocolate every night as a kind of simple pleasure, a reward to himself for his efforts throughout the day. As he eats, he savors the flavor of the chocolate and the luxury of the moment. He enjoys this nightly ritual for those brief moments and then does not tend to think of it again until the next evening when he has another piece. He is consuming primarily for pleasure-based reasons.

Person 3 consumes chocolate every night because she likes to learn about the flavors, growing practices, and chocolate-making traditions and innovations across the world. She thinks very critically about fair trade standards, the chocolatiers who stay in business thanks to her purchases, and the uniqueness of the flavors in the chocolate that she samples each night. Eating chocolate makes her feel like she is connected to other people and she loves to reflect on how much she has learned about chocolate and the chocolate community over the years. Sometimes she eats a piece and realizes that the flavors are not necessarily her favorite but it does not really dampen the overall experience. Person 3 is consuming primarily for meaning-based reasons.

These vignettes together provide an extreme example of the ways in which the same product might contribute to a person's overall well-being in very different ways. It is important to note that often a consumption experience, particularly in the domain of food, will provide a mix of utilitarian, pleasure-, and meaning-based benefits. Indeed, work on the experiential pleasure of food (EPF) defines EPF as "the enduring cognitive (satisfaction) and emotional (i.e., delight) value consumers gain from savoring in multisensory, communal, and cultural meaning in food experiences [34]." This definition, therefore is consistent with our conceptualization but focuses instead on foods that provide both meaning-based ("enduring cognitive") and pleasure-based (emotional delight) benefits, whereas, in this work, we examine the unique implications of each type of benefit. Nevertheless, consumers can have primary motivations even when multiple benefits are possible. We hope that these vignettes can also help to build the intuition for the unique role that information plays in supporting meaning making.

Meaning making is reliant on expertise and supplementary information in a way that pleasure is not. While most people can enjoy hedonic benefits from consuming a piece of cheese or dark chocolate, the ability to derive meaning from these and other experiences is enhanced by baseline expertise about the product category or supplementary information provided at the time of consumption [12]. This is because while pleasure is a momentary affective assessment (i.e., asking oneself "Am I enjoying myself?"), meaning making is a cognitive process (i.e., asking oneself "What does this mean or change in a meaningful way about me or the way that I think about myself in relation to the world?"). While product knowledge and expertise have little or no effect on a consumer's ability to determine whether she feels relatively good or bad in a moment, they do affect her ability to integrate the consumption experience into her self-concept and her subjective evaluation of whether the experience contributes to a purposeful and self-actualized life. Drawing on this insight, in this work, we examine consumers' lay beliefs [35] about the role of product information in facilitating pleasure and meaning from food, particularly specialty or artisan food produced by small-scale producers. Our focus on lay beliefs about well-being and their implications for consumption is consistent with recent work on meaningful consumption [11–13].

2.2. Unique Challenges of Small-Scale Farmer Producers

The agricultural sector in highly-developed nations has seen a dramatic shift in the composition of both the competitive landscape and the level of specialization necessary to compete in conventional agricultural markets over the past several decades. In both the US and Europe, the number of farms has dramatically decreased and the size of individual farms has increased as farmland and food processing has become increasingly industrialized and commodified [36,37]. Smaller-scale and diversified operations are being pushed out as the profitability of such operations decreases; in 2018, the median farm income in the US was a loss of $1840 [38].

Small-scale farms trying to make it in this climate find themselves faced with a choice: continue to sell their products into the commodity market where the primary competitive advantage is economies of scale or begin to market their products as non-commodity or value-added products [39–41]. While this strategy has the potential to increase profitability for small-scale producers who choose to remain small, it also demands different knowledge, resources, and abilities [42]. What commodity markets lack in flexibility and profitability, they make up for in terms of offering a simplified selling process. When a small-scale farmer shifts from streamlined, B2B, largely pre-determined sales processes to the much more abstract and nuanced B2C market for non-commodity and value-added products, she is forced to perform all of the functions of a major corporation with infinitesimal fractions of the time, resources, or experience. Recent research has demonstrated that this gap in knowledge, skills, and resources necessary to uncover and creatively meet consumer needs is a primary concern of small-scale farmers and craft and artisan food producers [43–45].

Small-scale farmers and craft and artisan food producers are uniquely suited to creatively approach their product design and positioning decisions. Smaller, more nimble producers are better able to ideate, prototype, and implement in quick succession. Small-scale producers who know both the production process and the customer intimately should be able to leverage this unique knowledge effectively. Finally, many small-scale agricultural and craft and artisan food producers were drawn to the work that they do for the same or similar reasons that consumers search for unconventional food products—a desire to live and consume more natural products, a desire to honor the past and preserve agricultural traditions, an urge to live more holistically and healthily. These producers may be able to better empathize with their target markets. Consumers increasingly pursue products not because of the way the product specifically performs but because of the experience that the consumer has during consumption [46,47].

3. Hypothesis Development

In formulating our hypotheses for this study, we drew on work demonstrating that consumers value and pursue meaningful consumption experiences and that these experiences are improved when consumers receive additional information about the product before the consumption experience [13]. Our goal in this study was to build on the observational insights of small-scale agricultural producers and craft and artisanal food makers to provide evidence and guidance on how consumers distinguish between their more specialized offerings and more typical food products, particularly with regard to effects on well-being. To do this, we examine consumer expectations about how specialty versus conventional food products affect well-being and how small, artisan producers can use that information to design better customer experiences. Thus, we offer the following hypotheses:

Hypothesis 1 (H1). *Consumers will report buying and consuming specialty products is more associated with the pursuit of meaning-based benefits than buying and consuming conventional products.*

We also wanted to provide insight that smaller-scale producers could use to design more empathetic consumption experiences that support the pursuit of meaning by examining consumer preferences for information about special versus typical food products. Thus, we offer the following three hypotheses:

Hypotheses 2 (H2). *Consumers will report that they use information on products labels more often when purchasing specialty products than conventional products.*

Hypotheses 3 (H3). *Consumers will report that they seek out additional information about a product from sources other than the product label more often when purchasing specialty products than conventional products.*

Hypotheses 4 (H4). *Consumers will report that their purchase and consumption experience would be improved with access to more information than is typically provided on the label to a greater degree for specialty than for conventional products.*

Finally, we tested whether our data support our hypothesis that expectations of meaning serve as the mechanism explaining differences in the value of product information for specialty and conventional products. We thus offer Figure 1 as a representation of our proposed process and the following mediational process hypothesis.

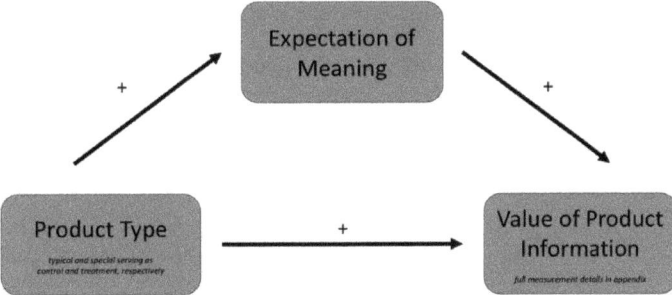

Figure 1. Conceptual model—consumer expectations of meaningful consumption experiences mediate the effect of product type on the value of product information.

Hypothesis 5 (H5). *The effect of specialty versus conventional product type on use of product label information will be mediated by expectations of meaning.*

Hypothesis 6 (H6). *The effect of specialty versus conventional product type on tendency to look for additional product information other than that included on the label will be mediated by expectations of meaning.*

Hypothesis 7 (H7). *The effect of specialty versus conventional product type on expectations of improved purchasing and consumption experiences resulting from more product information will be mediated by expectations of meaning.*

4. Materials and Methods

We surveyed high-involvement consumers of a food that is commonly consumed in more typical and more specialized forms and varieties: cheese. "Involvement" is a concept in the marketing literature which refers to an individual difference in the cognitive and emotional resources a consumer dedicates to thinking about and interacting with a given product class [48]. The literature often distinguished between enduring (more stable) and situational (affected by one's temporary environment) involvement [49]. We believe that consumers in our study were high in both enduring and situational involvement at the time of this study. Measuring within subject, we contrast consumers' expectations about the implications of consumption on well-being for typical and specialized versions of the same product and examine the role of information in designing more compelling consumption experiences for each product type.

The study materials and procedure were reviewed and approved by the authors' Institutional Review Board. This study was conducted at a food festival in the Northeastern United States focused on artisanal cheese produced in the state in which the festival was held. The event was sponsored by the formal guild of cheesemakers in the state and members of the guild comprised the majority of the vendors in attendance at the event. Tickets to attend the festival ranged from $20 to 35 and the festival included cheese-tasting events and informational classes, live music, and access to a variety of artisan foods including cow, goat, and sheep's milk cheeses and yogurts as well as goat's milk caramels, spices, beer, and wine. We believe that the festival attendees are representative of the target market that small-scale dairy farmers and artisan cheesemakers target with their products (we are assuming psychographic segmentation based on the value consumers place on purchasing local, sustainable, and artisanal goods and not a segmentation based on demographic characteristics). Participants (N = 150, 66% female, average age = 43 years, median annual household income $50,000–59,999) participated in this study in exchange for a chance to win one of five prizes (choice of either $25 in cash or $25 worth of cheese selected by participating cheesemakers, though many participants indicated that they completed this study simply to help the members of the guild of cheesemakers). Study participants completed this study using paper and in an area the authors coordinated with the event sponsor to have set aside specifically for data collection.

In this study, we asked participants to think about "typical" and "special" cheese purchases (scales were developed by the authors; full descriptions and instructions are available in Appendix A). We used this terminology to limit the extent to which participants might think that we might have favored particular varieties of cheese, farms, or production methods; our goal was to contrast the products of small-scale producers with the more typical products of large-scale producers. We wanted to allow participants to define what made small-scale producers' products special in whatever way felt most natural to them.

We conducted all analyses using R. All effects were measured within subject. We examined whether the order of presentation of the special and conventional measures had any effect on our results and found none. We thus report the results of the simplified models for each hypothesis; all results are consistent and there are no changes to the significance of results when we include order of presentation in the models.

5. Results

5.1. Analyses

We present our findings in three parts. First, we examine consumer beliefs about the effect of typical and special products on well-being.

We first tested H1, that consumers would report buying and consuming specialty products is more associated with the pursuit of meaning-based benefits than buying and consuming conventional products. We did not have a prediction about the effect of product type on experienced pleasure. Participants indicated that products they deemed special were more likely to lead to both pleasurable ($M_{Special}$ = 4.59, $M_{Typical}$ = 3.68, t (147) = -10.43, $p < 0.0001$) and meaningful ($M_{Special}$ = 4.00, $M_{Typical}$ = 2.87, t (145) = -9.96, $p < 0.0001$) consumption experiences. While we did not expect the significant difference in expectations of pleasure, it is not inconsistent with out conceptualization which treats pleasure and meaning as orthogonal constructs. Our vignettes described extreme consumer types motivated by the pursuit of utilitarian, pleasure, or meaning-based needs but consumers can be motivated to fulfill multiple needs in a single purchase.

Our further hypotheses, however, were developed with an expectation that the difference in consumer expectations about the psychological benefits of specialty and conventional products would be particularly strong for meaning. Thus, as a follow-up analyses, we tested the magnitude of the product type difference for pleasure and meaning. We expected to find that the magnitude of the difference for our predicted effect on meaning would be greater than the magnitude of the difference for the effect we did not predict on pleasure. In other words, we expected the difference in meaning for special versus typical products would be greater than the difference in pleasure for special versus typical products. We created new variables by subtracting the typical ratings from the special ratings for meaning and for pleasure for each participant and then compared the two difference measures. Consistent with our expectations, the difference between special and typical product experiences was significantly larger for meaning than for pleasure ($M_{meaning} = 1.13$, $M_{pleasure} = 0.90$, $t(145) = 2.42$, $p = 0.02$).

Building on this insight and recent research examining the role of knowledge and expertise in meaning making, we next examined how consumer needs for and reactions to product information differ for more typical versus special food products. We examined responses to items measuring consumer use of, search for, and value of information regarding special versus typical food. Paired samples t-tests comparing special versus typical food revealed that special foods were associated with greater use of ($M_{Special} = 3.71$, $M_{Typical} = 3.06$, $t(149) = -6.53$, $p < 0.0001$), search for ($M_{Special} = 3.01$, $M_{Typical} = 2.05$, $t(150) = -10.66$, $p < 0.0001$), and value of ($M_{Special} = 3.38$, $M_{Typical} = 2.82$, $t(148) = -7.21$, $p < 0.0001$) information than were typical foods. Note that degrees of freedom for the t-tests vary due to incomplete data on some items. All valid pairwise comparisons were utilized in each analysis. Removing participants with incomplete data from all analyses does not affect the pattern or significance of results. Across all three measures concerned with the utility of additional information about products, participants were more interested in information about special products.

Finally, we examine whether the expectation that a consumption experience will prove meaningful mediates the effect of product type on participants' use of, search for, and expected value of product information and find support for our theoretical model. We expected to find that the increased preference for information about special versus typical products would be driven at least in part by the expectation that special products were more likely to prove meaningful. However, we wanted to provide a conservative test of our theory given that the effect of product type on pleasure showed a similar pattern of results to the effect it had on meaning, the former of which we did not predict. Thus, we report our results below controlling for the effect of expectations of pleasure and examining only the unique contribution of expectations of meaning. The pattern and significance of all effects we report here is identical and the proportion of the total effect mediated is larger when we exclude pleasure from the models but we feel that this more conservative test is a better test of our proposed theoretical model.

We conducted our mediation analyses using the causal mediation analysis [50] which allowed us to account for the within-subjects random effects that result from our within-subject manipulation of product type and repeated measures of expected meaning, expected pleasure, and preference for information. The confidence intervals for each model were approximated utilizing a quasi-Bayesian method and 10,000 simulations of the data. The results of models predicting each measure of the value of product information supported our theoretical model and suggest a pattern of partial mediation; the indirect effect of product type (treating typical as control and special as treatment) through expectations of meaning was significant when predicting use of (indirect effect estimate = 0.163, 95% CI [0.08, 0.26], $p < 0.001$), search for (indirect effect estimate = 0.164, 95% CI [0.08, 0.27], $p < 0.001$), and value of (indirect effect estimate = 0.089, 95% CI [0.03, 0.17], $p = 0.004$) information. Importantly for our theory, the pattern of effects does not hold if we control for meaning and examine pleasure as a mediator (p's = 0.68, 0.18, and 0.66). Full results of all mediation analyses are available in Appendix B (Table A1).

5.2. Study Limitations

While we expect these results to generalize beyond these foci, it is important to note that our data used only a single product category (cheese) and collected data in one region (the Northeastern United States). It is also important to note that in our study, we sampled on a factor we expect is necessary for these effects to emerge: a moderate to high degree of involvement. While we expect that these results would generalize to consumers that place at least a moderate value on the pursuit of meaning and have an interest and appreciation for artisan and specialty foods, we similarly expect that the results would not emerge among consumers with less motivation to pursue meaning or involvement with specialty and artisan foods. Critically, our data collection was limited to participants for whom it can reasonably be argued, the product category is of greater than average personal interest and importance. Participants in this study travelled and paid a minimum of $20 to enter a festival specifically focused on cheese. While this sample is not representative of the average consumer, we pursued this narrow sampling because product involvement is a theoretically important moderator for the effects that we investigated; a sample containing lower involvement consumers would likely find moderation of the effects found here. However, studies focused specifically on the judgments and decisions of this type of niche market are extremely limited and we believe that our unique sample allows for a novel and valuable contribution to our understanding of diverse consumer preferences.

One potential moderator of the effects we investigated in this paper unrelated to consumer involvement that should be considered in future work is the interaction between package labeling and retail outlet. Many of the people in this study were completing this study in a context in which they had direct access not only to the cheesemaker but also to the dairy producer—in many cases, this was the same person. Yet, past research has shown that consumer reaction to label content for sustainable product claims (specifically organic labeling) varies by retail outlet [51]. Consumer trust in information that appears on a label may be moderated by retail outlet, by the attributed source of the information [52], or by perceived distance in the value chain between the cheesemaker and the final consumer.

6. Discussion

In this paper, we examine how consumer expectations of the effect of typical versus special food products on distinct aspects of well-being differ and how the value of information differs for special vs. typical food products. We hypothesized and found that special food products are more associated with meaning than typical products and that when consumers encounter special food products they are likely to search for, value, and use information at higher levels than they would for typical products in part to cultivate meaningful consumption experiences. For producers of special products (such as artisan cheesemakers), this is notable, suggesting that they need to carefully design and promote the information accompanying the product and consumption experience.

6.1. Theoretical Contribution

Our investigation joins the growing body of work examining how consumers think about the relationship between different types of consumption and distinct aspects of their own well-being. We believe this work also contributes to a growing body of work in food psychology, agroecology, and sustainability that has identified the unique barriers to entry and long-term profitability for smaller and artisan producers [53] and has appealed to the growing interest among consumers in sustainable products as an inevitable part of the solution [54]. As this body of work continues to develop, it is critical that researchers consider more than measures of intention to purchase as there is often a gap between intention and behavior [55]. Decades of research on consumer behavior has shown us that a deeper and richer understanding of the mechanisms driving consumer judgements, decisions, and behaviors allows us to better predict and influence purchase behavior. We believe this work is an interesting case study of theories developed and mechanisms identified in the consumer literature [11–13] and an extension of that work to the broader discussion in food psychology and sustainability.

6.2. Managerial Implications

We believe that the managerial implications for this work for small-scale and artisan producers are even more significant. So many of these producers find themselves overextended, trying to manage legal, operational, entrepreneurial, and growth concerns on a day to day basis, often with a single employee and lacking formal education on all relevant aspects of their businesses and support to address structural barriers to their success [56]. Even when producers find the time to consult the relevant and actionable literature on marketplace trends and best business practices, that literature is very rarely developed with businesses at their scale in mind; instead, much of the focus is on markets for commodity goods or on larger-scale businesses. Much of the work that is focused on smaller-scale and artisan producers is focused on identifying barriers and reasons to explain the slow extinction of small family farms and other small agricultural enterprises. Instead, we focus on the unique opportunity that small-scale and artisan producers have to use their unique and intimate access and control over the story of production to design distinctive and meaningful consumption experiences.

Determining how best to tell the story of artisan production requires careful thought and consideration. In her extensive examination of artisan cheese, Heather Paxson refers to the complex cross-section of economic values and social values that underlie small-batch cheesemaking, "economies of sentiment" and points out that these economies, "point to the cultural, emotional, ethical, and political dispositions that motivate people to assume the economic risk and backbreaking labor of making cheese in small batches using minimal technology. These sentiments are multifaceted [57]." Intrinsic to economies of sentiment is an understanding of the values motivating the producer to perform the work, and these values largely connect production to wider social and ecological spheres. Paxson for instance found that artisan cheesemakers pursue this work for reasons including producing high-quality products, preserving local markets, supporting the dairy industry, and creating environmentally, socially and financially sustainable businesses [57]. With inspiration, the producer's main aim is to creatively explore both these values, resultant ties to the product's attributes, and how the consumer's experience can be more fully realized through information.

Implications of this research for specialty and artisan producers are clear; make additional information that can bolster meaning making available to consumers interested in meaningful aspects of consumption. Our items in this study were written with impersonal information in mind (i.e., product labels, websites, etc.). This is in part for efficiency; once developed and tested, these materials are easily reproduced without further taxing the time of the small-scale producer. Similarly, other research has suggested that while fostering connection between producers and consumers of local food contributes to the perceived value of local food, those connections may be more effective when the connection is indirect [58]. Producers should consider introducing supplemental information through a variety of channels. For instance, cheesemakers could use cards near their cheese displays that relay the provenance of the cheese—its sourced animal, date of milking, process used, and aging—this channel could be augmented with personal stories told by the cheese monger (when the cheesemaker is not directly involved in sales). Cards and information could be printed as takeaways, or consumers could be directed to a website, podcast, video series, or social media account with more detailed information. The ability to report granular, personal information of this nature is unique to smaller-scale producers and these producers should take advantage of the unique opportunity to cultivate meaningful consumer experiences this provides. Both producers and future academic research should further investigate a variety messages and media to determine the most effective messaging and outlets. As consumers' preferences evolve, small-scale producers need to adopt approaches to managing their products that help them to identify these unique opportunities for differentiation and increased profitability.

Author Contributions: Both authors contributed substantially to this project. E.P.C. and S.W. conceptualized this study, collected data, contributed to the writing, and secured funding. E.P.C. conducted all analyses using R and wrote the first draft of the paper. All authors have read and agreed to the published version of the manuscript.

Funding: This research was supported by the Competitive Energy Services Sustainability Fund managed by the Maine School of Business at the University of Maine. The fund provides curriculum support to promote economic development while protecting ecosystem health and fostering community well-being.

Institutional Review Board Statement: The study was conducted according to the guidelines of the Declaration of Helsinki, and judged exempt by the Institutional Review Board of the University of Maine (approved on 5 September 2019).

Informed Consent Statement: Informed consent was obtained from all subjects involved in the study.

Data Availability Statement: Anonymized data are available upon request, consistent with the Institutional Review Board review and approval of this project.

Acknowledgments: The authors thank the Maine Farmland Trust, the Maine Cheese Guild, and the 2020 class of Business, Agriculture, and Rural Development (BARD) Sustainable Business Fellows participating in the Technical Assistance pipeline program through the University of Maine for inspiring and supporting this work.

Conflicts of Interest: The authors declare no conflict of interest.

Appendix A

Appendix A.1. Descriptions of "Typical" and "Special" Cheeses That Appeared Earlier in the Survey

For the next set of question, we'd like you to stop and think about a very typical cheese buying experience and what it means to you. Think about cheese that you purchase and eat frequently as well as where you typically purchase the cheese.

For the next set of question, we'd like you to stop and think about how you go about buying "special" cheese and what it means to you. Special means different things to different people; it might mean for a special occasion, it might mean something you've never had, it might mean something that you need for a specific recipe. There are no right or wrong definitions of special, we are interested in your opinions and experiences about what makes a particular cheese special to you.

Appendix A.2. Items Measuring Use of, Search for, and Value of Information

How often would you say that you closely read the labels to try to get more information when making typical and special cheese purchases?
→ Typical: five point scale from never to always
→ Special: five point scale from never to always

For each type of purchase, how often do you look for more information about the cheese from somewhere other than the label/packaging on the cheese itself (cheesemaker, cheesemonger, magazines, look online, ask friends, etc.)?
→ Typical: five point scale from never to always
→ Special: five point scale from never to always

To what extent would you say that your experience buying and consuming each type of cheese would be improved if you had more information about the cheese than what it typically provided on the package?
→ Typical: five point scale from never to always
→ Special: five point scale from never to always

Appendix A.3. Items Measuring Effect on Well-Being

To what extent would you say that your experience buying and consuming each type of cheese is pleasurable?

→ Typical: five point scale from "buying and consuming is not at all pleasurable" to "buying and consuming is extremely pleasurable"

To what extent would you say that your experience buying and consuming each type of cheese is meaningful?

→ Special: five point scale from "buying and consuming is not at all meaningful" to "buying and consuming is extremely meaningful"

Appendix B

Table A1. Full causal mediation results.

Testing for mediation of the effect of product type on use of information			
	Simple model with meaning as mediator	Conservative model with meaning as mediator controlling for pleasure	Pleasure as mediator controlling for meaning
Indirect Effect	0.394, CI = [0.27, 0.54], $p < 0.001$	0.163, CI = [0.08, 0.26], $p < 0.001$	−0.013, CI = [−0.08, 0.05], $p = 0.68$
Direct Effect	0.59, CI = [0.39, 0.79], $p < 0.001$	0.605, CI = [0.39, 0.81], $p < 0.001$	0.605, CI = [0.40, 0.81], $p < 0.001$
Total Effect	0.99, CI = [0.80, 1.17], $p < 0.001$	0.767, CI = [0.55, 0.98], $p < 0.001$	0.59, CI = [−0.14, 0.09], $p < 0.001$
Testing for mediation of the effect of product type on search for information			
	Simple model with meaning as mediator	Conservative model with meaning as mediator controlling for pleasure	Pleasure as mediator controlling for meaning
Indirect Effect	0.396, CI = [0.27, 0.54], $p < 0.001$	0.163, CI = [0.08, 0.27], $p < 0.001$	0.039, CI = [−0.02, 0.10], $p = 0.18$
Direct Effect	0.593, CI = [0.39, 0.80], $p < 0.001$	0.607, CI = [0.40, 0.82], $p < 0.001$	0.321, CI = [0.13, 0.51], $p = 0.001$
Total Effect	0.99, CI = [0.80, 1.18], $p < 0.001$	0.771, CI = [0.55, 0.99], $p < 0.001$	0.360, CI = [0.17, 0.55], $p < 0.001$
Testing for mediation of the effect of product type on value of information			
	Simple model with meaning as mediator	Conservative model with meaning as mediator controlling for pleasure	Pleasure as mediator controlling for meaning
Indirect Effect	0.395, CI = [0.27, 0.54], $p < 0.001$	0.089, CI = [0.02, 0.17], $p = 0.004$	−0.013, CI = [−0.08, 0.05], $p = 0.66$
Direct Effect	0.594, CI = [0.39, 0.80], $p < 0.001$	0.364, CI = [0.12, 0.60], $p = 0.003$	0.606, CI = [0.39, 0.81], $p < 0.001$
Total Effect	0.989, CI = [0.80, 1.17], $p < 0.001$	0.453, CI = [0.22, 0.69], $p < 0.001$	0.593, CI = [0.388, 0.79], $p < 0.001$

References

1. International Food Information Council. 2021 Food & Health Survey. Available online: https://foodinsight.org/2021-foodhealth-survey/ (accessed on 19 May 2021).
2. Rytkönen, P.; Bonow, M.; Girard, C.; Tunón, H. Bringing the Consumer Back in—The Motives, Perceptions, and Values behind Consumers and Rural Tourists' Decision to Buy Local and Localized Artisan Food—A Swedish Example. *Agriculture* 2018, *8*, 58. [CrossRef]
3. Feldmann, C.; Hamm, U. Consumers' perceptions and preferences for local food: A review. *Food Qual. Prefer.* 2015, *40*, 152–164. [CrossRef]
4. Zoll, F.; Specht, K.; Opitz, I.; Siebert, R.; Piorr, A.; Zasada, I. Individual choice or collective action? Exploring consumer motives for participating in alternative food networks. *Int. J. Consum. Stud.* 2017, *42*, 101–110. [CrossRef]
5. Batat, W.; Manna, V.; Ulusoy, E.; Peter, P.C.; Ulusoy, E.; Vicdan, H.; Hong, S. New paths in researching "alternative" consumption and well-being in marketing: Alternative food consumption/Alternative food consumption: What is "alternative"?/Rethinking "literacy" in the adoption of AFC/Social class dynamics in AFC. *Mark. Theory* 2016, *16*, 561. [CrossRef]
6. Conner, D.S.; Montri, A.D.; Montri, D.N.; Hamm, M.W. Consumer demand for local produce at extended season farmers' markets: Guiding farmer marketing strategies. *Renew. Agric. Food Syst.* 2009, *24*, 251–259. [CrossRef]
7. Pérez-Caselles, C.; Brugarolas, M.; Martínez-Carrasco, L. Traditional Varieties for Local Markets: A Sustainable Proposal for Agricultural SMEs. *Sustainability* 2020, *12*, 4517. [CrossRef]

8. Block, L.G.; Grier, S.A.; Childers, T.L.; Davis, B.; Ebert, J.E.; Kumanyika, S.; Laczniak, R.N.; Machin, J.E.; Motley, C.M.; Peracchio, L.; et al. From Nutrients to Nurturance: A Conceptual Introduction to Food Well-Being. *J. Public Policy Mark.* **2011**, *30*, 5–13. [CrossRef]
9. Ganglmair-Wooliscroft, A.; Wooliscroft, B. Well-Being and Everyday Ethical Consumption. *J. Happiness Stud.* **2017**, *20*, 141–163. [CrossRef]
10. Baumeister, R.F.; Vohs, K.D.; Aaker, J.L.; Garbinsky, E.N. Some key differences between a happy life and a meaningful life. *J. Posit. Psychol.* **2013**, *8*, 505–516. [CrossRef]
11. Percival Carter, E.L.; Williams, L. Meaningful Consumption Provides Long Lasting Benefits at a High Cost. *ACR North American Advances.* 2017. Available online: https://www.acrwebsite.org/volumes/1024806/volumes/v45/NA-45 (accessed on 21 June 2021).
12. Percival Carter, E.L.; Williams, L. Prolonging the Search for Meaning: How Hedonic Versus Eudaemonic Consumption Experiences Shape Preference for Variety. *ACR North American Advances.* 2014. Available online: https://www.acrwebsite.org/volumes/1017718/volumes/v42/NA-42 (accessed on 21 June 2021).
13. Williams, L.; Percival Carter, E.L. *Meaningful Consumption Connects (unpublished manuscript)*; University of Colorado Boulder: Boulder, CO, USA, 2020.
14. Mugel, O.; Gurviez, P.; Decrop, A. Eudaimonia around the Kitchen: A Hermeneutic Approach to Understanding Food Well-Being in Consumers' Lived Experiences. *J. Public Policy Mark.* **2019**, *38*, 280–295. [CrossRef]
15. Gupta, A. Meaningful consumption: A Eudaimonic Perspective on the Consumer Pursuit of Happiness and Well-Being. Ph.D. Thesis, The University of Nebraska-Lincoln, Lincoln, NE, USA, 2019.
16. Chandy, R.K.; Johar, G.V.; Moorman, C.; Roberts, J.H. Better Marketing for a Better World. *J. Mark.* **2021**, *85*, 1–9. [CrossRef]
17. Schwarz, N.; Clore, G.L. Mood, misattribution, and judgments of well-being: Informative and directive functions of affective states. *J. Pers. Soc. Psychol.* **1983**, *45*, 513–523. [CrossRef]
18. Kahneman, D.; Diener, E.; Schwarz, N. *Well-Being: Foundations of Hedonic Psychology*; Russell Sage Foundation: Manhattan, NY, USA, 1999; ISBN 978-0-87154-424-7.
19. Diener, E.; Wirtz, D.; Tov, W.; Kim-Prieto, C.; Choi, D.-W.; Oishi, S.; Biswas-Diener, R. New Well-being Measures: Short Scales to Assess Flourishing and Positive and Negative Feelings. *Soc. Indic. Res.* **2010**, *97*, 143–156. [CrossRef]
20. Kahneman, D.; Krueger, A.B. Developments in the Measurement of Subjective Well-Being. *J. Econ. Perspect.* **2006**, *20*, 3–24. [CrossRef]
21. Organization of Economic Cooperation and Development. OECD Guidelines on Measuring Subjective Well-Being. Available online: https://www.oecd.org/statistics/oecd-guidelines-on-measuring-subjective-well-being-9789264191655-en.htm (accessed on 21 June 2021).
22. Hirschman, E.C.; Holbrook, M.B. Hedonic Consumption: Emerging Concepts, Methods and Propositions. *J. Mark.* **1982**, *46*, 92. [CrossRef]
23. Batra, R.; Ahtola, O.T. Measuring the hedonic and utilitarian sources of consumer attitudes. *Mark. Lett.* **1991**, *2*, 159–170. [CrossRef]
24. Botti, S.; McGill, A.L. The Locus of Choice: Personal Causality and Satisfaction with Hedonic and Utilitarian Decisions. *J. Consum. Res.* **2011**, *37*, 1065–1078. [CrossRef]
25. Dhar, R.; Wertenbroch, K. Consumer Choice between Hedonic and Utilitarian Goods. *J. Mark. Res.* **2000**, *37*, 60–71. [CrossRef]
26. Okada, E.M. Justification Effects on Consumer Choice of Hedonic and Utilitarian Goods. *J. Mark. Res.* **2005**, *42*, 43–53. [CrossRef]
27. Voss, K.E.; Spangenberg, E.R.; Grohmann, B. Measuring the Hedonic and Utilitarian Dimensions of Consumer Attitude. *J. Mark. Res.* **2003**, *40*, 310–320. [CrossRef]
28. Deci, E.; Ryan, R.M. Self-determination theory: A macrotheory of human motivation, development, and health. *Can. Psychol. Can.* **2008**, *49*, 182–185. [CrossRef]
29. Huta, V.; Waterman, A.S. Eudaimonia and Its Distinction from Hedonia: Developing a Classification and Terminology for Understanding Conceptual and Operational Definitions. *J. Happiness Stud.* **2014**, *15*, 1425–1456. [CrossRef]
30. Ryan, R.M.; Deci, E.L. Self-determination theory and the facilitation of intrinsic motivation, social development, and well-being. *Am. Psychol.* **2000**, *55*, 68–78. [CrossRef] [PubMed]
31. Kashdan, T.B.; Biswas-Diener, R.; King, L.A. Reconsidering happiness: The costs of distinguishing between hedonics and eudaimonia. *J. Posit. Psychol.* **2008**, *3*, 219–233. [CrossRef]
32. Mascaro, N.; Rosen, D.H. The Role of Existential Meaning as a Buffer against Stress. *J. Humanist. Psychol.* **2006**, *46*, 168–190. [CrossRef]
33. Rudd, M.; Catapano, R.; Aaker, J. Making Time Matter: A Review of Research on Time and Meaning. *J. Consum. Psychol.* **2019**, *29*, 680–702. [CrossRef]
34. Batat, W.; Peter, P.C.; Moscato, E.M.; Castro, I.A.; Chan, S.; Chugani, S.; Muldrow, A. The experiential pleasure of food: A savoring journey to food well-being. *J. Bus. Res.* **2019**, *100*, 392–399. [CrossRef]
35. Friestad, M.; Wright, P. Persuasion Knowledge: Lay People's and Researchers' Beliefs about the Psychology of Advertising. *J. Consum. Res.* **1995**, *22*, 62–74. [CrossRef]
36. Eurostat. Agriculture, Forestry and Fishery Statistics—2018 Edition. Available online: https://ec.europa.eu/eurostat/web/products-statistical-books/-/ks-fk-18-001 (accessed on 21 June 2021).

37. Wang, S.L.; Nehring, R.; Mosheim, R. *Agricultural Productivity Growth in the United States: 1948–2015. Amber Waves:The Economics of Food, Farming, Natural Resources, and Rural America*; United States Department of Agriculture, Economic Research Service: Washington, DC, USA, 2018.
38. USDA Economic Research Service. Beginning Farmers and Age Distribution of Farmers. Available online: https://www.ers.usda.gov/topics/farm-economy/beginning-disadvantaged-farmers/beginning-farmers-and-age-distribution-of-farmers/ (accessed on 20 October 2019).
39. Alonso, A.D.; Northcote, J. Investigating farmers' involvement in value-added activities. *Br. Food J.* **2013**, *115*, 1407–1427. [CrossRef]
40. Clark, J. Entrepreneurship and diversification on English farms: Identifying business enterprise characteristics and change processes. *Entrep. Reg. Dev.* **2009**, *21*, 213–236. [CrossRef]
41. Morris, J.R.; Brady, P.L. The muscadine experience: Adding value to enhance profits. AAES Research Report 982, Arkansas Agricultural Experiment Station. Available online: http://arkansasagnews.uark.edu/Musc_Rev07_reduced4Web.pdf (accessed on 20 January 2011).
42. Carlisle, L.; De Wit, M.M.; DeLonge, M.S.; Iles, A.; Calo, A.; Getz, C.; Ory, J.; Munden-Dixon, K.; Galt, R.; Melone, B.; et al. Transitioning to Sustainable Agriculture Requires Growing and Sustaining an Ecologically Skilled Workforce. *Front. Sustain. Food Syst.* **2019**, *3*. [CrossRef]
43. Bieman, B. Growing Maine's Food Industry, Growing Maine. Available online: https://www.hks.harvard.edu/centers/mrcbg/publications/awp/awp50 (accessed on 21 June 2021).
44. Welcomer, S.; MacRae, J.; Davis, B.; Searles, J. Maine's Artisan Cheesemakers: The Opportunities and Challenges of Being an Artist, Scientist, Agriculturalist, Alchemist, and Entrepreneur. *Maine Policy Rev.* **2017**, *26*, 59–71.
45. Wilson, R.; Roberts, J. *Action Plan for Agriculture and Food System Development*; Northern Community Investment Corporation: St Johnsbury, VT, USA, 2014.
46. Lebergott, S. *Pursuing Happiness*; Princeton University Press: Princeton, NJ, USA, 1993.
47. Pine, B.J.; Gilmore, J.H. *The Experience Economy*; Harvard Business Press: Cambridge, MA, USA, 2011; ISBN 978-1-4221-6197-5.
48. Laurent, G.; Kapferer, J.-N. Measuring Consumer Involvement Profiles. *J. Mark. Res.* **1985**, *22*, 41. [CrossRef]
49. Houston, M.; Rothschild, M. Conceptual and Methodological Perspectives on Involvement, Educators Proceedings. *Res. Front. Mark. Dialogues Dir.* **1978**, *184*, 187.
50. Tingley, D.; Yamamoto, T.; Hirose, K.; Keele, L.; Imai, K. Mediation: R Package for Causal Mediation Analysis. *J. Stat. Softw.* **2014**, *59*, 1–38. [CrossRef]
51. Ellison, B.; Duff, B.R.; Wang, Z.; White, T.B. Putting the organic label in context: Examining the interactions between the organic label, product type, and retail outlet. *Food Qual. Prefer.* **2016**, *49*, 140–150. [CrossRef]
52. Oates, C.; McDonald, S.; Alevizou, P.; Hwang, K.; Young, W.; McMorland, L. Marketing sustainability: Use of information sources and degrees of voluntary simplicity. *J. Mark. Commun.* **2008**, *14*, 351–365. [CrossRef]
53. Kwil, I.; Piwowar-Sulej, K.; Krzywonos, M. Local Entrepreneurship in the Context of Food Production: A Review. *Sustainability* **2020**, *12*, 424. [CrossRef]
54. Ghadge, A.; Kara, M.E.; Mogale, D.G.; Choudhary, S.; Dani, S. Sustainability implementation challenges in food supply chains: A case of UK artisan cheese producers. *Prod. Plan. Control.* **2020**, 1–16. [CrossRef]
55. Wenzig, J.; Gruchmann, T. Consumer Preferences for Local Food: Testing an Extended Norm Taxonomy. *Sustainability* **2018**, *10*, 1313. [CrossRef]
56. Calo, A. How knowledge deficit interventions fail to resolve beginning farmer challenges. *Agric. Hum. Values* **2018**, *35*, 367–381. [CrossRef]
57. Paxson, H. *The Life of Cheese*, 1st ed.; University of California Press: Berkeley, CA, USA, 2013; ISBN 978-0-520-27018-3.
58. Albrecht, C.; Smithers, J. Reconnecting through local food initiatives? Purpose, practice and conceptions of 'value'. *Agric. Hum. Values* **2018**, *35*, 67–81. [CrossRef]

Article

A Typological Classification for Assessing Farm Sustainability in the Italian Bovine Dairy Sector

Margherita Masi [1], Yari Vecchio [1], Gregorio Pauselli [1], Jorgelina Di Pasquale [2,*] and Felice Adinolfi [1]

[1] Department of Veterinary Medical Science, University of Bologna—Alma Mater Studiorum, 40064 Ozzano dell'Emilia, Italy; margherita.masi4@unibo.it (M.M.); yari.vecchio@unibo.it (Y.V.); gregorio.pauselli2@unibo.it (G.P.); felice.adinolfi@unibo.it (F.A.)
[2] Faculty of Veterinary Medicine, University of Teramo, 64100 Teramo, Italy
* Correspondence: jdipasquale@unite.it

Abstract: Italy is among the most important countries in Europe for milk production. The new European policies encourage a transition towards sustainability and are leading European dairy farms to follow new trajectories to increase their economic efficiency, reduce their environmental impact, and ensure social sustainability. Few studies have attempted to classify dairy farms by analyzing the relationships between the structural profiles of farms and the social, environmental, and economic dimensions of sustainability. This work intends to pursue this aim through an exploratory analysis in the Italian production context. The cluster analysis technique made it possible to identify three types of dairy farms, which were characterized on the basis of indicators that represented the three dimensions of sustainability (environmental, social, and economic sustainability) and the emerging structural relationships based on the structural characteristics of the dairy farms. The classification made it possible to describe the state of the art of the Italian dairy sector in terms of sustainability and to understand how different types of farms can respond to the new European trajectories.

Keywords: sustainability; farm type; sustainable management; small-/medium-scale animal farms

1. Introduction

Italy was the fifth European country in terms of the quantity of milk delivered in 2019 with a total of 1,208,647 thousand tons [1]. This amount was the main portion of the total milk production of farms, which reached 1,330,010,000 tons when domestic consumption, direct sale, and cattle feed were included [2].

Within the country, four regions produced almost 80% of the total milk in 2020: Lombardy, Emilia-Romagna, Veneto, and Piedmont. These areas produce larger quantities and have better efficiency compared to the others [3]. In 2018, the sector recorded a production value in the agricultural sector of approximately EUR 4.68 billion [4], while the turnover of the dairy industry was around EUR 16.63 billion [5]. These data show how the dairy industry is a fundamental asset for the national agri-food sector, representing 10% of the value of agricultural production and 12% of the dairy industry. In the last decade, the number of farms in the sector has decreased by about 17,000 units, reaching just over 26,000 in the first half of 2020 [6]. This trend originated in the disadvantageous position of small farms relative to capital-intensive investors [7], which made farmers susceptible to the "get-big-or-get-out" syndrome, leading to a process of concentration [8]. This situation, together with many factors, such as the environmental and social context in which farmers operate, have led to the development of larger farms over time; these represent the production base of the entire sector and have a greater concentration in the north of Italy [9].

In recent years, the rapid expansion of intensive farming systems has been analyzed, as they have caught the attention of researchers and experts, who have questioned the sustainability of these farming methods in terms of the negative effects on the environment and animal welfare that they produce [3,10,11]. These factors have generated higher social costs than those of private systems, with negative consequences for the whole society. This type of agriculture generates negative externalities due to its linear perspective [12,13], which follows the "take–make–dispose" principle of linear economics [14]. This is especially true in the animal farming sector due to the large number of inputs used to feed intensive farming systems, as well as their wastes [15]. Today, the possibility of having a greater social acceptability of these practices and the need to make the sector more efficient and more environmentally sustainable represent future challenges. This has been requested by the new European trajectories, such as the Farm-to-Fork Strategy [16]. In fact, the sector is now facing a transition towards sustainability, and although it is progressing slowly, more sustainable production methods have been developed in recent years [17–19].

The studies on farm sustainability are very heterogeneous, but they can be enclosed within some dominant strands of research. The first strand concerns studies that have attempted to explore the sustainability performance of livestock farming systems [20] and strategies for improving their resilience by looking at the self-sufficiency of their inputs [21,22]. In fact, it is widely recognized that increasing input self-sufficiency can be a strategy for improving the sustainability of agricultural systems because those are less dependent on the variability of the input market [23]. Although one of the suggested strategies was to reduce the inputs, other views suggest enhancing the waste outputs, too. Another strand, in fact, concerns studies that have designed the possibility of transitioning to a more sustainable system through a circular economy approach [24,25], which is intended to be a system-wide approach to economic development that is designed for the benefit of businesses, society, and the environment [26]. The transition to a circular economy can result in a minimization of the use of external inputs and in the reduction of waste and emissions into the environment through the recycling and valorization of agricultural waste [27], thus avoiding the exploitation animals but using them "for what they know how to do" [28].

Finally, the last strand concerns farm sustainability classification studies, of which there have been very few attempts in the bovine sector [28]. The need for an objective classification of dairy farms that covers all aspects of the production systems has been underlined by researchers [29,30]. In his classification study on the goat sector, Gelasakis et al. [31] provided a framework of factors that are representative of the types of farms from a structural point of view and of the level of intensity of a production system, thus laying the foundations for secondary evaluations in terms of farm sustainability performance. The importance of associating a sustainability assessment with the type of farm has already been recognized for other sectors in several countries [29,30,32–36].

The assessment of sustainability must be calibrated on a farm's structural assets [30,32], which mainly include the area of the farm, the number of animals raised, the age and the education of the farmer, and the production methods (e.g., organic).

Lebacq et al. [37] provided a useful description of all three dimensions of sustainability: environmental, economic, and social. Environmental sustainability is described through the management of inputs, such as fertilizers and pesticides, and the use of resources. The consumption of water and the use of fertilizers and pesticides play a fundamental role in the environmental impact of agriculture and livestock [38–40]. The same authors described economic sustainability as the economic viability of agricultural systems, that is, their ability to be profitable in order to provide prosperity to the agricultural community. This can be described through profitability, self-sufficiency of inputs, diversification, and durability (family and employee labor costs). The inclusion of these concepts in the assessment of the economic sustainability of a farm is supported by the literature [22,41,42]. Social sustainability is defined at two levels: that of the farm and that of the society. At the farm level, it can be defined by the educational qualification, the working conditions, and the use of machinery [43]. On the social level, it is defined by multifunctionality (quality of rural areas, ecosystem services) and by extra-farm labor [44].

For the Italian context, there are no studies that have attempted to classify bovine dairy farms by assessing the relationships between the structural profile and the three dimensions of sustainability. Furthermore, while the environmental impact of bovine farming has been widely studied [3,45,46], the Italian literature lacks a characterization of farms that can also provide perspectives for improving farm management from an economic and social point of view.

This kind of classification studies in terms of sustainability-produced standardized procedures to better estimate input and output of livestock farming systems, evaluate changes and target information and policies to farmers [31,47].

Our study aims to address this gap. Furthermore, this study can be considered a novelty, as it can represent a preliminary strand with respect to studies that attempt to measure farms' performance in sustainability; only through a classification of bovine dairy farms that aggregates them according to homogeneous characteristics by calibrating the assessment of the sustainability of their structural assets will it be possible to proceed, in the second instance, to a more objective assessment of the social, environmental, and economic performance.

Indeed, to respond to this gap, this work carries out an exploratory analysis in order to identify a classification for Italian dairy farms by identifying emerging relations between the three dimensions of sustainability (environmental, social, and economic) and their structural profiles. This analysis is a novelty in that it cannot only allow us to understand how different types of farms might respond to the new European trajectories, but it also provides a general model that can be replicated across Europe, as the indicators and databases come from a European sample survey. To this end, an exploratory approach is used by applying multivariate statistical techniques. A principal component analysis was used to summarize the factors, and then a cluster analysis was run to gather the different livestock farms into homogeneous groups. This two-step approach is consistent with the statistical literature [48]. This analysis allows the creation of a typological grid that is representative of the aspects of structural polymorphism and sustainability [49,50] through a process of a posteriori interpretation [51].

The work is made up of four sections. First, the methodology and its theoretical basis are defined. The second part reports the descriptive analysis of the sample, followed by a description of the clusters obtained. Finally, a discussion and conclusions are provided based on the evidence from the analysis and the literature.

2. Materials and Methods

The empirical analysis was conducted on secondary data from the 2018 Farm Accountancy Data Network (FADN). The FADN was established in 1965 by the European Commission and is the official source of microeconomic data, as it is based on harmonized accounting standards. The Italian FADN (https://rica.crea.gov.it, accessed on 21 April 2021) sample consists of 11,000 annual farms, which are structured to represent the different production types, sizes, locations (e.g., region, altitude, etc.) that are present in the national territory. It allows a national average coverage of 95% of the utilized agricultural area (UAA), 97% of the value of standard production, 92% of the labor units (ULs), and 91% of the livestock units. For the present analysis, a filter was made to allow the isolation of farms in the dairy sector, thus leading to the extraction of 1216 farms.

The FADN consists of 25 tables that contain different types of information. In order to select useful indicators for the analysis, the indices were grouped according to their social, environmental, economic, and structural dimensions, as shown in the following graph (Figure 1). All of the following indicators were therefore selected in order to assess the farms' sustainability.

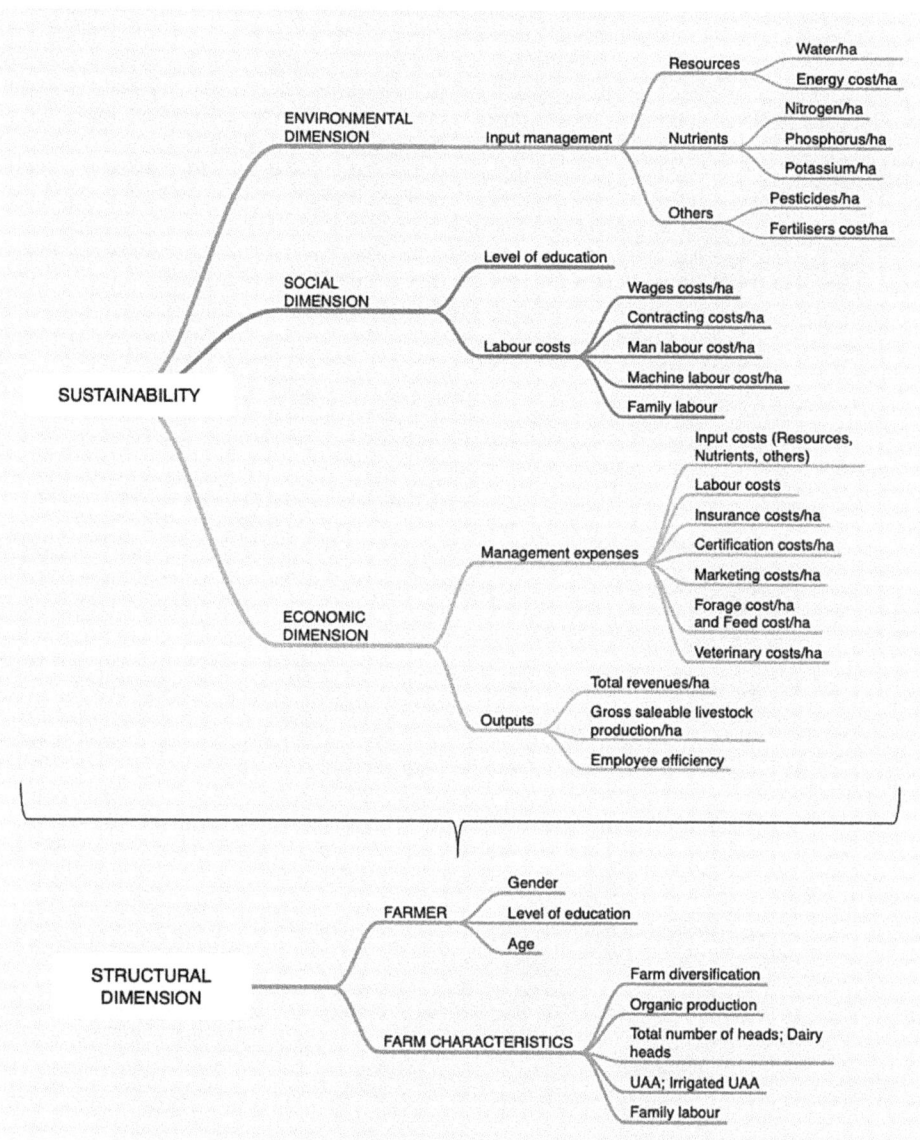

Figure 1. Authors' elaboration of the Italian FADN indicators.

In order to carry out this study, the selection of indicators was performed while considering the objective of the research and the targets of the farms examined. Furthermore, based on the available data, and the minimum representative sets of indicators were chosen for each dimension (Table 1). The literature confirms the need to contextualize the indicators of all three dimensions with respect to the reality of the production of dairy cows [37,52].

Table 1. Minimum representative sets of indicators for each dimension.

			Sustainability
Environmental dimension	Input management	Resources	Water/ha: volume of water distributed per hectare Energy cost/ha: costs incurred for the purchase of fuel, electricity and heating per hectare
		Nutrients	Nitrogen/ha: quantity of nitrogen distributed per hectare Phosphorus/ha: quantity of phosphorus distributed per hectare Potassium/ha: quantity of potassium distributed per hectare
		Others	Pesticides/ha: quantity distributed per hectare Fertilizer cost/ha: costs for fertilizers per hectare
Social dimension	Level of education		No title Primary school Secondary school Diploma Professional diploma Bachelor's degree Master's degree
	Labor costs		Wage costs/ha: expenses incurred for wages, social charges, and rent payable per hectare Contracting cost/ha: cost of agro-mechanical and technological services offered by external suppliers/ha Human labor cost/ha: cost of human labor per hectare Machine labor cost/ha: cost of machine labor per hectare Family labor: relationship between family UL and total UL
Economic dimension	Managerial expenses		Input costs, described in the previous section as "Resources, Nutrients, and others" Labor costs, described in the previous section as "Labor costs" Insurance costs/ha: insurance costs per hectare Certification costs/ha: costs for purchasing certifications per hectare Marketing costs/ha: marketing costs per hectare Forage costs/ha: expenses for the purchase of non-farm forage per hectare Feed costs/ha: expenses for the purchase of feed per hectare Veterinary costs/ha: costs of veterinary services and pharmaceutical costs per hectare
	Outputs		Total revenues/ha: total farm revenue Gross saleable livestock production/ha (Livestock GSP): revenues strictly related to the livestock activity Employee efficiency: relationship between added value and work unit
			Structural Dimension
Farmer	Gender		
	Level of education		No title, Primary school, Secondary school, Diploma, Professional diploma, Bachelor's degree, Master's Degree
	Age: age of the owner		
Farm characteristics			Farm diversification: presence or absence of other activities Organic production: presence or absence of organic production Dairy heads: number of heads in lactation Total number of heads UAA: in hectares Irrigated UAA: in hectares Family labor: relationship between family UL and total UL

The selected indicators were therefore the following:

The number of indicators identified was high, so in order to perform the analysis more efficiently, optimal scaling techniques were carried out to synthesize the information. The clustering process was articulated as follows:

- Given the highest number of continuous variables, a principal component analysis (PCA) was performed. PCA is a descriptive method that aims to summarize a data matrix in order to express its structure with a reduced number of dimensions. Thus, PCA is a method for identifying a particular transformation of the observed variables (a linear combination) and trying to explain a large part of the variance of the observed variables with a few components. In order to interpret the factorial weights more easily, it is possible to perform rotations of the factorial axes that maintain scale invariance by simplifying the structure of the weight system. The most commonly used solutions respect the orthogonality of the factors; in the present case, the Varimax rotation [53] was used, which is a useful method when there are several factors and a clear separation between the extracted factors is desired. Based on the rule of having an eigenvalue greater than 1 and on the interpretability of the data, the top 5 factors that explained 60% of the variance were chosen (Table 2).
- Subsequently, a cluster analysis was carried out with the aim of creating homogeneous groups of dairy farms based on the 5 previously extracted factors. Cluster analysis allows the generation of groups in which the points of the same group are more similar to each other than the points of the other clusters. Thus, the technique allows the formation of groups in which the internal inertia is minimal (within inertia), while the inertia between groups is maximal. The clustering technique used was the agglomerative hierarchical technique [54]. According to Ward's criterion, 10 successive iterations were performed.

Table 2. Results of the Varimax rotation.

Component	Total	% of Variance	Cumulative %
1	4.941	23.528	23.528
2	2.667	12.698	36.226
3	1.804	8.592	44.818
4	1.803	8.585	53.404
5	1.58	7.524	60.928

The data were processed through two types of software: SPSS version 26 and SPAD version 3.21.

With cluster analysis, it is possible to estimate the inputs and outputs of farms with different farming systems. Many studies in the literature (e.g., [52,55,56]) confirmed the methodological approach of this study. In particular, Micha et al. [52] calculated some performance indicators that were chosen to classify the dairy farms in Ireland, which permitted the identification of the farms that performed better or worse from the point of view of sustainability. In addition, by implementing a contextualization of the indicators for the Italian production context, the present study intends to find a classification of farms based on sustainability indicators and their socio-economic profiles in order to help in the understanding of how to act in order to improve the sustainability of the sector.

This process can lead to a standardization of the results, which is essential for providing policy guidelines for the improvement of these systems.

3. Results

Our sample consisted of 1211 farms, of which 95.5% were dairy farms and 4.5% were buffalo farms. The sample used, which included farms that were distributed throughout Italy, was representative of the Italian context and its distribution in the different areas (Figure 2).

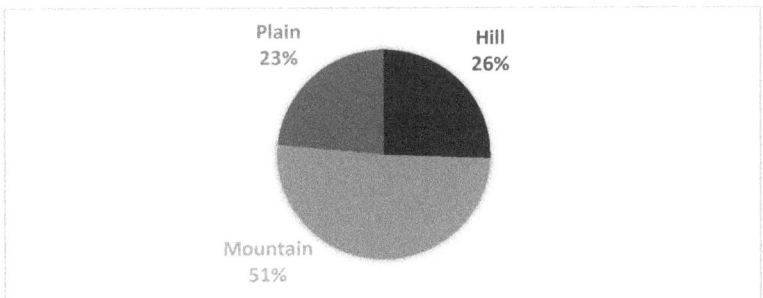

Figure 2. Distribution by altitude zone.

As already mentioned, dairy farming in Italy mainly developed in four regions (Lombardy, Emilia-Romagna, Veneto, and Piedmont); the following graph shows the distribution of the average number of animals of the analyzed farms on a provincial basis (Figure 3). As can be observed, the distribution was greater in the provinces of northern Italy, particularly in the area of the Po Valley.

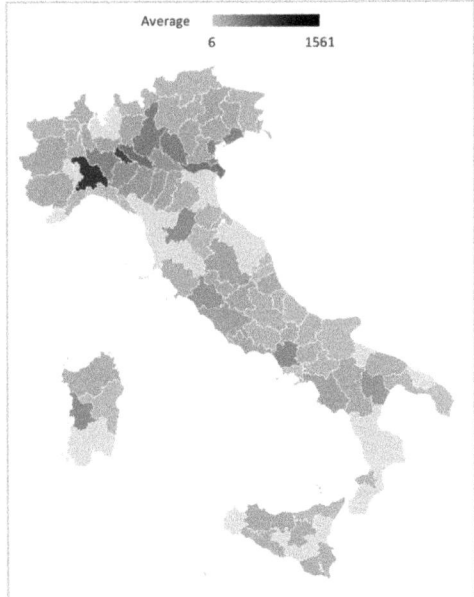

Figure 3. Distribution of the average number of animals per farm.

Table 3 shows the most important socio-demographic characteristics of the sample examined. In 82.66% of the cases, the farms were led by male entrepreneurs, while they were led by female ones in 17.34% of the cases. Furthermore, as shown in the following table, 13.29% of them were young farmers (under 40 years old). A total of 39.8% of the respondents attended secondary school, and 44.84% had a professional or high school diploma. Only 2.72% of the respondents had a university degree. In addition, 11.73% of the sample practiced organic production and 15.03% practiced activities complementary to livestock farming. Over 91.25% of the farms were family-run, while 8.75% had extra-family labor.

Table 3. Descriptive statistics.

	Variable	Count	%
Gender	Female	210	17.34
	Male	1001	82.66
Young	No	1050	86.71
	Yes	161	13.29
Educational Level	No school	35	2.89
	Primary school	118	9.75
	Secondary school	482	39.80
	High school diploma	264	21.80
	Professional diploma	279	23.04
	Bachelor's degree	9	0.74
	Master's degree	24	1.98
Organic production	No	1069	88.27
	Yes	142	11.73
Diversification	No	1029	84.97
	Yes	182	15.03
Type of management	Other type	1	0.08
	With employees	13	1.08
	Subcontracting only	1	0.08
	Direct with extra-family prevalence	91	7.51
	Direct with predominantly family members	534	44.10
	Direct with family members only	571	47.15

Empirical Analysis

In order to characterize the farms involved in the study based on aspects of their structure and sustainability, a cluster analysis was carried out, and three typological groups were identified. The variables used for the analysis were both the structural variables of the sample and those that characterized the farms from the environmental, social, and economic points of view. The highlighted clusters (CLs) are shown in Figure 4.

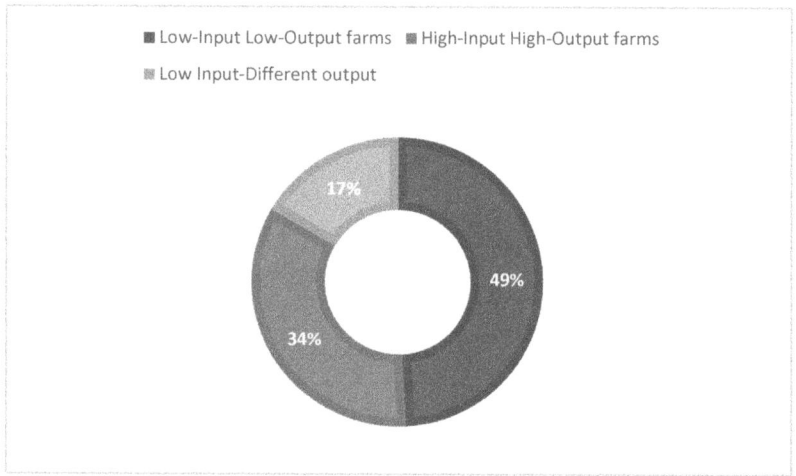

Figure 4. Distribution of the clusters.

- CLUSTER 1: Low-Input, Low-Output Farms that Are Attentive to the Environmental Dimension (49.1% of the Sample)

This CL was made up of the smallest farms in terms of consistency; they were characterized by an average of 79 total heads, of which 42 were in the lactation phase. As regards the overall size, however, the farms reported an average UAA of 52.8 hectares (of which 15% was irrigated). Compared to the others, in this cluster, the minimum values of the indicators of the environmental dimension were recorded. In fact, the lowest values were recorded for the use of "negative" inputs, such as non-renewable resources, fertilizers, nutrients, and pesticides. In support of this, 16% of the cluster's producers complied with the organic specifications. At the same time, this "low-input" profile was also associated with a "low-output" one. In fact, the indicators associated with the economic dimension tended to record the lowest values here, both among the cost items and for the total revenues (EUR 4626.6/ha), as well as in the added value per labor unit (EUR 38,638.7/ha). This last parameter highlighted the lowest economic efficiency per employee. For the social dimension, the lowest values were recorded for wages (EUR 337.88/ha), contracting (EUR 1.52/ha), human labor (EUR 777.27/ha), and machinery (EUR 239.93/ha). In the social dimension of this group, a higher amount of family labor was employed compared to in other groups (family work unit/total work unit equal to 0.88). This group also included the highest percentage of female leaders (20.3%). For the many aspects that were highlighted, these companies were distinguished from those of the other groups due to their greater attention to the environmental aspects.

- CLUSTER 2: High-Input, High-Output Farms that Are Attentive to the Economic Dimension (34.3% of the Sample)

This CL was made up of the largest farms, where, on average, there were 176 heads, of which 92 were in the lactating phase. However, the high number of animals was associated with an average UAA of only 45.99 hectares—the smallest among the groups—thus characterizing a production system with high inputs. A high degree of intensification of activities was also suggested by the high irrigated UAA and by the level of labor costs (mainly extra-family employees), which was the highest. The values of the environmental dimension, which represented the highest values per hectare among the clusters, confirmed that this was a type of farm that produced by using a high level of inputs. The high-cost values in the economic dimension were followed by the maximum values achieved for total revenues (EUR 10,952.88/ha), livestock GSP (EUR 7454.53/ha), and labor productivity. The latter, which was measured in added value per labor unit, was, on average, equal to EUR 60,639.96/ha. These then described the image of the farms based not only on the high inputs, but also on the economic sustainability. The maximum values were also reached for all indicators of the social dimension: wages (EUR 749.73/ha), contracting (EUR 22.09/ha), Human labor, and machine labor. The relationship between the latter two highlighted how this cluster was the one with the greatest use of machines compared to human labor. For the many aspects that were highlighted, these companies were distinguished from those of the other groups due to their greater attention to the economic aspects. Here, the cost per hectare found for insurance was the highest among the groups—to manage their business risk.

- CLUSTER 3: Low-Input, Different-Output Farms that Are Attentive to Social Sustainability (16.6% of the Sample)

This cluster was distinguished by an average of 115 animals that were reared, of which 52 were in the lactation phase. The average size of the farms belonging to this cluster was 82 hectares, which is the largest found among the groups, and this is where the greatest presence of organic production fell (16.5%). Furthermore, in this group, we found the highest percentage of tenants who had a degree (3%), confirmed by the fact that 18% of the cluster was made up of young entrepreneurs. This cluster showed the greatest attention to the social dimension among the clusters, as shown by the highest percentage of complementary activities that were added to the zootechnical activity (90.5%). In the economic dimension, the small amount spent on the purchase of non-farm fodder indicated that these were large farms practicing self-production. Here, the cost per hectare found for the certifications for enhancing the quality of the product was the highest among the groups. The indicators of the social dimension, such as human labor and machine labor, were found here to have the lowest values among the groups and, when compared to each other, highlighted a clear prevalence of the use of human work over machinery. These farms were therefore characterized by their greater attention to the social aspects, as they were more attentive to multifunctionality compared to those belonging to the other groups.

4. Discussion

Three different profiles emerged from the comparison among the clusters, which were distinguished according to the values achieved for the indicators used in the different dimensions to characterize the dairy farms: environmental, economic, social, and structural. In fact, on the basis of the structural characteristics detected within each cluster, different behavioral profiles were highlighted with respect to the farms' sustainability choices. The distribution by altitude zone was not significant in determining the clusters.

Although the region did not represent a net determinant of the entrepreneurs' behavior, most of the farms located in the areas of the four most important regions for the Italian bovine dairy sector fell within cluster 2.

CL1 represented small farms that stood out due to their performance in environmental sustainability compared to the others. In fact, they were characterized by the lowest level of input in the environmental dimension. This aligned well with the fact that this group included a high percentage of female leaders and of organic practices. The literature confirms the greater predisposition of women towards environmental issues and recognizes them as an active and essential part in environmental conservation [57]. In addition, CL1 highlighted the importance of family labor in the social dimension, which combined well with the high percentage of women in the cluster, as shown in the studies by Trauger [58] and Fairlie and Robb [59]. This group also recorded the lowest values in the economic dimension because, as the literature confirmed, it is precisely small companies that often demonstrate lower economic efficiency [60].

This group was clearly opposed by the CL2 farms, which directed their business choices toward the economic dimension. This was in line both with the medium–large structural assets of the farms and the fact that there was a high percentage of male leaders in the cluster [61]. In fact, these farms stood out not only by having the highest values in the economic dimension, but also by having the highest level of input in the environmental dimension. This showed a lack of attention to the environmental impacts of the practice of intensive farming, which was also indicated by the fact that the percentage of organic producers was almost zero. In fact, as reported in a study by Muller et al. [62], the organic method mandates a clear reduction of chemical inputs. The profile of this group also denoted a high degree of intensification of practices in the social dimension, with the highest costs for contracting and wages and the greatest value for the use of machines compared to human labor. The high values of these latter indicators agreed with the economic settings of the companies, as these also had the highest economic efficiency per employee.

CL3 was represented by farms that were very attentive to the social aspects. It was made up of farms with the largest farm size, which highlighted a production choice oriented towards multifunctionality from which the farmer obtained an important share of the company revenues. Here, there was a low livestock consistency, and a substantial percentage of organic livestock (16.5%) was found, which was in line with the findings highlighted by CL1. In fact, although these two clusters had similar consistencies and breeding methods, which were confirmed by the observed value of the livestock GSP, they differed in the values of their total revenues. This was due to both the diversification of the activities carried out in CL3 and the scarce quantity of investments of the farms in CL1 in activities that would allow them to enhance their farm products, such as certifications. In fact, the farms in CL3 sought different sources of income through complementary activities (90.5%), confirming that this group had the highest percentage of young people (18%). It is precisely the literature that confirms that young farmers may be more inclined to support a multifunctional approach on a farm [63]. Unlike in CL3, farms with a more limited surface area (CL1, CL2), regardless of the intensification of activities, were mainly oriented toward livestock production.

5. Conclusions

This work aimed to define the types of farms in the dairy cattle sector in order to provide a detailed picture of the Italian situation and, at the same time, to create a classification of homogeneous groups of farmers based on their structural profiles from the point of view of sustainability. This study provides a homogeneous and common basis on which it is possible to develop studies that measure the performance of farms in terms of sustainability through a more objective and comparable assessment of the social, environmental, and economic performance of farms. To obtain this result, multivariate statistical approaches were used, as these methods allowed a better exploration of the characteristics of management and the comparison of the profitability of the clusters that were obtained, as was already done in similar analyses [31,55].

Kelly et al. [64] highlighted the limitations of the FADN sample surveys, as they do not capture data from small farms that do not fall under the definition of "commercial", even though they receive subsidies as a result of the Common Agricultural Policies. Nevertheless, the FADN is the most reliable sample survey, as shown by the extensive literature [65–67]. If the sample base has limitations in its precise representation of the agricultural sectors, a further limitation of the study is the choice of indicators, which was based on a careful survey of the literature [52] and based on the characteristics of the sector, but it could be enriched with additional factors [31,32,55] and supplemented with indications that a primary survey could provide, such as a survey of farmers' perceptions and intentions [68].

Despite these limitations, the analysis carried out here allowed the set objectives to be reached and an identification of Italian farm types to be provided based on the emerging relations between the aspects of structure and sustainability. This classification potentially highlights the state of the art of the bovine dairy sector by classifying farms on the basis of sustainability indicators in order to understand how far the breeding strategies are from the European trajectories and how some types are already on the road to this sustainable transition.

Knowing how to combine the objectives of environmental sustainability with those of economic competitiveness will be important in complying with what is required by the European Commission through the Green Deal and related strategies (the Farm-to-Fork and Biodiversity strategies) [69,70].

Thus, this work is in the vein of research on policy support and is an attempt to animate discussions and future analyses to be replicated in other European countries so that a framework for sustainability can be provided. The different analyses could give hints to policymakers for making policies—such as the CAP or rural development policies—that are closer to the needs of different territorial realities.

Some farms have a high level of eco-compatibility; however, they should also be pushed to pay more attention to the economic dimension through a better use of socio-technical tools. Other farms are proving to be more attentive to the economic dimension, but need to improve their commitment to the environmental and social fields. The existence of the third group of farms demonstrates that a multifunctional management that includes sustainability in its three dimensions (environmental, social, and economic) is possible. The policy tools for supporting this sector will have to be defined in a different way to support the development of an articulated and complex sector.

Author Contributions: Conceptualization, M.M., Y.V., G.P., J.D.P., and F.A.; methodology, M.M., Y.V., G.P., J.D.P., and F.A.; software, M.M., Y.V., and G.P.; validation, M.M., Y.V., and G.P.; formal analysis, M.M., Y.V., and G.P.; investigation, M.M., Y.V., G.P., J.D.P., and F.A.; resources, M.M., Y.V., G.P., and F.A.; data curation, M.M., Y.V., and G.P.; writing—original draft preparation, M.M., Y.V., and G.P.; writing—review and editing, M.M., Y.V., G.P., and J.D.P.; visualization, M.M., Y.V., G.P., J.D.P., and F.A.; supervision, M.M., Y.V., G.P., J.D.P., and F.A.; project administration, M.M., Y.V., and G.P. All authors have read and agreed to the published version of the manuscript.

Funding: The present study was carried out in the framework of the Project "Demetra" (Dipartimenti di Eccellenza 2018–2022, CUP_C46C18000530001), funded by the Italian Ministry for Education, University, and Research.

Institutional Review Board Statement: Not applicable.

Informed Consent Statement: Not applicable.

Data Availability Statement: The data can be found at https://rica.crea.gov.it.

Conflicts of Interest: The authors declare no conflict of interest. The funding sponsors had no role in the design of the study; in the collection, analyses, or interpretation of data; in the writing of the manuscript, and in the decision to publish the results.

References

1. Eurostat 2019. Available online: https://ec.europa.eu/eurostat/product?code=APRO_MK_POBTA&mode=view (accessed on 22 June 2021).
2. Eurostat 2019. Production of Milk on Farms. Available online: https://ec.europa.eu/eurostat/product?code=APRO_MK_FARM&mode=view (accessed on 22 June 2021).
3. Battini, F.; Agostini, A.; Tabaglio, V.; Amaducci, S. Environmental impacts of different dairy farming systems in the Po Valley. *J. Clean. Prod.* **2016**, *112*, 91–102. [CrossRef]
4. Istat 2019. Available online: http://dati.istat.it (accessed on 28 May 2021).
5. Caroli, M.; Brunetta, F.; Valentino, A. 2019 L'Industria Alimentare in Italia. Sfide, Traiettorie Strategiche e Politiche di Sviluppo. Available online: www.federalimentare.it (accessed on 28 May 2021).
6. Banca Dati Nazionale dell'Anagrafe Zootecnica (BDN)—Istituto Zooprofilattico Sperimentale dell'Abruzzo e del Molise "G. Caporale". Available online: https://www.vetinfo.it (accessed on 28 May 2021).
7. Bunkus, R.; Theesfeld, I. Land Grabbing in Europe? Socio-Cultural Externalities of Large-Scale Land Acquisitions in East Germany. *Land* **2018**, *7*, 98. [CrossRef]
8. Maynard, H.; Nault, J. Big Farms, Small Farms. Strategies in Sustainable Agriculture to Fit All Sizes. 2005. Available online: https://foodsecurecanada.org/sites/foodsecurecanada.org/files/AIC_discussion_paper_Final_ENG1.pdf (accessed on 22 June 2021).
9. Adinolfi, F.; Russo, C.; Sabbatini, M. L'evoluzione della Struttura Delle Aziende Agricole Negli Anni '90: Un'analisi Alla Luce dei Dati Censuari. *Econ. Dirit. Agroaliment.* **2006**, *11*, 55–74.
10. Flaten, O.; Koesling, M.; Hansen, S.; Veidal, A. Links between profitability, nitrogen surplus, greenhouse gas emissions, and energy intensity on organic and conventional dairy farms. *Agroecol. Sustain. Food Syst.* **2019**, *43*, 957–983. [CrossRef]
11. Lovarelli, D.; Bava, L.; Zucali, M.; D'Imporzano, G.; Adani, F.; Tamburini, A.; Sandrucci, A. Improvements to dairy farms for environmental sustainability in Grana Padano and Parmigiano Reggiano production systems. *Ital. J. Anim. Sci.* **2019**, *18*, 1035–1048. [CrossRef]
12. Murray, A.; Skene, K.; Haynes, K. The Circular Economy: An interdisciplinary exploration of the concept and application in a global context. *J. Bus. Ethics* **2017**, *140*, 369–380. [CrossRef]
13. Barros, M.V.; Salvador, R.; de Francisco, A.C.; Piekarski, C.M. Mapping of research lines on circular economy practices in agriculture: From waste to energy. *Renew. Sustain. Energy Rev.* **2020**, *131*, 109958. [CrossRef]
14. Sariatli, F. Linear Economy Versus Circular Economy: A Comparative and Analyzer Study for Optimization of Economy for Sustainability. *Visegr. J. Bioecon. Sustain. Dev.* **2017**, *6*, 31–34. [CrossRef]
15. Kılkış, Ş.; Kılkış, B. Integrated circular economy and education model to address aspects of an energy-water-food nexus in a dairy facility and local contexts. *J. Clean. Prod.* **2017**, *167*, 1084–1098. [CrossRef]
16. Mohammed, S.; Alsafadi, K.; Takács, I.; Harsányi, E. Contemporary changes of greenhouse gases emission from the agricultural sector in the EU-27. *Geol. Ecol. Landsc.* **2020**, *4*, 282–287. [CrossRef]
17. Neethirajan, S. The role of sensors, big data and machine learning in modern animal farming. *Sens. Bio-Sens. Res.* **2020**, *29*, 100367. [CrossRef]
18. Yunan, X.; Weixin, L.; Yujie, Y.; Hui, W. Evolutionary game for the stakeholders in livestock pollution control based on circular economy. *J. Clean. Prod.* **2021**, *282*, 125403. [CrossRef]
19. Segerkvist, K.A.; Hansson, H.; Sonesson, U.; Gunnarsson, S. Research on Environmental, Economic, and Social Sustainability in Dairy Farming: A Systematic Mapping of Current Literature. *Sustainability* **2020**, *12*, 5502. [CrossRef]

20. Gaudino, S.; Reidsma, P.; Kanellopoulos, A.; Sacco, D.; Van Ittersum, M.K. Integrated Assessment of the EU's Greening Reform and Feed Self-Sufficiency Scenarios on Dairy Farms in Piemonte, Italy. *Agriculture* **2018**, *8*, 137. [CrossRef]
21. Perrin, A.; Cristobal, M.S.; Milestad, R.; Martin, G. Identification of resilience factors of organic dairy cattle farms. *Agric. Syst.* **2020**, *183*, 102875. [CrossRef]
22. Lebacq, T.; Baret, P.; Stilmant, D. Role of input self-sufficiency in the economic and environmental sustainability of specialised dairy farms. *Animal* **2014**, *9*, 544–552. [CrossRef] [PubMed]
23. Ward, S.M.; Holden, N.M.; White, E.P.; Oldfield, T.L. The 'circular economy' applied to the agriculture (livestock production) sector—Discussion paper. In Proceedings of the Workshop on the Sustainability of the EU's Livestock Production Systems, European Commission, DG Agriculture and Rural Development, Brussels, Belgium, 14–15 September 2016.
24. Ellen McArthur Foundation. 2020. Available online: https://www.ellenmacarthurfoundation.org/explore/the-circular-economy-in-detail (accessed on 22 June 2021).
25. Verduna, T.; Blanc, S.; Merlino, V.M.; Cornale, P.; Battaglini, L.M. Sustainability of four dairy farming scenarios in an Alpine environment: The case study of Toma di Lanzo cheese. *Front. Vet. Sci.* **2020**, *7*. [CrossRef]
26. Ghisellini, P.; Protano, G.; Viglia, S.; Gaworski, M.; Setti, M.; Ulgiati, S. Integrated Agricultural and Dairy Production within a Circular Economy Framework. A Comparison of Italian and Polish Farming Systems. *J. Environ. Account. Manag.* **2014**, *2*, 367–384. [CrossRef]
27. Van Zanten, H.H.; Van Ittersum, M.K.; De Boer, I.J. The role of farm animals in a circular food system. *Glob. Food Secur.* **2019**, *21*, 18–22. [CrossRef]
28. Bánkuti, F.I.; Prizon, R.C.; Damasceno, J.C.; De Brito, M.M.; Pozza, M.S.S.; Lima, P.G.L. Farmers' actions toward sustainability: A typology of dairy farms according to sustainability indicators. *Animal* **2020**, *14*, s417–s423. [CrossRef]
29. Gelasakis, A.; Valergakis, G.; Arsenos, G.; Banos, G. Description and typology of intensive Chios dairy sheep farms in Greece. *J. Dairy Sci.* **2012**, *95*, 3070–3079. [CrossRef] [PubMed]
30. Usai, M.G.; Casu, S.; Molle, G.; Decandia, M.; Ligios, S.; Carta, A. Using cluster analysis to characterize the goat farming system in Sardinia. *Livest. Sci.* **2006**, *104*, 63–76. [CrossRef]
31. Gelasakis, A.I.; Rose, G.; Giannakou, R.; Valergakis, G.E.; Theodoridis, A.; Fortomaris, P.; Arsenos, G. Typology and characteristics of dairy goat production systems in Greece. *Livest. Sci.* **2017**, *197*, 22–29. [CrossRef]
32. Castel, J.M.; Mena, Y.; Delgado-Pertñez, M.; Camúñez, J.; Basulto, J.; Caravaca, F.; Guzmán-Guerrero, J.L.; Alcalde, M.J. Characterization of semi-extensive goat production systems in southern Spain. *Small Rumin. Res.* **2003**, *47*, 133–143. [CrossRef]
33. Toro-Mujica, P.; García, A.; Gómez-Castro, A.; Perea, J.; Rodríguez-Estévez, V.; Angón, E.; Barba, C. Organic dairy sheep farms in south-central Spain: Typologies according to livestock management and economic variables. *Small Rumin. Res.* **2012**, *104*, 28–36. [CrossRef]
34. Ruiz, F.A.; Vázquez, M.; Camuñez, J.A.; Castel, J.M.; Mena, Y. Characterization and challenges of livestock farming in Mediterranean protected mountain areas (Sierra Nevada, Spain). *Span. J. Agric. Res.* **2020**, *18*, e0601. [CrossRef]
35. Stylianou, A.; Sdrali, D.; Apostolopoulos, C.D. Capturing the diversity of Mediterranean farming systems prior to their sustainability assessment: The case of Cyprus. *Land Use Policy* **2020**, *96*, 104722. [CrossRef]
36. Scarpato, D.; Civero, G.; Rusciano, V.; Risitano, M. Sustainable strategies and corporate social responsibility in the Italian fisheries companies. *Corp. Soc. Responsib. Environ. Manag.* **2020**, *27*, 2983–2990. [CrossRef]
37. Lebacq, T.; Baret, P.V.; Stilmant, D. Sustainability indicators for livestock farming. A review. *Agron. Sustain. Dev.* **2013**, *33*, 311–327. [CrossRef]
38. Leip, A.; Billen, G.; Garnier, J.; Grizzetti, B.; Lassaletta, L.; Reis, S.; Simpson, D.; Sutton, M.A.; De Vries, W.; Weiss, F.; et al. Impacts of European livestock production: Nitrogen, sulphur, phosphorus and greenhouse gas emissions, land-use, water eutrophication and biodiversity. *Environ. Res. Lett.* **2015**, *10*, 115004. [CrossRef]
39. Mubareka, S.; Maes, J.; LaValle, C.; De Roo, A. Estimation of water requirements by livestock in Europe. *Ecosyst. Serv.* **2013**, *4*, 139–145. [CrossRef]
40. Legesse, G.; Ominski, K.H.; Beauchemin, K.A.; Pfister, S.; Martel, M.; McGeough, E.J.; Hoekstra, A.Y.; Kroebel, R.; Cordeiro, M.R.C.; McAllister, T.A. BOARD-INVITED REVIEW: Quantifying water use in ruminant production. *J. Anim. Sci.* **2017**, *95*, 2001–2018. [CrossRef] [PubMed]
41. Guth, M.; Smędzik-Ambroży, K.; Czyżewski, B.; Stępień, S. The Economic Sustainability of Farms under Common Agricultural Policy in the European Union Countries. *Agriculture* **2020**, *10*, 34. [CrossRef]
42. O'Donoghue, C.; Devisme, S.; Ryan, M.; Conneely, R.; Gillespie, P.; Vrolijk, H. Farm economic sustainability in the European Union: A pilot study. *Stud. Agric. Econ.* **2016**, *118*, 163–171. [CrossRef]
43. Gaviglio, A.; Bertocchi, M.; Marescotti, M.E.; DeMartini, E.; Pirani, A. The social pillar of sustainability: A quantitative approach at the farm level. *Agric. Food Econ.* **2016**, *4*, 21. [CrossRef]
44. Barnaud, C.; Couix, N. The multifunctionality of mountain farming: Social constructions and local negotiations behind an apparent consensus. *J. Rural Stud.* **2020**, *73*, 34–45. [CrossRef]
45. Fantin, V.; Buttol, P.; Pergreffi, R.; Masoni, P. Life cycle assessment of Italian high quality milk production. A comparison with an EPD study. *J. Clean. Prod.* **2012**, *28*, 150–159. [CrossRef]
46. Gaudino, S.; Goia, I.; Grignani, C.; Monaco, S.; Sacco, D. Assessing agro-environmental performance of dairy farms in northwest Italy based on aggregated results from indicators. *J. Environ. Manag.* **2014**, *140*, 120–134. [CrossRef]

47. Madry, W.; Mena, Y.; Roszkowska-Mądra, B.; Gozdowski, D.; Hryniewski, R.; Castel, J.M. An overview of farming system typology methodologies and its use in the study of pasture-based farming system: A review. *Span. J. Agric. Res.* **2013**, *11*, 316–326. [CrossRef]
48. Romesburg, C.; Marshall, K. *User's Manual for CLUSTER/CLUSTID Computer Programs for Hierarchical Cluster Analysis*; Lifetime Learning Publications: Belmont, CA, USA, 1984.
49. Eboli, M.G. Aziende e famiglie in due differenti contesti del Lazio meridionale: Latina e Frosinone. In *Agricoltura Familiare in Transizione*; De Bendictis, M., Ed.; Studi e Ricerche, Inea: Rome, Italy, 1995.
50. Perito, M.A.; De Rosa, M.; Bartoli, L.; Chiodo, E.; Martino, G. Heterogeneous Organizational Arrangements in Agrifood Chains: A Governance Value Analysis Perspective on the Sheep and Goat Meat Sector of Italy. *Agriculture* **2017**, *7*, 47. [CrossRef]
51. Giovannini, E.; Sabbatini, M.; Turri, E. *Le Statistiche Agrarie verso il 2000—Contributi di Ricerca all'Analisi Strutturale e Socioeconomica delle Aziende*; ISTAT Collana Argomenti: Rome, Italy, 1999.
52. Feil, A.A.; Schreiber, D.; Haetinger, C.; Haberkamp, Â.M.; Kist, J.I.; Rempel, C.; Maehler, A.E.; Gomes, M.C.; Da Silva, G.R. Sustainability in the dairy industry: A systematic literature review. *Environ. Sci. Pollut. Res.* **2020**, *27*, 33527–33542. [CrossRef] [PubMed]
53. Kaiser, H.F. The varimax criterion for analytic rotation in factor analysis. *Psychometrika* **1958**, *23*, 187–200. [CrossRef]
54. Contreras, P.; Murtagh, F. Herarchical Clustering. In *Handbook of Cluster Analysis*; Henning, C., Meila, M., Murtagh, F., Rocci, R., Eds.; CRC Press: Boca Raton, FL, USA, 2016.
55. Micha, E.; Heanue, K.; Hyland, J.J.; Hennessy, T.; Dillon, E.J.; Buckley, C. Sustainability levels in Irish dairy farming: A farm typology according to sustainable performance indicators. *Stud. Agric. Econ.* **2017**, *119*, 62–69. [CrossRef]
56. Vitali, G.; Cardillo, C.; Albertazzi, S.; Della Chiara, M.; Baldoni, G.; Signorotti, C.; Trisorio, A.; Canavari, M. Classification of Italian Farms in the FADN Database Combining Climate and Structural Information. *Cartogr. Int. J. Geogr. Inf. Geovis.* **2012**, *47*, 228–236. [CrossRef]
57. Ghasemi, M.; Badsar, M.; Falahati, L.; Karamidehkordi, E. The mediation effect of rural women empowerment between social factors and environment conservation (combination of empowerment and ecofeminist theories). *Environ. Dev. Sustain.* **2021**, 1–23. [CrossRef]
58. Trauger, A. Because they can do the work: Women farmers insustainable agriculture in Pennsylvania, USA. *Gend. Place Cult.* **2004**, *11*, 289–307. [CrossRef]
59. Fairlie, R.W.; Robb, A. Gender differences in business performance: Evidence from the characteristics of business owners survey. *Small Bus. Econ.* **2009**, *33*, 375–395. [CrossRef]
60. Galluzzo, N. An analysis of the efficiency in a sample of small Italian farms part of the FADN dataset. *Agric. Econ.* **2016**, *62*, 62–70. [CrossRef]
61. Adinolfi, F.; Capitanio, F.; De Rosa, M.; Vecchio, Y. Gender differences in farm entrepreneurship: Comparing farming performance of women and men in Italy. *New Medit* **2020**, *19*, 69–82. [CrossRef]
62. Muller, A.; Schader, C.; Scialabba, N.E.-H.; Brüggemann, J.; Isensee, A.; Erb, K.-H.; Smith, P.; Klocke, P.; Leiber, F.; Stolze, M.; et al. Strategies for feeding the world more sustainably with organic agriculture. *Nat. Commun.* **2017**, *8*, 1290. [CrossRef]
63. Gullino, P.; Battisti, L.; Larcher, F. Linking Multifunctionality and Sustainability for Valuing Peri-Urban Farming: A Case Study in the Turin Metropolitan Area (Italy). *Sustainability* **2018**, *10*, 1625. [CrossRef]
64. Kelly, E.; Latruffe, L.; Desjeux, Y.; Ryan, M.; Uthes, S.; Diazabakana, A.; Dillon, E.; Finn, J. Sustainability indicators for improved assessment of the effects of agricultural policy across the EU: Is FADN the answer? *Ecol. Indic.* **2018**, *89*, 903–911. [CrossRef]
65. Porrini, D.; Fusco, G.; Miglietta, P.P. Post-Adversities Recovery and Profitability: The Case of Italian Farmers. *Int. J. Environ. Res. Public Health* **2019**, *16*, 3189. [CrossRef] [PubMed]
66. Bassi, I.; Iseppi, L.; Nassivera, F.; Peccol, E.; Cisilino, F. Alpine agriculture today: Evidence from the italian alps: Acces la success. *Calitatea* **2020**, *21*, 122–127.
67. Baležentis, T.; Galnaitytė, A.; Kriščiukaitienė, I.; Namiotko, V.; Novickytė, L.; Streimikiene, D.; Melnikiene, R. Decomposing Dynamics in the Farm Profitability: An Application of Index Decomposition Analysis to Lithuanian FADN Sample. *Sustainability* **2019**, *11*, 2861. [CrossRef]
68. Vecchio, Y.; Pauselli, G.; Adinolfi, F. Exploring Attitudes toward Animal Welfare through the Lens of Subjectivity—An Application of Q-Methodology. *Animals* **2020**, *10*, 1364. [CrossRef] [PubMed]
69. European Commission. Communication from the Commission to the European Parliament, the European Council, the Council, the European Economic and Social Committee and the Committee of the Regions. The European Green Deal. Com/2019/640 Final. 2019. Available online: https://eur-lex.europa.eu/resource.html?uri=cellar:b828d165-1c22-11ea-8c1f-01aa75ed71a1.0002.02/DOC_1&format=PDF (accessed on 23 June 2021).
70. European Commission. Communication from the Commission to the European Parliament, the Council, the European Economic and Social Committee and the Committee of the Regions. A Farm to Fork Strategy for a Fair, Healthy and Environmentally-Friendly Food System. Com/2020/381 Final. 2020. Available online: https://eur-lex.europa.eu/resource.html?uri=cellar:ea0f9f73-9ab2-11ea-9d2d-01aa75ed71a1.0001.02/DOC_1&format=PDF (accessed on 23 June 2021).

Article

Investigating the Suitability of a Heat Pump Water-Heater as a Method to Reduce Agricultural Emissions in Dairy Farms

Patrick S. Byrne *, James G. Carton and Brian Corcoran

School of Mechanical and Manufacturing Engineering, Faculty of Engineering and Computing, Dublin City University, Whitehall, Dublin 9, Ireland; james.carton@dcu.ie (J.G.C.); brian.corcoran@dcu.ie (B.C.)
* Correspondence: patrick.byrne49@mail.dcu.ie; Tel.: +353-851094981

Abstract: The performance of an air-source heat pump water-heater (ASHPWH) system manufactured by Kronoterm was benchmarked in this study for the application of dairy farming in Ireland. The COP of the system was calculated to be 2.27 under normal operating conditions. The device was able to supply water at 80 °C, however a full tank at this temperature was not achieved or deemed necessary for the dairy application. Litres per kWh was used as a performance metric for the device and the usable water per unit of energy for the system was found to be 397 L when using both electric heaters and 220 L when using just the top heater both in conjunction with the heat pump. The performance of the heat pump system in terms of its cost to run and efficiency was also compared with five other water heaters. The heat pump is seen to be very efficient, however due to the carbon intensity of the Irish grid electricity and high water temperatures required, the solar water heater with gas backup was found to be the best performing under energy efficiency and carbon emissions per litre of usable water. In conclusion, although the heat pump was not the best-performing system under these metrics, the cost and complexity of the solar-gas system may be a deterrent for dairy farmers and for this reason, the heat pump is considered a cost-effective, efficient and viable option for dairy farmers trying to reduce their carbon footprint and energy bills.

Keywords: heat pump; dairy farming; water heater; ASHP; ASHPWH; usable water; emissions

Citation: Byrne, P.S.; Carton, J.G.; Corcoran, B. Investigating the Suitability of a Heat Pump Water-Heater as a Method to Reduce Agricultural Emissions in Dairy Farms. *Sustainability* **2021**, *13*, 5736. https://doi.org/10.3390/su13105736

Academic Editor: Rajeev Bhat

Received: 5 April 2021
Accepted: 6 May 2021
Published: 20 May 2021

Publisher's Note: MDPI stays neutral with regard to jurisdictional claims in published maps and institutional affiliations.

Copyright: © 2021 by the authors. Licensee MDPI, Basel, Switzerland. This article is an open access article distributed under the terms and conditions of the Creative Commons Attribution (CC BY) license (https://creativecommons.org/licenses/by/4.0/).

1. Introduction

One of the main drivers of environmental damage in the form of climate change is the emission of greenhouse gases (GHGs) in the production of energy, and in particular electricity. Due to the increase in the burning of fossil fuels since the industrial revolution, the increase in concentration of these GHGs in our atmosphere has resulted in a warming effect on the planet [1]. There are many ways in which emissions can be reduced and the implementation of all of the possible strategies improves the chances of meeting the targets set out by the EU. Figure 1, below, shows how our climate targets interact with one another. By improving the energy efficiency, the demand for energy decreases, because less energy is required to perform the same basic task, e.g., electricity for agricultural water-heating. By decreasing the demand for energy, the proportion of energy that is generated through renewable sources increases, because the total energy has decreased. In the same way, by reducing the demand for energy through improvements in energy efficiency, the amount of fossil fuels burned is reduced. The same works for increasing the quantity of energy produced from renewable energy. For this reason, all avenues to reducing energy demand and increasing renewable energy production must be explored. For example, in thermal power plants using fossil fuels to produce electricity, only about 55% of the energy stored within the fuel is converted to electricity as remaining energy is lost as heat [2]. Other losses which occur in the production of electricity are distribution and parasitic losses. The net supply efficiency of electricity in Ireland in 2017 was 49%, which means that 51% of the energy produced was lost [2].

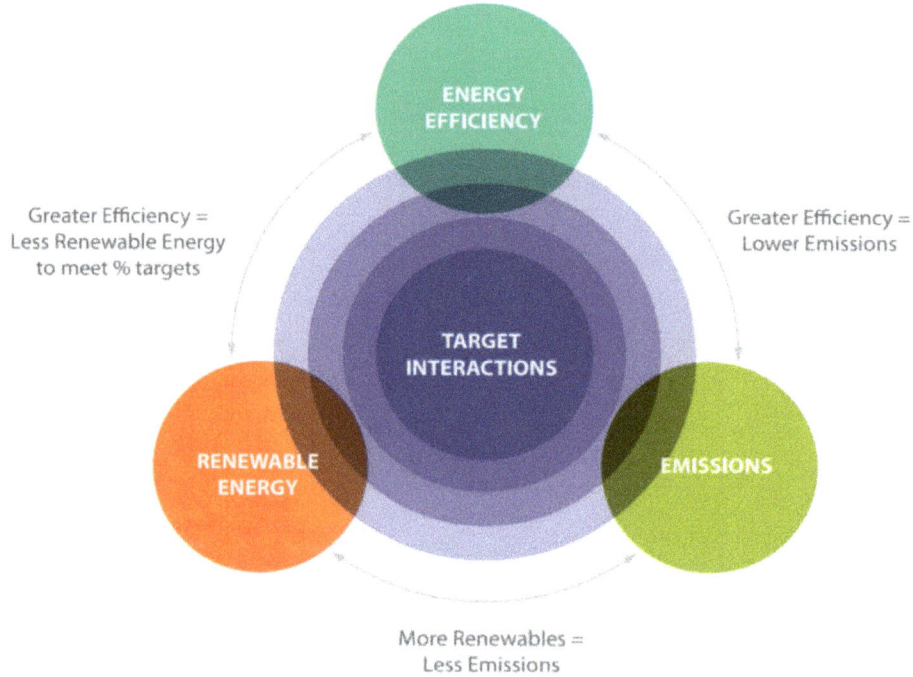

Figure 1. Interaction of climate actions [3].

Contrasting to renewables the inefficiency of fuel sources such as coal and peat, as well as the higher carbon intensity of those fuels highlight the importance of investment in the renewable energy space in order to reduce GHG emissions in the production of electricity, as electricity accounted for 27% of the energy used in Ireland in 2018 [4]. An important point to note is that in the production of electricity from non-combustible renewable sources such as wind, solar and hydro (not pumped hydro storage), there are no transformation losses incurred, as there is no energy lost in the production of electricity from these sources. There are still some losses that occur in the transport of the electricity generated, and there may be some energy required to produce the electricity from these sources also. These losses are much smaller than the transformation losses incurred when producing electricity.

Reducing electricity consumption by way of increasing the efficiency of appliances and of electrical water-heating systems have a large impact on reducing Ireland's GHG emissions [5]. This is because as we reduce the demand for electricity the fuel mix becomes more renewable as fewer fossil fuels need to be burned to make up the demand.

Currently, the agriculture sector in Ireland is responsible for 34% of Ireland's total GHG emissions, equating to 45% of Ireland's non-Emissions Trade Scheme (ETS) emissions [6]. The contribution of each industry to Ireland's GHG emissions is shown in Figure 2. It can be seen that agriculture is the largest contributor to Ireland's GHG emissions. Due to the high proportion of Ireland's non-ETS emissions coming from this sector, it is imperative that energy emissions in this sector are reduced if the targets set by the EU are going to be met.

In the agriculture sector, emissions have increased year on year since 2014 according to the Sustainable Energy Authority Ireland (SEAI) and the Environmental Protection Agency (EPA) [4,7]. This contrasts greatly with the general trend in emissions in Ireland which have been reducing, see Figure 3. The main reason suggested by the EPA for this is an increase in dairy cow numbers, up 27% over the past five years, which equates to an increase of over 300,000 dairy cows [7]. The increase in dairy cows is the result of the milk quota's abolition in 2015 [8]. Although abolition of the quotas has resulted in increased incomes for Irish dairy farmers, failure to improve efficiency in practices has resulted in an increase in emissions over this time [8]. Most of the emissions in agriculture are non-energy related, and a large proportion of these non-energy-related emissions are difficult to avoid without the reduction of the herd numbers [9]. Therefore, in order to see any significant reduction in emissions from agriculture, dairy farmers will have to be much more efficient in terms of their energy use.

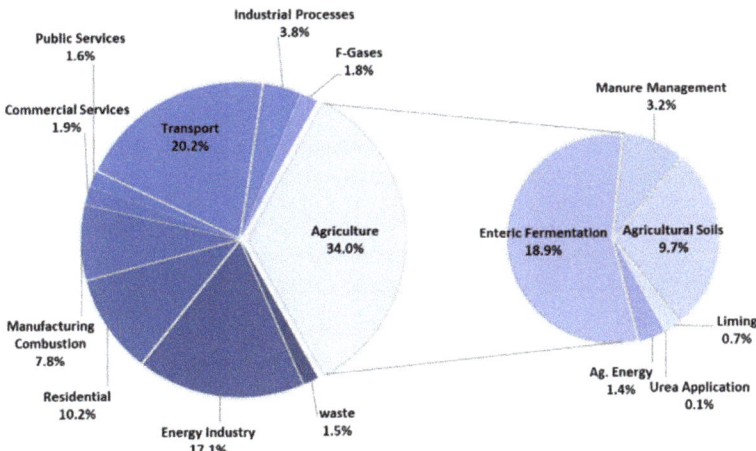

Figure 2. Breakdown of 2018 emissions (data from [6,10]).

Figure 3. Agriculture emissions from 2000–2018 (data from [10]).

Maintaining low total bacteria count and a low thermoduric count is important for maintaining milk quality on dairy farms, and one way this is achieved is through using the correct cleaning procedures. Hot water is required for washing milking equipment and the bulk tank on dairy farms. Temperatures of 65 °C to 75 °C are required for cleaning the milking equipment, and 60 °C to 70 °C for cleaning the bulk tank [11]. Water temperatures will drop during the washdown procedure, so according to Glanbia, starting temperatures of 70–75 °C are essential [12]. The quantity of hot water required on dairy farms fluctuates based on the frequency of milking and the size of the farm, so it is difficult to quantify the hot water required for an average farm on a daily basis. For the milking machines, roughly 10 litres of hot water are required per milking unit, and an average farm has 16 milking units [11]. An 8000 L bulk tank, which is adequate for an average sized farm requires about 160 L of hot water for cleaning.

A method which has been used to reduce energy consumption in an industrial and residential setting has been the implementation of heat pumps for space and water heating. A study carried out by Hong and Howarth found that in 2009, 18% of energy used in the home was for water heating [13]. They implemented a scenario to estimate the energy savings that could be incurred by using heat pumps instead of the more popular gas and electric heaters, for domestic hot water in residential and commercial setting. The findings were that if heat pumps were used instead, total emissions from fossil fuels in residential and commercial use would be 26% lower [13]. It was concluded following the study that heat pumps are an effective method of reducing GHG emissions in water heating [13].

Upton et al. analysed the electricity consumption of 22 Irish dairy farms along with their daily and seasonal trends to identify the strategies which could reduce electricity usage on dairy farms and also maximise the use of electricity during off peak hours to reduce costs [14]. An interesting result found that 20 of the 22 dairy farms studied were using electric water heaters [14]. The study also found that 60% of the direct energy used in the farm was electricity, and of that electricity used, 80% was used in the milking parlour [14]; the two practices that used the most energy were milk cooling (31%) and water heating (23%) [14].

Gas boilers and liquefied petroleum gas (LPG) instant boilers are also common methods of heating water in dairy farms [15]. Both of these are lower in emissions in comparison to the electric storage heaters due to the lower carbon intensities of the fuels. However, as the production of electricity is based more on renewables all the time, this may not be the case for the entire life cycle of a new system which is installed, which should be considered when purchasing a new water heater. Oil boilers were a popular choice, but they have higher installation costs and cause greater emissions due to the higher carbon intensity of the fuel. In terms of renewable water-heating technologies, solar thermal is an option, but according to Upton et al., it can only supply around 40% of the total hot water demand [15]. Heat pumps are another renewable technology that can be used for water heating. Heat pump water heaters work on the basis of a reverse refrigeration cycle, which means they can transfer low grade heat into useable heat to increase the temperature of the water. There are three main heat pump technologies: water source heat pumps (WSHPs), ground source heat pumps (GSHPs) and air-source heat pumps (ASHPs).

Heat pumps have advantages compared to other systems because of their ability to heat water efficiently due to their high coefficient of performance (COP). The COP of a heat pump refers to the amount of heat energy supplied by the heat pump, divided by the amount of energy consumed by the heat pump. Figure 4 shows each type of water heater with emissions (excluding solar as it has no emissions). Table 1 shows the information in Figure 4 in a tabular format. It can be seen from Figure 4 that the COP of the heat pump seems to make it much more efficient in terms of carbon emissions produced per unit of useful heat. Due to the lack of education in heat pump technology, higher initial cost and a history of limited availability, the uptake of heat pumps in Irish dairy farming has been limited. However, there are now grants available in Ireland by the SEAI for businesses which reduce the initial purchase cost of heat pumps [16].

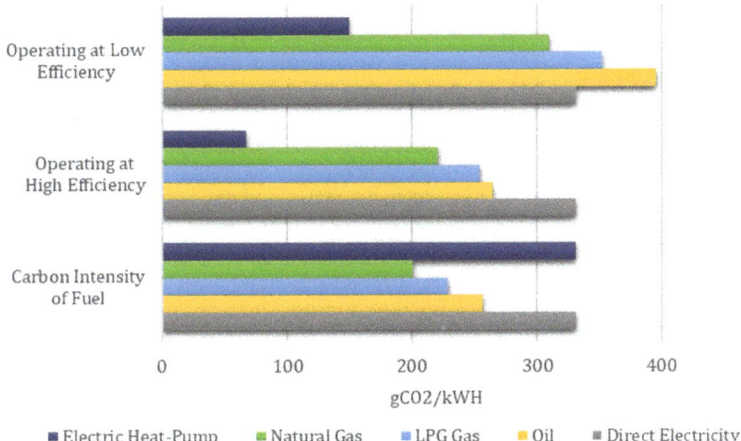

Figure 4. Heat pumps vs. water heaters (adapted from SEAI [17]).

Table 1. Comparison of water heater efficiency.

	Direct Electricity	Oil	LPG Gas	Natural Gas	Electric Heat-Pump
High Efficiency Conversion Rate	100%	97%	91%	91%	560%
Low Efficiency Conversion Rate	100%	65%	65%	65%	250%
Carbon Intensity of Fuel (gCO_2/kWh)	331	265	229	201	331

There are many different types of heat pump which could be used in the application of dairy farming, but the type which would be most applicable to all farms would be the air-source heat pump water-heater (ASHPWH) [17]. Other heat pumps are more expensive to install and may not be possible to install in certain farms, and therefore this study will deal with ASHPWHs in particular. In this study the performance of an ASHPWH system is of interest for the application of dairy farming in Ireland. A study by Upton et al. [18] investigated the performance of an ASHPWH with an electric heating element for the dairy farm application. The heat pump was used to heat the water to 55 °C, and the electric heating element was then used to heat the water to 80 °C. The study used usable water as a metric for the performance of the heat pump and it defined usable water as water between the temperatures of 60 °C and 80 °C [18]. This is the temperature range for which the water is deemed usable for washing down the bulk tank and the temperature range for cleaning the milking machines is 65 °C to 75 °C. The temperature of the water in the tank may not be the temperature that the water will be at when it reaches the bulk tank and milking machines, so a higher threshold for useable water temperature is used in this study to account for this. The metric of L/kWh was used to compare the performance of the heat pump water heater with an electric emersion element only water heater. The ASHPWH was able to produce 15.12 L/kWh in comparison to 9.66 L/kWh with the electric emersion heater [18].

Upton's study found that the maximum amount of usable water which could be drawn out of the system was found when the water was mixed during the heating process. This study aims to use the same metrics to compare an ASHPWH with some of the other options available to a dairy farmer looking to improve the energy efficiency of their farm, in particular the performance of an air-source heat pump water-heater (ASHPWH) system, manufactured by Kronoterm, is analysed for its ability to reduce emissions in the application of dairy farming in Ireland. For this study usable water is in line with Glanbia's recommendation of greater than 70 °C [12].

ASHPWHs are very sensitive to their ambient conditions, but ambient temperature has the largest effect as the amount of energy available to the heat pump from the air decreases as temperature decreases. In general terms, as the ambient temperature increases, the COP of the heat pump increases and as the ambient temperature decreases, so does the COP. The true relationship though is slightly more complex, the COP of the heat pump is related to the difference between the ambient temperature and the temperature of the heat sink. This is why as the temperature of the water increases within the ASHPWH, the COP decreases. As the difference in temperature between the ambient conditions and the heat sink increases, it becomes more difficult for the system to draw useful heat from the air, and therefore it uses more power per unit of useful heat which it supplies to the heat sink. This relationship has been shown in multiple studies, which aimed to compare the performance of ASHPs during different periods of the year. For example, in a study carried out by Ji et al. which was carried out in China [19], the COP increased from 2 at 4.5 °C average temperature to 3.42 at 31 °C average temperature. A similar study carried out by Zhang et al. investigated the COP of a heat pump over each season, in order to find the seasonal performance factor (SPF) of an air-source heat pump [20]. The study found that the COP during the winter, at an average ambient temperature of 0 °C was 2.61, which increased to 4.82 during spring and autumn with an average temperature of 25 °C and increased further during summer to 5.66 for an average temperature of 35 °C [20]. These studies show how great an effect the ambient temperature of the environment can have on the performance of air-source heat pumps.

2. Materials and Methods

2.1. Heat Pump System

The heat pump system used in this study is an air-source heat pump water heater (ASHPWH) made by Kronoterm and supplied by Glenergy Ltd. (Sydney, Australia) [21], which can be seen in Figure 5. The condenser in the heat pump is a wrap-around condenser. The heat pump incorporates two electric water heaters which are used to heat the tank above 65 °C or to heat just the top portion of the tank when only a small amount of hot water is needed. For the purpose of this study, the comfort setting and the external heater settings on the device were used. The comfort setting uses the heat pump to heat water to 65 °C and then uses the electric heaters to heat the water to 80 °C, and the external heater setting allows one to choose for either the top heater or both heaters to be used. The two configurations that the heat pump was tested in are shown in Table 2.

Parameter	Units	Value
Tank Volume	L	450
Electric Heaters	No.	2
Power of Top Heater	W	1920
Power of Bottom Heater	W	1920
Nominal Electrical Power	W	970
Heat Pump Max. Power	W	1940

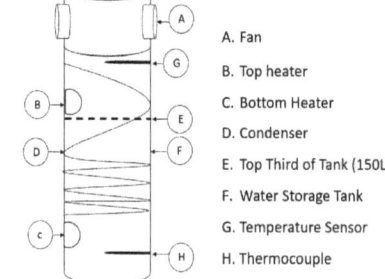

A. Fan
B. Top heater
C. Bottom Heater
D. Condenser
E. Top Third of Tank (150L)
F. Water Storage Tank
G. Temperature Sensor
H. Thermocouple

Figure 5. Heat pump schematic and statistics.

Table 2. Heat pump system configurations.

Configuration	Heat Pump Status	Immersion Element Status
1	On from 20–65 °C	Top Heater Only from 65–80 °C
2	On from 20–65 °C	Both Heaters On from 65–80 °C

2.2. Measurement Equipment

The measurement equipment used in this study included a Fluke 435 Series II Power Quality and Energy Analyser, a Pico data logger with thermocouples, the Kronoterm cloud system and an EL-USB II data logger. Information such as temperatures, relative humidity and power were collected by these systems in order to characterise the performance of the system under the operating conditions of the given test, as ambient conditions can have a large effect on the performance of heat pumps.

2.3. Testing Setup and Procedure

The setup for the tests was consistent for each of the tests performed, and the heat pump settings were used to vary the test parameters. In order to measure the power and energy, the loops and clips on the Fluke system were connected to the relevant wires. The EL-USB II data logger was attached to the air intake to ensure that the ambient conditions were measured accurately. The Pico data logger was connected to k type thermocouples which were used to measure the temperature of the water as it exited the tank, in order to measure the useable water.

2.4. Equations

This formula was used to calculate the energy increase in the water so that the COP of the heat pump could be calculated. The equation used for this was:

$$E = mc_p(\Delta T) \tag{1}$$

where m is the mass of the water, c_p is the specific heat capacity and ΔT is the temperature difference between the start and end temperature of the water.

The COP of an ASHPWH has been defined as the amount of energy the system can transfer to the water divided by the energy consumed by the system.

$$COP = \frac{E}{Q} \tag{2}$$

The carbon emissions per cycle in kg (C) of each water heater was calculated using this formula.

$$C = \frac{EF}{\varepsilon} \times \frac{m}{1000} \tag{3}$$

where EF is the emissions factor for the fuel, ε is the L/kWh value calculated for the water heater and m is the mass of the water heated.

The cost per cycle (P) of each water heater was calculated using the following formula.

$$P = \frac{c_f}{\varepsilon} \times m \tag{4}$$

where c_f is the cost of the fuel per kWh. For electricity, there are day and night rates, so the formula for the electricity calculations is shown below.

$$P = \left(\frac{c_{fD}}{\varepsilon} \times m_D\right) + \left(\frac{c_{fN}}{\varepsilon} \times m_N\right) \tag{5}$$

where c_{fD} is the cost per kWh of electricity on the day rate, m_D is the mass of the water heated on the day rate. The subscript N denotes the same for the night rate.

The area of the solar panels needed for the system, A_s, can be found using the following formula.

$$A_s = \frac{E}{c_s} \times 0.7 \tag{6}$$

where c_s is the solar collection of one square metre of solar panel, and 0.7 represents the solar fraction used in this study.

The average energy generated by the solar installation is found by multiplying the panel area by the average collection per square metre as shown below. This value was then subtracted from the total energy needed to heat the water to find the total energy needed from the backup system, E_B.

$$E_s = A_s \times c_s \tag{7}$$

$$E_B = E - E_s \tag{8}$$

2.5. Comparison Section

The systems which have been chosen to be compared to the heat pump are as follows: electric, natural gas boiler, LPG instant water heater and a solar water heater with an electric or natural gas backup. These systems have been chosen as they all offer different benefits in terms of either fuel source, convenience or efficiency. Solar was chosen as it is generally considered to be one of the most efficient methods of water heating, and the flat plate collector in particular was chosen as flat plates systems are generally cheaper than evacuated tube collectors, which make the cost of the system slightly more comparable to the price of a heat pump water heater. In the literature, the electric backup tends to be seen more than a gas backup for solar water heaters, but both were considered in this study, as the author felt that a natural gas backup would be less carbon intensive than an electric backup. Natural gas was chosen as gas boilers are less carbon intense than oil boilers and have a lower cost. The LPG system was chosen as these systems offer instant hot water which could be a good option for smaller farms with less water-heating requirements. The electric option will also be considered as it is the most common type of water heater used on dairy farms according to Upton's report [14].

The comparison was performed on the basis of one full cycle of the heat pump to a maximum of 80 °C. The amount of usable water from the system from that cycle was used as a comparison metric, and the other values for the heat pump were calculated using the L/kWh figure. The other water heaters were compared to the heat pump in terms of how many L/kWh those systems can produce in a full heating cycle. This method was chosen, as it removes variables such as heat loss which cannot be estimated for the solar or natural gas systems. The water was heated from 20 °C to an average water temperature of 70.7 °C. By using the average temperature, the amount of energy and therefore energy-related emissions of the other water heaters which those heaters would require to achieve with the same amount of water at that average temperature will be calculated. The other metric used to compare the systems is the cost per day, the formula for which can be seen in Equation (5).

2.5.1. Gas Heating Option

The gas heating option considered is an instant water heater made by LPG Flogas. In order to calculate the number of kilowatt hours (kWh) of gas burned to heat the water to the required temperature, the energy formula in equation 1 was used, and an efficiency of 90% was assigned to the gas heater, as this is the high end in terms of efficiency for gas water heaters.

2.5.2. Solar with Electric or Gas Backup

In order to calculate the efficiency of a solar system with a backup heat source, first the sizing of the solar system needed to be calculated. In order to calculate the size of the system, the first step was to find the solar radiation for the summer period, as this is the period where the solar radiation will be highest. By sizing the system based on the maximum solar radiation the system will not be oversized, which in solar installations could be very costly. This is because in oversized solar water-heating systems, at peak solar radiation, stagnation occurs in the system resulting in irreparable damage to the solar collectors.

By finding the average solar radiation per metre squared on a day during the summer, the size of the system can be calculated by dividing the energy required for heating water on an average day by the average solar radiation per meter squared during summer. This gave the area of panels required to supply a 100% solar fraction. Solar water heaters however are not designed to heat water to 80 °C, so to accommodate for this the solar fraction of the installation should be 70%. The remainder of the heating was done by a backup source. Equation (6) shows the mathematical formulation of the calculations described above. The area was rounded to the nearest whole number to give the number of panels required.

To calculate an average day of heating with the designed solar system, the average solar radiation for the year was divided by 365 to find the average solar radiation per metre squared for a day in the year. This solar radiation value was multiplied by the efficiency values for the collector and for the system which were taken from a paper by Duffy and Ayompe [22], which was conducted in Dublin using flat plate collectors with an external storage tank with an electric backup. The solar collection of one panel was then multiplied by the calculated area of the solar installation to calculate the kWh of heat energy made up by the solar system (see Equation (7)). The result of this calculation is a value for the total amount of energy which will be stored in the water over a day by the solar installation. By subtracting this value from the total amount of energy required to heat the water for the given day, the amount of energy required from the backup source on that day can be calculated (see Equation (8)). The amount of energy produced by the solar system is considered free energy in terms of both cost and carbon emissions, so only the energy used by the backup source is considered in the calculations for comparing the efficiency of the system with the ASHPWH.

2.5.3. Calculation Methodology

The two metrics to be used to compare the systems are the carbon emissions and cost per day, assuming daily water use of 397 L. By dividing the carbon intensity of the fuel for each system (gCO_2eq/kWh) by the efficiency of that system (L/kWh), the carbon intensity of each system (gCO_2eq/L) can be found. This is shown in a mathematical formulation in Equation (3).

By dividing the cost per kWh of fuel (€/kWh) by the efficiency of each system (L/kWh), the cost per litre can be found. This can be multiplied by the quantity of useable water needed to find the cost per day of operating the system. The mathematical formulation of this can be seen in Equation (4). To take into account the different cost rates of electricity, Equation (5) should be used.

For the heat pump, it is possible that for some farmers using this system they could operate using only the night rate of electricity, as heat loss from the system during the day is minimal. To account for topping up the heat during the day with the electric heaters, it is assumed that 25% of the electricity use will be during the day, meaning 75% will be on the night rate. The rates of electricity cost used in the calculations are shown in Table 3.

Table 3. Conversion units and cost values.

Fuel Source	Carbon Intensity (gCO_2eq/kWh)	Cost Per kWh (Day)	Cost Per kWh (Night)
Electricity	331	0.15	0.07
Gas	201	0.066	0.066
LPG	229	0.12	0.12

It would be expected that most of the solar radiation will be absorbed by the system after the first milking, and therefore most of the energy used by the electric backup would be used in the night rate. To account for this, 75% of the energy used by the solar with electric backup has been assigned to the night rate and 25% to the day rate.

In order to extrapolate the results to see the comparative yearly performance of the systems, for the sake of simplicity the daily water use will be assumed to be the usable water number calculated for the heat pump for a full cycle.

Electricity pricing comes from a report from J. Upton et al. which was presented at the Teagasc National Dairy Conference [23]. The carbon intensity of the fuel sources comes from the SEAI website which has the most up to date information which is from 2020 [2]. The price of natural gas and LPG comes from a domestic fuel cost comparison published by SEAI [24].

3. Results

The tests carried out on the heat pump were done to benchmark its performance under normal operating conditions. The results in Figure 6 show that the heat pump used in this test had high efficiency at lower water temperatures with a COP of 4.3 for the temperature range of 20–30 °C. The efficiency of the heat pump reduced significantly as the temperature of the water increased. This was expected, as the COP for heat pumps decreases as the discharge temperature increases. Although the heat pump is more efficient than other water heater technologies even at higher water temperatures, it is not as efficient as some of the heat pumps which were seen in the literature such as one in the study by Ibrahim et al. [25] where the COP of the heat pump was still greater than 2.5 at temperatures around 55 °C. The COP of 2.27 for a full cycle is also at the lower end of what was seen in the literature, although most of the studies did not heat the water as high as 65 °C, which did have an impact on the results of the experiment. If the heat pump had been run to 55 °C the COP of the cycle would have been 2.66 which would be comparable with some of the studies in the review of heat pump water heaters by Hepbasli et al. [26]. The performance of an ASHPWH is dependent on ambient temperature and humidity and therefore the COP may change throughout the year. For this reason, in cooler climates it has been seen that locating an ASHPWH indoors can improve the average COP over the year and does not affect space heating bills significantly [27]. Some of the studies cited in that review of water heaters also began with lower initial water temperatures, which would result in a higher COP [26]. The temperature varies by month and by storage method, but a normal range for Ireland is between 8 and 14 °C [28]. In practice, the initial water temperature may be lower than that used in this study but do not affect the overall findings. Although the COP of the heat pump decreases significantly at higher water temperatures (see Table 4) it would still be worth using the heat pump at these higher temperatures as the efficiency of the system is still higher than it would be with an electric water heater.

Table 4. COP of heat pump.

Temperature Range	COP
20–25	4.28
25–30	4.36
30–35	3.45
35–40	2.91
40–45	2.55
45–50	2.13
50–55	1.66
55–60	1.54
60–65	1.33
20–65	2.27

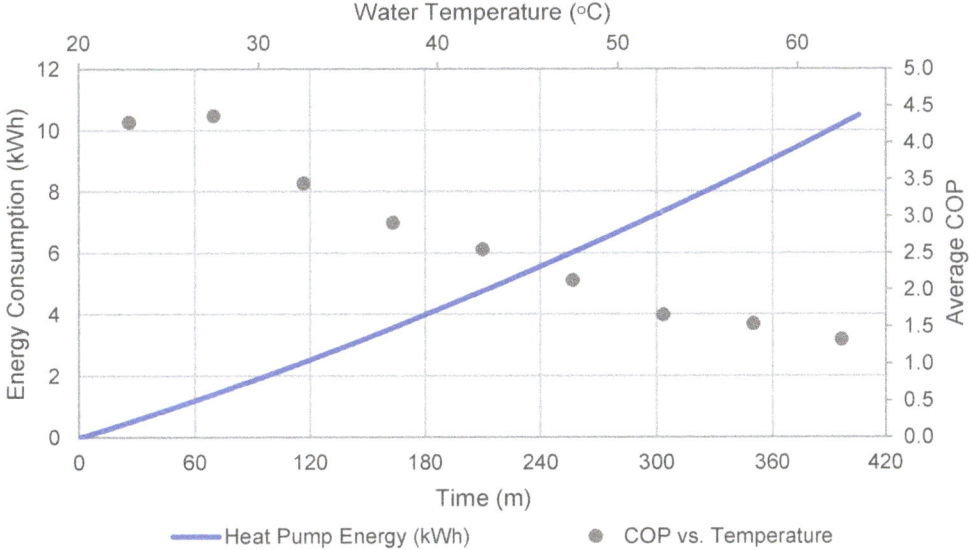

Figure 6. Energy performance of heat pump.

Another key performance indicator of any water-heating system is the time taken to heat the water to a given temperature. The time taken to heat the water from 20–65 °C was 6 h and 46 min. This is a significant amount of time, but it should be noted that the amount of water being heated is also quite large at 450 L. The heat pump is significantly slower to heat water than a typical gas boiler and the instant LPG instant water heater, and therefore the farmer should be aware of this and heat water in advance. The heat pump has a slow rate of temperature loss, which can be seen from the orange line in Figure 7, and therefore in order to save energy costs, the heat pump system should be run to at least 65 °C at the night rate of electricity. The electric heaters can then be used to heat the water to useable temperatures shortly before it is needed. In comparison to the other types of water heater, a gas system would have greater heat losses as gas water-heater tanks are more difficult to insulate due to the internal combustion. The instant water heater has no standby losses, which could mean that for small farms with little water usage, such a system may be ideal. A key performance metric for energy use in water heaters is sizing. If the system is too large it usually leads to excess energy use over time. As more water is heated than needed, heat loss from the system can be quite significant and therefore this is another important factor for a farmer to consider.

Despite the heat pump using more energy towards the end of the cycle, the rate of temperature increase falls, this means that at higher temperatures, the heat pump uses more energy and takes longer to heat the water. It should be mentioned that when the heat pump is used to heat the water, the water heats relatively uniformly. This is because the system uses a wrap-around condenser which can be seen in Figure 5. The uniform heating of the tank was checked with a thermocouple which was placed near the bottom of the tank as seen in Figure 5.

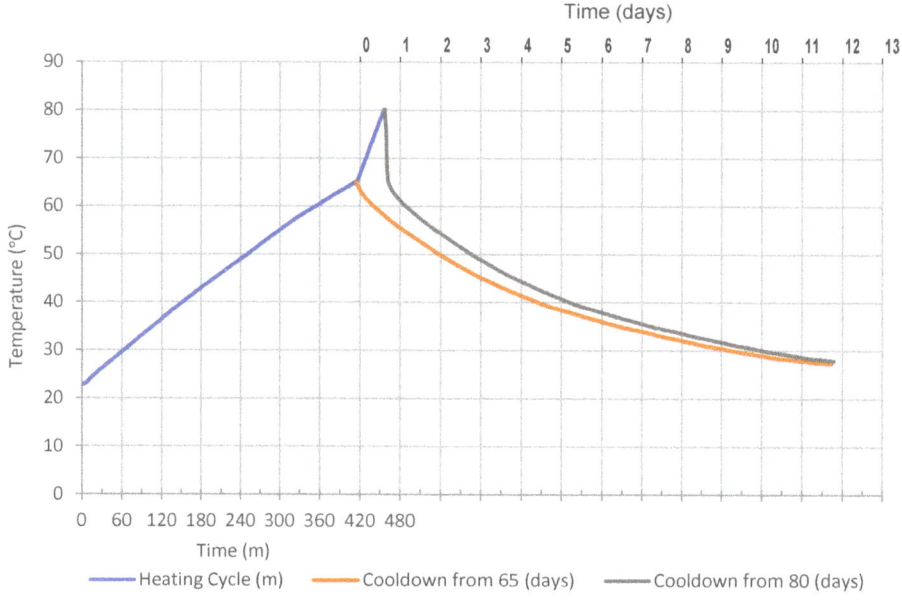

Figure 7. Heat pump heating and cooling cycles.

With regards to the energy used for the full cycle, the total energy when only the top heater was used was 12.4 kWh and using both heaters the energy used was 13.82 kWh, see Table 5. This shows one potential issue with this system, which is that it will not heat the full tank to 80 °C. The top heater heats only roughly the top third of the tank due to thermal stratification, which means that 1.91 kWh of energy is required to heat the top third of the tank from 65–80 °C, see Table 6. The energy consumption over time for both configurations is shown in Figure 8. The system has a built-in thermostat which shuts off the system when it reaches 80 °C. This is located near the top of the tank as shown in Figure 5 and therefore, due to the thermal stratification in the tank caused by the top heater, the whole tank is not heated to 80 °C. Thermal stratification is common in water heaters containing multiple heating elements, particularly when the tank has a large aspect ratio and is an important tool used in designing hot water tanks to reduce the need to reheat the whole tank [29]. One way to solve this issue which would also improve the performance of the heat pump system would be to mix the water while the emersion heaters are being used, as this was noted to improve the L/kWh value in Upton's study [18]. This improves the performance of the system in terms of L/kWh as it ensures that the full tank is heated to the set-point temperature. The difference in quantity of useable water is large, especially when considering that only 1.42 kWh more electricity is used to heat the remaining portion of the tank. The reason for this is that the useable water temperature in this study is considered above 70 °C, and therefore only the top portion of the tank can be considered as usable water when only the top heater is used between 65 and 80 °C. The remaining energy used to heat the water to this temperature is not wasted, as the tank has a very low rate of temperature decline. As cold water enters the system, it will push the hotter water to the top of the tank, and therefore the top heater can be used again to heat this portion of the tank.

Table 5. Full heating cycle energy performance.

Configuration	Time (h)	Air In (°C)	Energy Consumed (kWh)	Useable Water (L)	L/kWh
1	7.45	19.34	12.40	220	17.74
2	7.47	19.34	13.83	397	28.70

Table 6. Electric heater energy use.

Configuration	Power (W)	Time (m)	Total Energy (J)	Total Energy (kWh)
1	2860	40	6,864,000	1.91
2	4810	41.33	11,927,838	3.31

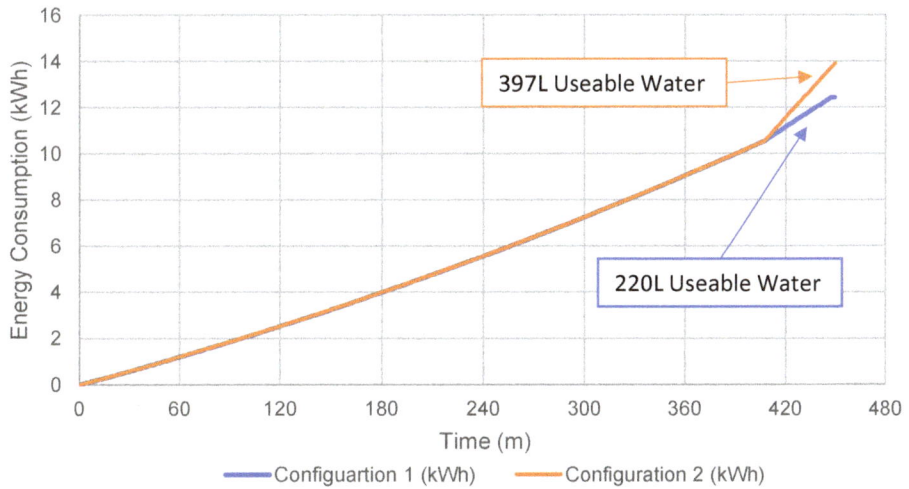

Figure 8. System energy performance.

Comparison to Other Water Heaters

A system, capable of supplying up to 120 L of water at 85 °C in 10 min, is an instant water heater by LPG Flogas [30]. The system has lower efficiency than the ASHPWH. There is no information available regarding the efficiency of the system, but as it is a gas system, an estimate of 90% efficiency has been used. A 90% efficiency will be assumed for the gas boiler also. The heat pump meanwhile has a COP of 2.27 up to 65 °C and then an electric heater of ~99% efficiency is used to heat the water to the set point temperature. The gas system however uses a less carbon intensive fuel, which is a clear advantage over the heat pump system in the current climate. One of the most obvious advantages of the gas system is that it can produce water much faster than the heat pump, which takes 7.5 h to heat 397 L of usable water from a starting temperature of 20 °C.

The other water heater compared to the heat pump system was a solar flat plate collector of 9 m². The method used to size the solar collector was described in the methodology section and the solar radiation information was taken from the Met Eireann website in the sunshine section [31]. Solar systems, similar to the heat pump system used in this study do not heat the water to useable temperatures. The majority of the heating is done by the solar collector when the sun is shining but a backup heater is used to heat the water up to the desired temperature. Because a lot of the water heating will be done by the backup heater, especially in winter and in overcast conditions, it is important that the backup heater is efficient in terms of both cost and carbon emissions. Two backup heating options will be considered in this comparison. The more common option is the electric backup heater, but perhaps the more efficient and cost-effective option is a gas backup heater. Gas is cheaper per kWh and is also a cleaner source of energy than electricity. In terms of a comparison with the heat pump, both systems offer a renewable method of heating water, they both take a relatively longer time to heat water unless the backup source is used and they are both relatively expensive in comparison to the gas system. Both the heat pump and the solar system could be more cost effective using the available government grants, which would offset some of the difference in price.

First, comparing the performance of the systems with reference to their energy efficiency and their carbon intensities—Table 7 shows the results of the calculations detailed in the comparison section of the methodology section of the report. It can be seen from Table 7 that regarding efficiency, the heat pump has the highest efficiency in terms of L/kWh, followed by the solar with electric backup. However, the solar with gas backup is more efficient than the heat pump with respect to its carbon emissions. It contributes to 25% less emissions when compared to the heat pump.

Table 7. Comparison of efficiency and carbon intensity and cost.

System	L/kWh	kgCO$_2$eq Per Day	% Dif. V. HP	Running Cost Per Day (€)	% Dif. V. HP
Heat Pump	28.70	4.57	-	1.24	-
Electric Heater	16.74	7.85	72%	2.13	72%
LPG Instant Water Heater	15.22	5.98	31%	3.05	146%
Gas Boiler	15.22	5.24	15%	1.91	54%
Solar with Gas Backup	23.18	3.44	−25%	1.13	−9%
Solar with Electric Backup	26.91	4.88	7%	1.33	7%

The carbon intensity of natural gas is much less than that of the carbon intensity of electricity, which is why the solar with gas backup is less carbon intensive than the solar with electric backup. The solar with electric backup is more carbon intensive and less fuel efficient than the heat pump, which is perhaps why it was deemed an unsuitable investment option by Upton et al. [32]. The solar with electric backup was pointed to as the most efficient option by Ibrahim et al. [25], but this is not the case in Ireland as Table 7 shows. In a country where the carbon intensity of electricity is lower and the solar radiation is more constant throughout the year, it is possible that the solar with electric backup could be the most efficient option.

It appears from the analysis of the results in Table 7 that the best overall option by efficiency and carbon emissions is the solar with gas backup. The installation of this sort of solar system may be unfeasible for a lot of farmers due to high initial costs, and if this is the case the heat pump is the next best option. Because the heat pump can be installed almost anywhere, and as it has a lower initial cost it would still be a very good option. The gas system is also a good option as it compares very closely in terms of emissions with the heat pump, which may not have been expected prior to the analysis. It should be pointed out that this comparison is a high-level overview and does not go into the details that may arise, like heat loss from the tanks and variability in terms of ambient temperature and solar radiation. The heat pump is still likely to be related to less emissions over its life cycle due to the ever-decreasing carbon intensity of electricity, which fell from 635 gCO$_2$/kWh to 331 gCO$_2$/kWh from 2005 to 2020.

In terms of comparing the cost of the systems, Table 7 shows the results of the calculations discussed. All running costs shown in Table 7 are in euros. The cost of operating the systems per litre and per day assuming that the amount of usable water heated is 397 L per day is shown and this is extrapolated out to show the yearly cost under these conditions. It can be seen that again the solar with gas backup performs better than the heat pump, but the heat pump outperforms the other systems. If it was possible to operate the heat pump on only night rate electricity it would be 17.5% cheaper to operate than the solar with gas backup, but it is unlikely that this would be possible for dairy farmers unless there is only one milking carried out per day. Because the solar with gas backup system would be more expensive to purchase and install than the heat pump, it is possible that the heat pump would yield a better return on investment over its life cycle.

The solar with electric backup and the gas system are more expensive to operate than the heat pump and they are also less efficient and more carbon intensive than the heat pump, which means that they are probably not as good an investment idea as the heat pump. It should also be pointed out that for small farms where hot water use is less than around 200 L per day the instant gas heater could be a very good option. The overall cost difference over the year period will decrease, and as the heat pump is more expensive to purchase than the gas system it could be a better investment for the farmer.

The heat pump system offers some advantages over the water heaters traditionally used in dairy farms. Its greater efficiency and cheaper running costs should serve as motivation to implement these systems in dairy farms in Ireland and in other countries with cool temperate climates. The higher efficiency of heat pumps is such that if 25% of Irish dairy farmers switched from electric emersion heaters to heat pumps as much as 4.7 ktCO$_2$eq. could be saved per year. This would be a saving of roughly 3.2% of total emissions from electricity use in dairy farms in Ireland. There are also some disadvantages to the system such as long heating cycle times, but the inconvenience of this can be negated by setting the system to heat water on a specific schedule to suit the milking times in the dairy farm.

4. Conclusions

There are many ways to reduce the total carbon emissions that are released for the production of energy, and all avenues to reducing these emissions should be explored. Jevons paradox is apparent in our energy systems today, as we improve the efficiency of each of the energy consuming systems we use, but the number of systems that we use continues to increase. This is why it is important to reduce our emissions in all ways including the production and use of electricity. The agriculture sector is the largest contributor to Ireland's GHG emissions, and although only 4% of these emissions are energy related, an effort should be made to reduce these energy-related emissions where possible. Within the dairy farming sector, which is the most carbon-intense sector within agriculture, the third largest user of energy is water heating. The two larger energy users, milk cooling and milk pumping have been made more efficient by using up-to-date technology such as variable speed drives and plate pre-coolers. The same effort should be made to reduce the carbon emissions related to water heating, and it appears that an ASHPWH is a viable option to do this for all dairy farmers. The ASHPWH tested in this study had a COP of 2.27, which was comparable with what was seen in the literature, considering the water was heated to 65 °C in this study. The ASHPWH when completing a full cycle in configuration 1 was able to produce 28.7 litres of useable water per kWh of energy used, while configuration 2 was able to produce 17.74 L/kWh. It was also noted that adding a mixer to the tank which would be used to de-stratify the tank could be used as a method to heat the full tank to the set-point temperature. The efficiency of heat pumps varies based on many external factors such as ambient temperature and humidity, and therefore its performance will change seasonally.

The heat pump system was compared to some of the current options used in dairy farms and some of the other potential renewable energy methods which could be used to reduce carbon emissions from water heating. The heat pump system performed better in terms of cost and emissions than the current methods used in dairy farms. One system that stood out by its comparative performance versus the heat pump was the solar water heater with a gas backup. This system was marginally cheaper to run than the heat pump but was 25% less carbon intense due to its less carbon intense fuel and benefit of using solar which has zero emissions. This system would be more expensive to implement than the ASHPWH which could serve as a barrier to entry for dairy farmers. It could be the topic of further research to investigate using a life-cycle analysis which system would be best out of ASHPWH and solar with gas backup in regard to cost and carbon emissions, as these seem to be the best of the options considered in the literature.

Supplementary Materials: The following are available online at https://www.mdpi.com/article/10.3390/su13105736/s1.

Author Contributions: Conceptualization, J.G.C. and B.C.; methodology, P.S.B.; software, P.S.B.; validation, P.S.B. and J.G.C.; formal analysis, P.S.B.; investigation, P.S.B.; resources, J.G.C.; data curation, P.S.B.; writing—original draft preparation, P.S.B.; writing—review and editing, J.G.C., B.C.; visualization, P.S.B.; supervision, J.G.C.; project administration, B.C.; funding acquisition, B.C., J.G.C. All authors have read and agreed to the published version of the manuscript.

Funding: This research was funded by Enterprise Ireland and Glenergy Ltd.

Data Availability Statement: The data presented in this study are available in the supplementary material.

Conflicts of Interest: The funders had no role in the design of the study; in the collection, analyses, or interpretation of data; in the writing of the manuscript, or in the decision to publish the results.

References

1. National Geographic. Greenhouse Gases. 13 May 2019. Available online: https://www.nationalgeographic.com/environment/global-warming/greenhouse-gases/ (accessed on 2 November 2019).
2. SEAI. Electricity. 2018. Available online: https://www.seai.ie/data-and-insights/seai-statistics/key-statistics/electricity/ (accessed on 1 November 2019).
3. SEAI. *Ireland's Energy Targets*; SEAI: Dublin, Ireland, 2016.
4. SEAI. CO_2 Emissions. SEAI, 2018. Available online: https://www.seai.ie/data-and-insights/seai-statistics/key-statistics/co2/ (accessed on 7 November 2019).
5. SEAI. *Energy Efficiency in Ireland*; SEAI: Dublin, Ireland, 2016.
6. SEAI. *Energy Related CO2 Emissions in Ireland 2005–2018*; SEAI: Dublin, Ireland, 2020.
7. Environmental Protection Agency. Ireland Exceeds Its Emissions Budget by 5 Million Tonnes and Moves Further from Climate Commitments. Environmental Protection Agency. 23 October 2019. Available online: http://www.epa.ie/newsandevents/news/name,67153,en.html (accessed on 3 November 2019).
8. Teagasc. Teagasc Roadmap for Dairying. 2011. Available online: https://www.teagasc.ie/media/website/animals/dairy/Chapter9Roadmapfordairying.pdf (accessed on 7 November 2019).
9. Jantke, K.; Hartmann, M.J.; Rasche, L. Agricultural Greenhouse Gas Emissions: Knowledge. *Land* **2020**, *9*, 130. [CrossRef]
10. Environmental Protection Agengy. Ireland's National Inventory Submissions 2020. 2020. Available online: http://www.epa.ie/pubs/reports/air/airemissions/ghg/nir2020/ (accessed on 14 October 2020).
11. Teagasc. Milking Machines Washing Routines. Available online: https://www.teagasc.ie/media/website/animals/dairy/milking-machine-cleaning-routines.pdf (accessed on 17 November 2019).
12. Glanbia. Dairy Wash Routine—Which One Is Right for You? Glanbia. Available online: https://www.glanbiaconnect.com/farm-management/detail/article/dairy-wash-routine (accessed on 29 November 2019).
13. Hong, B.; Howarth, R.W. Greenhouse gas emissions from domestic hot water: Heat pumps compared to most commonly used systems. *Energy Sci. Eng.* **2016**, *4*, 123–133. [CrossRef]
14. Upton, J. Energy demand on dairy farms in Ireland. *J. Dairy Sci.* **2013**, *96*, 6489–6498. [CrossRef] [PubMed]
15. Upton, J. *Water Heating Options for Dairy Farms*; Teagasc: Cork, Ireland, 2019.
16. SEAI. EXEED Certified Program. SEAI, 2019. Available online: https://www.seai.ie/business-and-public-sector/standards/exeed-certified-program/ (accessed on 22 September 2020).
17. Teagasc. Energy Fact Sheet No: 8—Heat Pumps in Agriculture. August 2018. Available online: https://www.teagasc.ie/media/website/publications/2016/08.-Heat-Pumps.pdf (accessed on 10 November 2019).
18. Upton, J.; Murphy, M.; O'Mahony, M. Suitability of Air-Source Heat pumps for water heating on Irish Dairy Farms. In *Proceedings of the Teagasc National Dairy Conference*; Teagasc: Cork, Ireland, Unpublished work.
19. Ji, J.; Chow, T.-T.; Pei, G.; Jun, D.; He, W. Domestic air-conditioner and integrated water heater for subtropical climate. *Appl. Therm. Eng.* **2003**, *23*, 581–592. [CrossRef]
20. Zhang, J.; Wang, R.; Wu, J. System optimization and experimental research on air source heat pump water heater. *Appl. Therm. Eng.* **2007**, *27*, 1029–1335. [CrossRef]
21. Glenergy Ltd. Heat Pumps. 2020. Available online: http://glenergy.ie/heat-pumps/ (accessed on 15 October 2020).
22. Ayompe, L.; Duffy, A. Analysis of the thermal performance of a solar water heating system with flat plate collectors in a temperate climate. *Appl. Therm. Eng.* **2013**, *58*, 447–454. [CrossRef]
23. Upton, J. Dairy Farm Energy Consumption. In Proceedings of the Teagasc National Dairy Conference, Mullingar, Ireland, 17–18 November 2010.

24. SEAI. *Domestic Fuel Cost Comparison*; SEAI: Dublin, Ireland, 2019.
25. Ibrahim; Fardoun, F.; Younes, F.; Louahlia-Gualous, H. Review of water-heating systems: General selection approach based on energy and environmental aspects. *Build. Environ.* **2014**, *72*, 259–286. [CrossRef]
26. Hepbasli, A.; Kalinci, Y. A review of heat pump water heating systems. *Renew. Sustain. Energy Rev.* **2009**, *13*, 1211–1229. [CrossRef]
27. Amirirad; Kumar, R.; Fung, A.S.; Leong, W.H. Experimental and simulation studies on air source heat pump water heater for year-round applications in Canada. *Energy Build.* **2018**, *165*, 141–149. [CrossRef]
28. Weather and Climate. Average Water Temperature in Dublin (Dublin County) in Celsius. 2020. Available online: https://weather-and-climate.com/average-monthly-water-Temperature,Dublin,Ireland (accessed on 21 October 2020).
29. Chandra, Y.P.; Matuska, T. Stratification analysis of domestic hot water storage tanks: A Comprehensive Review. *Energy Build.* **2019**, *187*, 110–131. [CrossRef]
30. LPG FLogas. Dairy Hot Water. 2020. Available online: https://www.flogas.ie/dairy-hot-water.html?gclid=Cj0KCQjwybD0BRDyARIsACyS8mu46eKm2bQ_rGHrlphNjPnUJJ5EDVGnTPZvPzTC8614lqopvfOe69kaArl3EALw_wcB (accessed on 27 March 2020).
31. Met Eireann. Sunshine and Solar Radiation. 2020. Available online: https://www.met.ie/climate/what-we-measure/sunshine (accessed on 25 March 2020).
32. Upton, J.; Murphy, M.; de Boer, J.M.; Koerkamp, P.W.G.G.; Berentsen, P.B.M.; Shalloo, L. Investment appraisal of technology innovations. *J. Dairy Sci.* **2015**, *98*, 898–909. [CrossRef] [PubMed]

Article

Manure Flushing vs. Scraping in Dairy Freestall Lanes Reduces Gaseous Emissions

Elizabeth G. Ross [1], Carlyn B. Peterson [1], Yongjing Zhao [2], Yuee Pan [1] and Frank M. Mitloehner [1,*]

[1] Department of Animal Science, University of California, Davis, 1 Shields Ave, Davis, CA 95616-8521, USA; eghumphreys@ucdavis.edu (E.G.R.); cbpeterson@ucdavis.edu (C.B.P.); yepan@ucdavis.edu (Y.P.)
[2] Air Quality Research Center, University of California, Davis, 1 Shields Ave, Davis, CA 95616-8521, USA; yjzhao@ucdavis.edu
* Correspondence: fmmitloehner@ucdavis.edu

Abstract: The objective of the present study was to mitigate ammonia (NH_3), greenhouse gases (GHGs), and other air pollutants from lactating dairy cattle waste using different freestall management techniques. For the present study, cows were housed in an environmental chamber from which waste was removed by either flushing or scraping at two different frequencies. The four treatments used were (1) flushing three times a day (F3), (2) flushing six times a day (F6), (3) scraping three times a day (S3), and (4) scraping six times a day (S6). Flushing freestall lanes to remove manure while cows are out of the barn during milking is an industry standard in California. Gas emissions were measured with a mobile agricultural air quality lab connected to the environmental chamber. Ammonia and hydrogen sulfide (H_2S) emissions were decreased ($p < 0.001$ and $p < 0.05$) in the flushing vs. scraping treatments, respectively. Scraping increased NH_3 emissions by 175 and 152% for S3 and S6, respectively vs. F3. Ethanol (EtOH) emissions were increased ($p < 0.001$) when the frequency of either scraping or flushing was increased from 3 to 6 times but were similar between scraping and flushing treatments. Methane emissions for the F3 vs. other treatments, were decreased ($p < 0.001$). Removal of dairy manure by scraping has the potential to increase gaseous emissions such as NH_3 and GHGs.

Keywords: ammonia emissions; dairy cow; flushing; freestall barn; scraping

1. Introduction

The United States has more than 14 million dairy animals that produce approximately 145,000,000 ton of manure per year and 1,663,735 ton of nitrogen per year [1]. California is the leading producer of fluid milk and produces 20% of all dairy products in the U.S., with the majority of production concentrated in the San Joaquin Valley. This large concentration of dairy cattle contributes to one of the worst air quality regions in the U.S. [2–4]. Dairies are a source of air pollutants such as NH_3, a precursor to particulate matter formation and smog forming volatile organic compounds (VOC) [5].

In 2016, the California Air Resources Board (CARB) released the Proposed Short Lived Climate Pollutant (SLCP) Reduction Strategy to reduce CH_4 emissions from dairy manure (i.e., urine and feces) management. In response to the SLCP reduction strategy, Senate Bill 1383 was passed in 2016, which requires a reduction of CH_4 emissions by 40% below 2013 levels by 2030. The majority of these reduction strategies consist of dairies converting from current liquid manure storage systems such as lagoons to dry manure storage, which would utilize scraping rather than flushing of freestall barns where the cows are housed. This conversion of manure management would eliminate so-called manure storage lagoons, which are considered to be a large CH_4 contributor, and encourages the use of anaerobic digesters to handle scraped manure [6].

The plan of action proposed by CARB to reduce CH_4 emissions from manure may have unintended consequences affecting other criteria pollutants, such as NH_3 emissions from dairies. Ammonia is a precursor to the formation of $PM_{2.5}$, which is a small aerosol that can subsist in the atmosphere for as long as 15 days. When $PM_{2.5}$ is inhaled it can carry pathogens that infiltrate the alveoli of the lungs and enter the blood stream, causing illness and respiratory disease [7,8]. This is problematic for the San Joaquin Valley as it exceeds the regulatory limits for $PM_{2.5}$ and ozone (O_3) and is classified as a serious nonattainment area by the California Air Resources Board [9].

Within livestock production, dairies were identified as the largest source of NH_3 emissions in California [10]. Sheppard et al. (2011) produced a model simulation that suggests up to 53% of the excreted total ammonia nitrogen (TAN) from a lactating cow will be emitted to the atmosphere as NH_3 during the housing, storage, and land spreading of manure [11]. Harper et al. (2009) reported estimates of excreted nitrogen (N) to be 7.6 ± 1.5% of input feed N based on data from three different dairies in Wisconsin from barns, manure treatment, and storage [12].

Further research is needed to better understand the full impacts of CH_4 mitigation strategies to comply with public policy including a better understanding of the variables effecting NH_3 emissions. It was hypothesized that scraping dairy freestall lanes would increase NH_3 emissions compared to flushing. The objective of the present study was to quantify NH_3 emissions, greenhouse gases, and other air pollutants as a result of scraping versus flushing manure removal strategies commonly utilized in dairy freestall barns.

2. Materials and Methods

2.1. Environmental Chamber Design

The study was conducted in an environmental chamber (4.4 m × 2.8 m × 10.5 m) under an IACUC approved protocol (#18818) at the University of California, Davis, Swine Teaching and Research Center. The environmental chamber, which is designed to work for various livestock species, was equipped to house 3 dairy cows under freestall conditions. The environmental chamber has a total volume of 142 m^3, a chamber residence time of approximately 3 min at the continuous ventilation rate of 51,848 L/min, and an air exchange rate of 20 times per hour. The chamber was air conditioned and set to 20 °C to maintain cow comfort. Industry standard freestall stanchions were assembled on the west end of the chamber to allow for the animals to maintain normal resting behaviors. Feed bunks and water troughs were located on the east end of the chamber that allowed for ad libitum access. The environmental chamber was certified by the Association for Assessment and Accreditation of Laboratory Animal Care International (AAALAC). Cows were housed at the University of California, Davis's Dairy Teaching and Research Facility when emissions measurements were not being collected in the environmental chamber. Emission measurements were collected on Monday, Wednesday, and Friday over 5 and a half weeks from the three cows assigned to the chamber on each testing day. Animals were milked at 04:00 h at the dairy facility and immediately transported to the environmental chamber for the 11 h data collection followed by transportation back to the dairy for the evening milking at 16:30 h. Cows were monitored in the chamber from approximately 05:00 to 16:00 h.

2.2. Animals and Diets

Twelve multiparous lactating Holstein cows were blocked by days in milk, milk production, parity, and pregnancy status before being randomly assigned to one of four groups (n = 4). Cows were fed the standard UC Davis dairy ration ad libitum during the testing period, upon arrival at the environmental chamber. While animals were housed at the UC Davis dairy, they were fed at: 04:00, 12:00, 16:00, and 22:00 h. The diet was analyzed by Cumberland Valley Analytical Services, Inc. (Hagerstown, MD, USA) for dry matter (DM), crude protein (CP), ash, acid detergent fiber (ADF), and neutral detergent fiber (aNDF). The chemical composition and ingredients of the total mixed ration (TMR) are shown in Tables 1 and 2, respectively. During gas emissions monitoring days, feed refusals were removed at the end of the day to assess group daily feed intake while in the chamber. Cows were milked twice daily at 04:00 and 16:30 h. Milk yield records for all animals were maintained for the duration of the study. Average feed intake across treatments was 32.70 ± 7.94 kg.

Table 1. Chemical composition of the total mixed ration.

Measures	Total Mixed Ration (% DM) [1]
Crude Protein	20.4
Ash	6.85
Neutral Detergent Fiber	31.8
Acid Detergent Fiber	23.7

[1] DM = dry matter.

Table 2. Ingredients of basal total mixed ration.

Feed Ingredients	As Fed (kg/d/cow)
Grain [1]	11.91
Alfalfa Hay	11.34
Whole Cotton Seed	2.27
Almond Hulls	2.27
Strata [2]	0.1
Milk Mineral	0.34
EnerGII [3]	0.29
Salt	0.07
Wheat Hay	0.91

[1] Grain mix contained: 20.50% rolled barley, 20.50% rolled corn, 21.03% dried distillers grains, 21.96% wheat mill run, 14.48% beet pulp, and 1.53% canola meal. [2] A calcium salt of fatty acids containing a blend of palmitic, stearic, and oleic fatty acids with a 16% eicosapentaenoic acid (EPA)/docosahexaenoic acid (DHA) omega-3 fatty acids (Virtus Nutrition, Corcoran, CA, USA). [3] A calcium salt of fatty acids containing 50% palmitic and 35% oleic fatty acids (Virtus Nutrition, Corcoran, CA, USA).

2.3. Treatments

The present manure removal study was designed as a Latin square with four treatments including: (1) flushing 3 times a day (F3, Control), (2) flushing 6 times a day (F6), (3) scraping 3 times a day (S3), and (4) scraping 6 times a day (S6). Each of the treatments occurred on different data collection days for a total of 16 days. The treatments were applied three times a day, at 08:30, 12:00, and 15:30 h, or six times a day at 06:45, 08:30, 10:15, 12:00, 13:45, and 15:30 h. Flushing consisted of spraying water on the concrete floor until all of the visible manure was flushed down the drain. The scraping treatment used metal scrapers to manually clear the manure into the drain. Each manure removal treatment took approximately ten minutes to complete and clean the pen. The drain in the chamber was plugged to keep urine and feces in the chamber and sewage gases from entering the chamber during the testing period. The drain plug was removed for cleanings and replaced after.

2.4. Equipment

A mobile agricultural air quality laboratory (MAAQL) was used to measure all emissions from the environmental chamber. This MAAQL contained gas analyzers, an air sampling system, and a data acquiring system to collect real-time air emission data from the environmental chamber. The environmental chamber had one incoming and one outgoing air duct. Teflon tubing (12.7 mm ID) transported air from inside the chamber through the air duct immediately above the ceiling and into the MAAQL. The Thermo 17i $NO/NO_x/NH_3$ analyzer (Thermo Scientific, Waltham, MA, USA) was used to measure NH_3, nitric oxide (NO), and oxides of nitrogen (NO_x). Methane was measured using the Thermo 55C CH_4 analyzer (Thermo Scientific, Waltham, MA, USA). Hydrogen sulfide (H_2S) was measured using the Thermo 450i sulfur dioxide (SO_2)/H_2S analyzer (Thermo Scientific, Waltham, MA, USA). Nitrous oxide (N_2O) was measured using the Thermo 46i

N₂O analyzer (Thermo Scientific, Waltham, MA, USA). Ethanol (EtOH), carbon dioxide (CO_2), NH_3, and methanol (MeOH) were measured with the INNOVA model 1412 Photoacoustic Gas Monitor (INNOVA AirTech Instrument, Ballerup, Denmark). Table 3 shows detection limits and upper monitoring ranges of all gas analyzers. Samples were analyzed for 15 min each, beginning with the inlet air duct, and then the outlet air duct, and were repeated for the 11 h testing period.

The concentrations of N_2O, NO, NO_x, SO_2, and methanol (MeOH) were detectable, but the inlet and outlet values were too close to derive meaningful emission rates of these gases. Therefore, their results therefore were not reported.

Table 3. Gas analyzers, gases monitored, detection limits, and detection ranges of the Mobile Agricultural Air Quality Laboratory (MAAQL) used to measure emissions from the environmental chamber.

Gas Analyzer	Gases [3]	Detection Limits	Upper Range
Thermo 17i	NO	1.25 ng/L	24.96 µg/L
$NO/NO_x/NH_3$	NO_x	1.54 ng/L	30.78 µg/L
analyzer [1]	NH_3	0.71 ng/L	14.14 µg/L
Thermo 55C CH_4 analyzer [1]	CH_4	13.31 ng/L	665.56 µg/L
Thermo 450i	SO_2	3.99 ng/L	26.62 µg/L
SO_2/H_2S analyzer [1]	H_2S	2.12 ng/L	14.14 µg/L
Thermo 46i N_2O analyzer [1]	N_2O	0.04 µg/L	36.61 µg/L
	CO_2	2.75 µg/L	1.83 g/L
Innova 1412	EtOH	0.15 µg/L	1.91 g/L
photo-acoustic	NH_3	0.71 µg/L	0.71 g/L
multi-gas analyzer [2]	MeOH	0.11 µg/L	1.33 g/L
	N_2O	0.05 µg/L	1.83 g/L

[1] Analyzers by Thermo Scientific, Waltham, MA, USA. [2] Analyzer by INNOVA AirTech Instrument, Ballerup, Denmark. [3] NO = nitric oxide; NO_x = oxides of nitrogen; NH_3 = ammonia; CH_4 = methane; SO_2 = sulfur dioxide; H_2S = hydrogen sulfide; CO_2 = carbon dioxide, EtOH = ethanol; MeOH = methanol; N_2O = nitrous oxide.

2.5. Emissions Calculations

Concentration data of the air samples from the environmental chamber over each 15 min period were truncated to remove the first five minutes and last two minutes of the sample to prevent carry over. The following equation was used to calculate emission rate mg/h of gases from the environmental chamber:

$$\text{Emission Rate (mg/h/head)} = \{[(MIX) \times (FL) \times (60)]/MV\} \times (MW) \times (Conv)/\text{Head} \quad (1)$$

where *MIX* is the net concentration (inlet concentration—outlet concentration) in either ppm (parts per million) or ppb (parts per billion), *FL* is the continuous ventilation rate of 51,848 L/min, 60 is the conversion from minute to hour, *MW* is the molecular weight of the gas in grams per mole, *Conv* is a conversion factor of 10^{-3} for concentration in ppm and 10^{-6} for concentration in ppb, and *V* is the volume of one molar gas at temperature *T* in liter/mole and is calculated as:

$$V = [(V_s) \times T)]/T_s \quad (2)$$

where V_s is the standard volume 22.4 L at 0 °C, T_s is the standard temperature 0 °C that equals to 273.15 K, *T* is the air temperature in K equaling to *T* in °C +273.15.

2.6. Data Analysis

Emission rates from the different manure removal methods were compared to evaluate their respective environmental impacts. All emissions data were analyzed using the lmerTest package in R [13]. The model used to evaluate emissions data is:

$$Y_{ijkl} = \mu + B_i + R_j + F_k + H_l + e_{ijkl} \tag{3}$$

where Y_{ijkl} is the dependent variable, μ is the overall mean, B_i is the block, R_j is the method of removal (scraping versus flushing), F_k is frequency of removal (three times versus six times), H_l is the hour of measurement, and e_{ijkl} is the error term associated with the model. Block was a random effect, with all other variables as fixed effects. The interaction of the main effects of method of removal * frequency of removal, was originally evaluated but removed from the model as this interaction was not significant. The milk data was analyzed using the lmerTest package in R [13]. The model for the milk data is:

$$Y_{ijkl} = \mu + C_i + B_j + D_k + T_l + e_{ijkl} \tag{4}$$

where Y_{ijkl} is the dependent variable, μ is the overall mean, C_i is the cow, B_j is the block, D_k is the date of milking, T_l is the time of milking, and e_{ijkl} is the error term associated with the model. Cow was a random effect, with all other variables as fixed effects. B_i, R_j, F_k, and H_l were categorical variables. Means are presented as least squares means (LSM) and were determined using the lsmeans package in R [14]. Pairwise comparisons of treatment LSM were determined by a Tukey test using the multcompView package in R [15]. Differences were declared significant at $p \leq 0.05$ and showed a trend at $0.05 \leq p \leq 0.10$.

3. Results and Discussion

3.1. Milk Production

Least squares means for milk yield were 42.4, 46.2, 42.7, and 39.9 kg (± 6.14 kg; $p = 0.91$) for each of the four groups of cows (blocks) during the duration of the study. Dry matter intake (DMI) for the 11 h period animals were housed in the environmental chamber was similar across treatments with group DMIs of 30.7, 31.1, 34.5, and 34.5 kg (± 4.31 kg; $p = 0.25$). A difference in milk yield could lead to differences in feed intake, affecting both manure output and gaseous emissions from manure and enteric sources [16,17].

3.2. Ammonia Emissions

Total NH_3 emissions from scraping were greater than flushing treatments ($p < 0.001$; Table 4). Scraping increased NH_3 emissions by 175 and 152% for S3 and S6, respectively, as compared to the control (F3; Table 4). The most common California industry practice of clearing freestall lanes is by flushing 2 to 3 times a day, which occurs while the cows are in the milking parlor. Scraping treatments left behind a film of manure that coated the concrete freestall lane in the environmental chamber. The urea being excreted in the animal's urine comes in contact with this manure film and is rapidly converted by the urease naturally present in the manure to NH_3 and volatized [18]. In contrast, flushing does not allow for the manure to create a film, which reduces the opportunity for urea to come into contact with urease. The presence of water with the flush treatment may also affect the amount of NH_3 that is volatilized. In the presence of water, NH_3 and ammonium (NH_4^+) exist at an equilibrium in solution that is dependent on pH and temperature [19]. An increase in the concentration of NH_4^+/NH_3 in the manure, an increase of temperature, or a disturbance to the manure, such as wind speed, can increase the volatilization of NH_3 [19–21]. The scraping treatments cause a physical disturbance to the manure and do not dilute the manure, which likely led to the greater NH_3 emission seen in these treatments. Flushing results in a lowering of urea and TAN concentrations in slurry by diluting and removing urine from the floor surface which reduces NH_3 emissions [22]. Kroodsma et al. (1993) found that scraping manure from a concrete stall did not reduce NH_3 emissions while flushing reduced NH_3 by 70% [23]. Urea is usually hydrolyzed within 2 h after urine is excreted on floors, but can continue to volatize for 15 h if left undisturbed [22,24]. Flushing more frequently dilutes the urine and removes it before the majority of the urea is hydrolyzed to NH_3. However, since flushing on commercial dairies occurs primarily when the cows are in the milking parlor, increasing the frequency of flushing would be difficult to implement with current management practices.

Table 4. Least squares means, pooled standard errors (SEM), and *p*-values for the 3 versus 6 times flushing or scraping treatments, respectively, for ammonia, methane, hydrogen sulfide, ethanol, and carbon dioxide emissions ($n = 4$). Emission measurements reported are on a per cow basis in either mg or g/h. A negative reduction potential equates to an increase in emissions.

	Treatment LSM [1]				SEM	*p*-Value		
	F3 [2]	F6	S3	S6		S vs. F [3]	3 vs. 6 [4]	Time
Ammonia								
Emission Rate (mg/h)	622.42 [a]	479.33 [a]	1712.90 [b]	1569.80 [b]	154.70	<0.001	0.12	<0.001
Reduction Potential (%)		23%	−175%	−152%				
Methane								
Emission Rate (g/h)	23.32 [a]	26.29 [b]	26.60 [b]	29.56 [c]	1.69	<0.001	<0.001	0.97
Reduction Potential (%)		−13%	−14%	−27%				
Hydrogen Sulfide								
Emission Rate (mg/h)	2.26	6.16	6.94	7.84	1.56	0.0496	0.29	0.13
Reduction Potential (%)		−173%	−207%	−247%				
Ethanol								
Emission Rate (g/h)	1.65 [ab]	2.17 [c]	1.55 [a]	2.07 [bc]	0.13	0.426	<0.001	<0.001
Reduction Potential (%)		−31%	6%	−25%				
Carbon Dioxide								
Emission Rate (g/h)	920.94 [a]	1056.81 [b]	1072.79 [b]	1208.79 [c]	58.42	<0.001	<0.001	0.98
Reduction Potential (%)		−15%	−16%	−31%				

[1] F3 = flush 3 times; F6 = flush 6 times; S3 = scrape 3 times; S6 = scrape 6 times. [2] F3 treatment is industry standard and considered the Control. [3] S = scrape; F = flush. [4] The difference between increasing the frequency of flush or scrape from 3 to 6 times. Means with the same letter ([abc]) are not significantly different ($p > 0.05$).

Ammonia emissions also changed over the 11 h period the cows were in the environmental chamber ($p < 0.001$; Figure 1). Both frequencies of the scraping treatments showed increased NH_3 emissions over the 11 h treatment period showing a compounding effect even after scraping occurred (Figure 1). Comparatively, the flushing treatments show a decrease in NH_3 emissions directly after flushing treatments occurred (Figure 1). The increase in concentration of NH_3 emissions over time for the scraping treatment is consistent with the literature [25,26].

Rotz et al. (2014) measured NH_3 emissions from dairies using both scrape and flushing systems in New York, Wisconsin, and Indiana [27]. Both the model simulation and the measured annual average NH_3 emissions were lower in flush barns compared with scrape barns. Vaddella et al. (2011) compared NH_3 emissions from simulated flushed manure storage with scraped manure storage and controlled for surface-area to volume ratio and found greater NH_3 emissions in the scraped versus flushed manure.

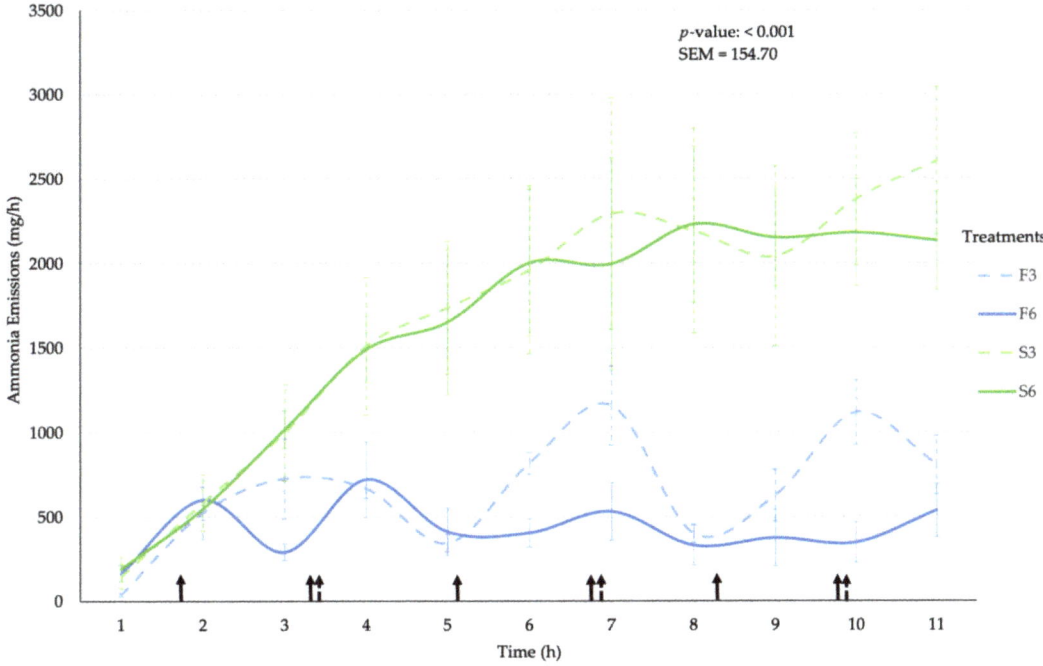

Figure 1. Average ammonia (NH$_3$) emissions for the four treatments F3 = flush 3 times; S3 = scrape 3 times; F6 = flush 6 times; S6 = scrape 6 times; over time in hours (n = 4). Emission measurements reported are on a per cow basis mg/h. Error bars represent the standard error for each point. The arrows correspond to the frequency of the four treatments applications (number of flushing or scraping events) that occurred in the environmental chamber. Solid arrows correspond to times the F6 and S6 treatments were applied, and dashed arrows correspond to times F3 and S3 treatments were applied.

3.3. Methane Emissions

In the present study, all treatments compared with the Control (F3) had negative reduction potentials (increased emissions) for CH$_4$ emissions ($p < 0.001$; Table 4). Both F6 and S3 treatments were similar. Surprisingly, the S6 treatment had larger ($p < 0.001$) CH$_4$ emissions as compared to other treatments. The primary source of CH$_4$ emissions during the testing period would be from enteric sources, which cannot be differentiated from manure CH$_4$ emissions inside the environmental chamber.

Sun et al. (2008) conducted a similar study in the same facility as the present study and found that minimal CH$_4$ emissions were attributed to fresh manure sources [28]. Sun et al. (2008) measured emissions in two phases: first with a cow plus manure, followed by manure only. Methane emissions substantially increased with the addition of the cows to the chamber and subsequently returned to near empty chamber concentrations when the cows were removed [28].

Methane emissions from livestock waste are produced by the decomposition of volatile solids in manure primarily from systems that promote an anaerobic environment, such as lagoons [29]. The production of CH$_4$ is dependent on methanogens, which thrive in anaerobic environments. Under aerobic conditions such as in an environmental chamber, there is little to no CH$_4$ production [29].

In the present study, cows were blocked in order to decrease variability in enteric CH_4 emissions. Blocking for milk yield groups cows with a similar dry matter intake, which has a linear relationship with CH_4 emissions [30]. Future research should remove the interference of enteric CH_4 emissions to determine if scraping fresh manure increases CH_4 emissions.

3.4. Hydrogen Sulfide Emissions

In the present study, individual treatments and frequency of treatments showed similar H_2S emissions; however, scraping resulted in higher H_2S emissions ($p < 0.05$) than flushing. Hydrogen sulfide emissions can be particularly dangerous in enclosed animal facilities. Without proper ventilation, a buildup of H_2S can cause mild eye irritation, and in large enough concentrations cause respiratory failure and death [31]. The majority of California dairy freestall barns are open air so health concerns from H_2S exposure are minimal. However, H_2S emissions should be carefully monitored in enclosed animal facilities particularly when the manure is disturbed for cleaning. Mixing or disturbing the surface of manure will lead to an increase in H_2S emissions because H_2S is contained in gas bubbles suspended in the manure, which burst when mixed [25,32,33]. Maasikmets et al. (2015) measured a farm with solid manure storage compared to a farm with liquid manure storage and found the solid manure storage to have a higher concentration of H_2S [34]. The concentration of H_2S was highest in the morning when there was little air movement inside the barns. With less air flow there is less dilution of the air pollutants, allowing for measurements at higher levels. Animal diets also play a key role in the production of H_2S in the manure. Cattle fed a higher concentrate diet, or a diet containing more sulfur substrate, will have manure that produces greater H_2S emissions than cattle fed a high forage diet or low sulfur substrate diet [32].

3.5. Ethanol Emissions

Ethanol emissions were similar for flushing vs scraping; however, flushing and scraping six times versus three times increased EtOH emissions ($p < 0.001$; Table 4). Ethanol emissions changed over the 11 h period in the environmental chamber ($p < 0.001$; Figure 2).

Ethanol is the primary VOC produced on dairies and is a precursor for O_3 formation [35]. Previous studies showed some enteric emissions of ethanol [28]; however, the majority of ethanol emissions comes from the manure. Another possible source of EtOH emission is the total mixed ration (TMR). Chung et al. (2004) quantified non-enteric VOC emissions sources from dairies and showed the highest emissions sources for EtOH to be from silage and silage-based TMR piles [36]. The TMR fed in the present study did not contain silage or other fermented feedstuffs; therefore, VOC emissions from the TMR is expected to be negligible as compared to a silage based TMR. The most common VOCs associated specifically with flush lanes are 2-butanone and toluene, with EtOH being a lesser source [36]. The highest emissions rates from feed occur during the feed out phase, due to greater oxygen exposure [37]. Ethanol is the major contributor of VOCs from animal feed at >70% of the total VOCs [35]. It is likely that the EtOH measured in the chamber during the current study was a combination of enteric processes, TMR, and manure. Further research should determine the effect of agitation on manure EtOH emissions to confirm the findings of the present study, that an increase in frequency of manure removal (i.e., agitation by either flush or scraping) increases EtOH emissions.

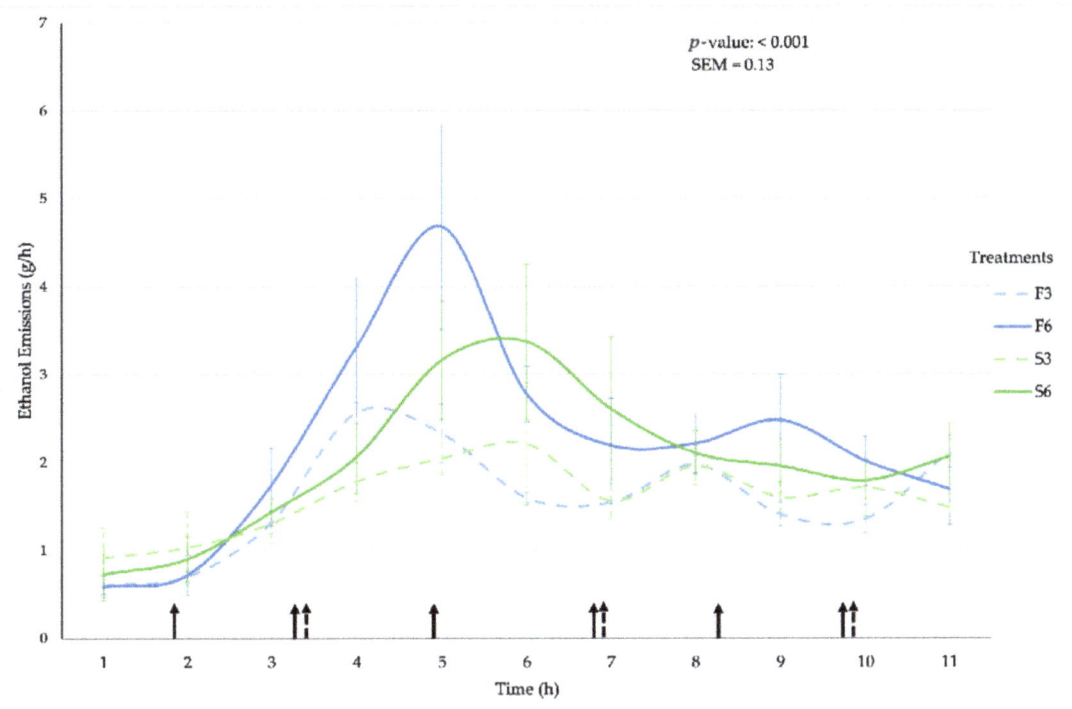

Figure 2. Ethanol (EtOH) emissions for the four treatments F3 = flush 3 times; S3 = scrape 3 times; F6 = flush 6 times; S6 = scrape 6 times; over time in hours (n = 4). Emission measurements reported are on a per cow basis g/h. Error bars represent the standard error for each point. The arrows correspond to the frequency of the four treatments applications (number of flushing or scraping events) that occurred in the environmental chamber. Solid arrows correspond to times the F6 and S6 treatments were applied, and dashed arrows correspond to times F3 and S3 treatments were applied.

3.6. Carbon Dioxide Emissions

In the present study, CO_2 increased with scraping and increased frequency of treatments ($p < 0.001$; Table 4). Carbon dioxide emissions can be from animal manure as products of microbial degradation, and from respiratory and enteric emissions [38]. Carbon dioxide from animal manure is a release of carbon sequestered by photosynthesis and is part of the cycling of carbon from the atmosphere to plants to animals and back to the atmosphere over a short period of time. For this reason, the USEPA does not consider CO_2 from animal feeding operations a contributor to the buildup of GHG in the atmosphere [38].

3.7. Relation to Manure Management

The present study has shown that converting from a flush to a scrape manure removal system can result in the unintended consequences of increasing NH_3 emissions. However, manure removal in the housing portion of the dairy is just one portion of the whole manure management system and a lifecycle assessment of the entire manure management train should be conducted.

Covered lagoon anaerobic digesters fit the current manure management system for California dairies better than a higher solids content digester, as this allows farmers to continue using a flush system. However, in either anaerobic digester system, the total ammoniacal nitrogen (TAN) is increased as well as the pH in the digested manure, which results in potentially higher NH_3 volatilization from digested manure [39]. Anaerobic digesters reduce the amount of easily degradable carbon in the manure through fermentation, which reduces CH_4 production from the effluent as well as the potential of N_2O emissions during soil application [40–42]. Montes et al. (2013) determined that N_2O emissions could be reduced by up to 70% from soil applied digested manure compared to fresh manure [43]. Given the changes in composition and gaseous emissions from digested manure, it is imperative to do a whole gaseous emissions balance for the varying manure management strategies to determine the most sustainable option for dairy farmers.

4. Conclusions

Flushing versus scraping manure from dairy freestall lanes was found to be advantageous for reducing both NH_3 and H_2S emissions. The control (F3) treatment showed the lowest CH_4 emissions compared with other treatments, which should be researched further to determine if increased agitation of manure or flushing of manure increases CH_4 emissions. Mitigation of CH_4 from dairy manure sources is vitally important, particularly in California where legislation requires it. For future research, we suggest conducting lifecycle assessments to predict emissions from the entire manure management train, from removal in the barn to storage/treatment, and finally land application. Mitigation strategies such as switching from flush to scrape to reduce CH_4 emissions in dairy housing have the potential to increase other important pollutants such as NH_3. The present study shows that NH_3 emissions are lower in flush vs. scrape systems. To align manure removal with storage, a flush system followed by anaerobic digestion should be considered. This combination would optimize mitigation of two of the most important gases emitted from dairies, CH_4 and NH_3. Potentially, the use of covered lagoon anaerobic digesters would best fit this combination. This is particularly important for areas like the San Joaquin Valley where NH_3 and $PM_{2.5}$ emissions are consistently above the regulatory limits. This research is directly related to the development of a sustainable dairy system, as manure management directly contributes to the environment and economic pillars of sustainability through mitigation of gaseous emission and economic feasibility for dairy producers.

Author Contributions: Conceptualization, E.G.R. and F.M.M.; methodology, execution, data collection, and curation, E.G.R., Y.Z., and Y.P.; data analysis, E.G.R. and C.B.P.; writing—original draft preparation, E.G.R.; results, discussion, and editing, E.G.R., C.B.P., Y.Z., Y.P., and F.M.M.; supervision and project administration, E.G.R., Y.Z., and F.M.M.; funding acquisition, F.M.M.; All authors have read and agreed to the published version of the manuscript.

Funding: This research received no external funding.

Institutional Review Board Statement: This study was conducted according to the guidelines of the Declaration of Helsinki, and approved by the Institutional Animal Care and Use Committee of the University of California Davis (protocol number 18818).

Informed Consent Statement: Not applicable.

Data Availability Statement: The data presented in this study is openly available in FigShare at doi:10.6084/m9.figshare.14569020.

Acknowledgments: We kindly thank: Doug Gisi the manager of the UC Davis Dairy and his staff for their help and support with this trial and the undergraduate students that were instrumental in the day to day.

Conflicts of Interest: The authors declare no conflict of interest.

References

1. Pagliari, P.; Wilson, M.; He, Z. Animal manure production and utilization: Impact of modern concentrated animal feeding operations. *Anim. Manure: Prod. Charact. Environ. Concerns Manag.* **2020**, *67*, 1–14.
2. Hatamiya, L. The Economic Importance of the California Dairy Quota Program. Available online: https://www.ams.usda.gov/sites/default/files/media/Exhibit%2054%20-%20Testimony%20of%20Lon%20Hatamiya.pdf (accessed on 26 February 2017).
3. USDA. Dairy Situation at a Glance. 2021. Available online: https://www.ers.usda.gov/topics/animal-products/dairy/ (accessed on 1 January 2021).

4. USDA. California Dairy Statistics and Trends Mid-Year Review January-June 2016 Data. Available online: https://www.cdfa.ca.gov/dairy/pdf/Annual/2016/MidYear2016.pdf (accessed on 27 April 2017).
5. CARB. Ambient Air Qualtiy Standards (AAQS) for Particulate Matter. Available online: https://www.arb.ca.gov/research/aaqs/pm/pm.htm (accessed on 16 February 2017).
6. CARB. Revised Proposed Short-Lived Climate Pollutant Reduction Strategy. 2017. Available online: https://ww2.arb.ca.gov/sites/default/files/2020-07/final_SLCP_strategy.pdf (accessed on 1 January 2021).
7. Samet, J.M.; Dominici, F.; Curriero, F.C.; Coursac, I.; Zeger, S.L. Fine particulate air pollution and mortality in 20 US cities, 1987–1994. *N. Engl. J. Med.* **2000**, *343*, 1742–1749. [CrossRef] [PubMed]
8. Aneja, V.P.; Blunden, J.; James, K.; Schlesinger, W.H.; Knighton, R.; Gilliam, W.; Jennings, G.; Niyogi, D.; Cole, S. Ammonia assessment from agriculture: US status and needs. *J. Environ. Qual.* **2008**, *37*, 515–520. [CrossRef] [PubMed]
9. USEPA. Air Actions in the San Joaquin Valley-PM2.5. 2015. Available online: https://19january2017snapshot.epa.gov/www3/region9/air/sjv-pm25/index.html (accessed on 1 January 2019).
10. SJVDMTFAP. San Joaquin Valley Dairy Manure Technology Feasibility Assessment Panel. An Assessment of Technologies for Management and Treatment of Dairy Manure in California's San Joaquin Valley. CARB. 2005. Available online: https://www.arb.ca.gov/ag/caf/dairypnl/dmtfaprprt.pdf (accessed on 1 January 2021).
11. Sheppard, S.; Bittman, S.; Swift, M.; Tait, J. Modelling monthly NH3 emissions from dairy in 12 Ecoregions of Canada. *Can. J. Anim. Sci.* **2011**, *91*, 649–661. [CrossRef]
12. Harper, L.A.; Flesch, T.K.; Powell, J.M.; Coblentz, W.K.; Jokela, W.E.; Martin, N.P. Ammonia emissions from dairy production in Wisconsin. *J. Dairy Sci.* **2009**, *92*, 2326–2337. [CrossRef]
13. Kuznetsova, A.; Brockhoff, P.B.; Christensen, R.H.B. lemerTest: Tests in Linear Mixed Effects Models; R package version 2.0-30. *J. Stat. Softw.* **2017**, *82*, 1–26. [CrossRef]
14. Lenth, R. Least-squares means: The R package lsmeans. *J. Stat. Sofw.* **2016**, *69*, 1–33.
15. Graves, S.; Piepho, H.-P.; Selzer, L.; Dorai-Raj, S. Multcompview: Visualizations of Paired Comparisons; R Package version 0.1-7. 2017. Available online: https//github.com/rvlenth/emmeans/issues (accessed on 18 June 2018).
16. Hassanat, F.; Gervais, R.; Benchaar, C. Methane production, ruminal fermentation characteristics, nutrient digestibility, nitrogen excretion, and milk production of dairy cows fed conventional or brown midrib corn silage. *J. Dairy Sci.* **2017**, *100*, 2625–2636. [CrossRef]
17. Sutter, F.; Schwarm, A.; Kreuzer, M. Development of nitrogen and methane losses in the first eight weeks of lactation in Holstein cows subjected to deficiency of utilisable crude protein under restrictive feeding conditions. *Arch. Anim. Nutr.* **2017**, *71*, 1–20. [CrossRef]
18. Bouwman, A.; Lee, D.; Asman, W.; Dentener, F.; Van Der Hoek, K.; Olivier, J. A global high-resolution emission inventory for ammonia. *Glob. Biogeochem. Cycles* **1997**, *11*, 561–587.
19. Hristov, A.; Hanigan, M.; Cole, A.; Todd, R.; McAllister, T.; Ndegwa, P.; Rotz, A. Review: Ammonia emissions from dairy farms and beef feedlots 1. *Can. J. Anim. Sci.* **2011**, *91*, 1–35. [CrossRef]
20. Sommer, S.G.; Petersen, S.O.; Søgaard, H.T. Greenhouse gas emission from stored livestock slurry. *J. Environ. Qual.* **2000**, *29*, 744–751. [CrossRef]
21. Teye, F.K.; Hautala, M. A comparative assessment of four methods for estimating ammonia emissions at microclimatic locations in a dairy building. *Int. J. Biometeorol.* **2010**, *54*, 63–74. [CrossRef]
22. Monteny, G.; Erisman, J. Ammonia emission from dairy cow buildings: A review of measurement techniques, influencing factors and possibilities for reduction. *Njas Wagening. J. Life Sci.* **1998**, *46*, 225–247. [CrossRef]
23. Kroodsma, W.; in't Veld, J.H.; Scholtens, R. Ammonia emission and its reduction from cubicle houses by flushing. *Livest. Prod. Sci.* **1993**, *35*, 293–302. [CrossRef]
24. Elzing, A.; Monteny, G. Modeling and experimental determination of ammonia emissions rates from a scale model dairy-cow house. *Trans. ASAE* **1997**, *40*, 721–726. [CrossRef]
25. Grant, R.H.; Boehm, M.T. Manure Ammonia and Hydrogen Sulfide Emissions from a Western Dairy Storage Basin. *J. Environ. Qual.* **2015**, *44*, 127–136. [CrossRef] [PubMed]
26. McGinn, S.; Janzen, H.; Coates, T.; Beauchemin, K.; Flesch, T. Ammonia emission from a beef cattle feedlot and its local dry deposition and re-emission. *J. Environ. Qual.* **2016**, *45*, 1178–1185. [CrossRef]
27. Rotz, C.A.; Montes, F.; Hafner, S.D.; Heber, A.J.; Grant, R.H. Ammonia Emission Model for Whole Farm Evaluation of Dairy Production Systems. *J. Environ. Qual.* **2014**, *43*, 1143–1158. [CrossRef]
28. Sun, H.; Trabue, S.L.; Scoggin, K.; Jackson, W.A.; Pan, Y.; Zhao, Y.; Malkina, I.L.; Koziel, J.A.; Mitloehner, F.M. Alcohol, volatile fatty acid, phenol, and methane emissions from dairy cows and fresh manure. *J. Environ. Qual.* **2008**, *37*, 615–622. [CrossRef]
29. USEPA. *Inventory of U.S. Greenhouse Gas Emissions and Sinks: 1990–2003*; EPA-430-R-05-003; USEPA: Washington, DC, USA, 2005.
30. Hindrichsen, I.K.; Wettstein, H.R.; Machmüller, A.; Kreuzer, M. Methane emission, nutrient degradation and nitrogen turnover in dairy cows and their slurry at different milk production scenarios with and without concentrate supplementation. *Agric. Ecosyst. Environ.* **2006**, *113*, 150–161. [CrossRef]
31. He, Z.; Pagliari, P.; Waldrip, H.M. Advances and outlook of manure production and management. *Anim. Manure: Prod. Charact. Environ. Concerns Manag.* **2020**, *67*, 373–383.

32. Andriamanohiarisoamanana, F.J.; Sakamoto, Y.; Yamashiro, T.; Yasui, S.; Iwasaki, M.; Ihara, I.; Tsuji, O.; Umetsu, K. Effects of handling parameters on hydrogen sulfide emission from stored dairy manure. *J. Environ. Manag.* **2015**, *154*, 110–116. [CrossRef]
33. Ni, J.-Q.; Heber, A.; Sutton, A.; Kelly, D. Mechanisms of gas releases from swine wastes. *Trans. ASABE* **2009**, *52*, 2013–2025.
34. Maasikmets, M.; Teinemaa, E.; Kaasik, A.; Kimmel, V. Measurement and analysis of ammonia, hydrogen sulphide and odour emissions from the cattle farming in Estonia. *Biosyst. Eng.* **2015**, *139*, 48–59. [CrossRef]
35. El-Mashad, H.M.; Zhang, R.; Rumsey, T.; Hafner, S.; Montes, F.; Rotz, C.A.; Arteaga, V.; Zhao, Y.; Mitloehner, F.M. A mass transfer model of ethanol emission from thin layers of corn silage. *Trans. ASABE* **2010**, *53*, 1903–1909. [CrossRef]
36. Chung, M.Y.; Beene, M.; Ashkan, S.; Krauter, C.; Hasson, A.S. Evaluation of non-enteric sources of non-methane volatile organic compound (NMVOC) emissions from dairies. *Atmos. Environ.* **2010**, *44*, 786–794. [CrossRef]
37. Bonifacio, H.; Rotz, C.; Hafner, S.; Montes, F.; Cohen, M.; Mitloehner, F. A process-based emission model of volatile organic compounds from silage sources on farms. *Atmos. Environ.* **2017**, *152*, 85–97. [CrossRef]
38. USEPA. Emissions from Animal Feeding Operations. Contract No. 68-D6-0011. 2001. Available online: https://www.epa.gov/sites/production/files/2020-10/documents/draftanimalfeed.pdf (accessed on 4 January 2018).
39. Petersen, S.O.; Sommer, S.G. Ammonia and nitrous oxide interactions: Roles of manure organic matter management. *Anim. Feed Sci. Technol.* **2011**, *166*, 503–513. [CrossRef]
40. Bertora, C.; Alluvione, F.; Zavattaro, L.; van Groenigen, J.W.; Velthof, G.; Grignani, C. Pig slurry treatment modifies slurry composition, N2O, and CO2 emissions after soil incorporation. *Soil Biol. Biochem.* **2008**, *40*, 1999–2006. [CrossRef]
41. Harrison, J.; Ndegwa, P. Anaerobic digestion of dairy and swine waste. *Anim. Manure: Prod. Charact. Environ. Concerns Manag.* **2020**, *67*, 115–127.
42. Petersen, S.O. Nitrous Oxide Emissions from Manure and Inorganic Fertilizers Applied to Spring Barley. *J. Environ. Qual.* **1999**, *28*, 1610–1618. [CrossRef]
43. Montes, F.; Meinen, R.; Dell, C.; Rotz, A.; Hristov, A.N.; Oh, J.; Waghorn, G.; Gerber, P.J.; Henderson, B.; Makkar, H. SPECIAL TOPICS—Mitigation of methane and nitrous oxide emissions from animal operations: II. A review of manure management mitigation options. *J. Anim. Sci.* **2013**, *91*, 5070–5094. [CrossRef] [PubMed]

Review

Examining the Variables Leading to Apparent Incongruity between Antimethanogenic Potential of Tannins and Their Observed Effects in Ruminants—A Review

Supriya Verma [1,*], Friedhelm Taube [1,2] and Carsten S. Malisch [1]

1 Grass and Forage Science/Organic Agriculture, Institute of Crop Science and Plant Breeding, Christian-Albrechts-Universität zu Kiel, Hermann-Rodewald Str. 9, DE-24118 Kiel, Germany; ftaube@gfo.uni-kiel.de (F.T.); cmalisch@gfo.uni-kiel.de (C.S.M.)
2 Grass Based Dairy Systems, Animal Production Systems Group, Wageningen University (WUR), 6708 PB Wageningen, The Netherlands
* Correspondence: sverma@gfo.uni-kiel.de; Tel.: +49-431-880-2137

Abstract: In recent years, several secondary plant metabolites have been identified that possess antimethanogenic properties. Tannin-rich forages have the potential to reduce methane emissions in ruminants while also increasing their nutrient use efficiency and promoting overall animal health. However, results have been highly inconclusive to date, with their antimethanogenic potential and effects on both animal performance and nutrition being highly variable even within a plant species. This variability is attributed to the structural characteristics of the tannins, many of which have been linked to an increased antimethanogenic potential. However, these characteristics are seldom considered in ruminant nutrition studies—often because the analytical techniques are inadequate to identify tannin structure and the focus is mostly on total tannin concentrations. Hence, in this article, we (i) review previous research that illustrate the variability of the antimethanogenic potential of forages; (ii) identify the source of inconsistencies behind these results; and (iii) discuss how these could be optimized to generate comparable and repeatable results. By adhering to this roadmap, we propose that there are clear links between plant metabolome and physiology and their antimethanogenic potential that can be established with the ultimate goal of improving the sustainable intensification of livestock.

Keywords: proanthocyanidins; condensed tannins; secondary plant metabolites; methane; ruminants; climate change

1. Introduction

Intensification and global expansion of livestock production systems have led to significant increased emissions of agricultural carbon dioxide (CO_2), nitrous oxide (N_2O), and methane (CH_4), with agriculture contributing to almost 15 % of the total anthropogenic greenhouse gas (GHG) emissions [1,2]. A major part of these emissions is in the form of CH_4 (44%), while the rest is divided between N_2O (29%) and CO_2 (27%) (proportions expressed in terms of CO_2-equivalent (CO_2-e)). From 1990 to 2012, global CH_4 emissions have increased by 11% from 1869 million tonnes to 2080 million tonnes CO_2-e. Methane has a shorter atmospheric lifespan (12 years) compared to N_2O (114 years) and CO_2 (up to thousands of years), and developing mitigation strategies for CH_4 abatement will help reach the global GHG-reduction targets and temperature stabilization goals [2,3]. In addition to this potential for temperature stabilization, a reduction in CH_4 emissions could further allow a reduction in existing atmospheric CH_4, as the remaining CH_4 emissions are naturally removed from the atmosphere within a short timespan [4–6]. Methane is released as a product of microbial degradation of feed macromolecules in the digestive tract of ruminants [7]. Ruminal methane emissions are the result of an inefficient pathway in ruminant digestion of feed and reducing these emissions would also be efficacious in preventing metabolizable energy losses; these comprise between 2 and 15% of the digestible energy intake depending on the forage quality [8–10]. Hence, the development and adoption of strategies and approaches to reduce CH_4 emissions from livestock systems would have both environmental benefits and lead to improved feed utilization and animal productivity. Since CH_4 production cannot be eliminated entirely without the ruminant losing its ability to digest fibre, the focus should be on increasing nutrient use efficiency in ruminant livestock [11].

One strategy with promising potential is the use of tannin-rich forages (TRFs). Tannins are polyphenolic plant secondary metabolites, which can precipitate or crosslink the proteins, thus making them less prone to proteolysis [12,13]. While several TRFs have been investigated for their antimethanogenic potential in numerous in vivo and in vitro trials, the results have so far been highly inconsistent. One such TRF is sainfoin (*Onobrychis viciifolia*). A study by Chung [14] indicated no difference between methane emissions from sainfoin and alfalfa hay in terms of dry matter intake (DMI), but identified 25% emission reductions from sainfoin based on the organic matter digested. In contrast, Huyen [15] showed that sainfoin silage diets decreased CH_4 emissions (per unit DMI) by 5.8% compared to grass and maize silage. On the other hand, other studies reported increments in CH_4 emissions when diets of sainfoin hay [16] and sainfoin silage [17] were fed. Similar discrepancies can be found in their effect on reducing bloat [18,19] and shifting nitrogen (N) excretion from urine to faeces [14,15,20,21]. These inconsistencies are still difficult to explain, although they may be partly explained by the lack of precise structural characterizations of tannins. Some studies have shown the intraspecies variation of tannin concentration and structures in sainfoin, indicating the complexity of tannin composition in forages [22,23], as well as the impact of the structures on antimethanogenic properties [24,25]. The co-presence of other secondary plant metabolites such as flavanols and saponins can also exert potential mutualistic or antagonistic effects [10,26]. Variations can also arise as a result of the growth conditions of the tested plants, which differed greatly across the reported experiments, and these can affect their secondary metabolite synthesis [27–30].

In this review, we (a) identify the potential of TRFs, specifically those containing condensed tannins (CTs, *syn.* proanthocyanidins), to affect rumen productivity and methanogenesis; (b) illustrate how the structural diversity within CTs is likely to contribute to explaining the inconsistencies observed; and (c) provide a roadmap to assess the bioactive potential of CT in livestock production systems. We aim to integrate the research on TRFs' potential to reduce methane emissions by understanding tannin synthesis, their mode of action in the animal and thereby, to indicate suitable analyses to improve their interpretability. If applied in practice, following this roadmap will increase the potential to extrapolate findings of antimethanogenic potential of forages.

2. Understanding Tannins and Their Functional Attributes

Previously, the sole function of tannins was regarded to be a part of a plant's defence mechanism against herbivory [13,31–33]. This trait conferred antiherbivory effects through (a) the ability of CT to precipitate proteins, thus rendering them unavailable for animal nutrition, and (b) they can have oxidative activities, which create oxidative stress in the herbivore gut [34].

In terms of their role in herbivore diets, plant tannins have surpassed their reputation of being purely antinutritional compounds and several of their beneficial functions have been identified. Tannins have been shown to possess the potential to reduce the impact of drought by acting as antioxidants and detoxifying reactive oxygen species produced as a result of drought stress [28,35]. Additionally, tannins and other polyphenols have been found to reduce the carbon and N mineralization rates in soil, by inhibiting the activity of soil microorganisms and enzymes [29,36]. At an individual plant level, this can result in long-term nutrient availability due to slower litter decomposition [29], while at the plant community level, this will enable better adaption of microorganisms to adapt to TRFs, thus generating a general "home field advantage" for one species [36], as well as increasing soil carbon stocks [37–40]. With the discovery of the additional functions of tannins, TRFs have emerged as a promising solution to help reduce CH_4 emissions in ruminants, while concomitantly providing a series of additional environmental or animal health benefits. A selection of relevant properties will subsequently be discussed in more detail.

2.1. What Are Tannins?

Tannins are the end products of energy demanding and extensive biosynthetic pathways, indicating that they play an important role in plant metabolism. They can be broadly divided into two groups—hydrolysable tannins (HTs) and CTs—depending on their structure [31,41]. Hydrolysable tannins contain central polyol esterified with gallic acid molecules [12,42]. They can be further divided into three groups: simple gallic acid derivatives, gallotannins (GTs), and ellagitannins (ETs). The two first classes contain only galloyl groups attached to the central core (glucose/polyol): simple gallic acid derivatives having only monogalloyls groups, but GTs having digalloyl or even trigalloyl groups in series attached to the polyol. In ETs, two of the galloyls are C-C linked to make the characteristic hexahydroxydiphenoyl (HHDP) group that can be modified even further [12,42,43]. Condensed tannins are the second most abundant polyphenols after lignins, and consist of two or more flavan-3-ol monomeric units. The most common flavan-3-ol subunits of CTs are characterized based on the number of hydroxyl groups on the A and B rings, and the relative stereochemistry between the B and C rings (Figure 1).

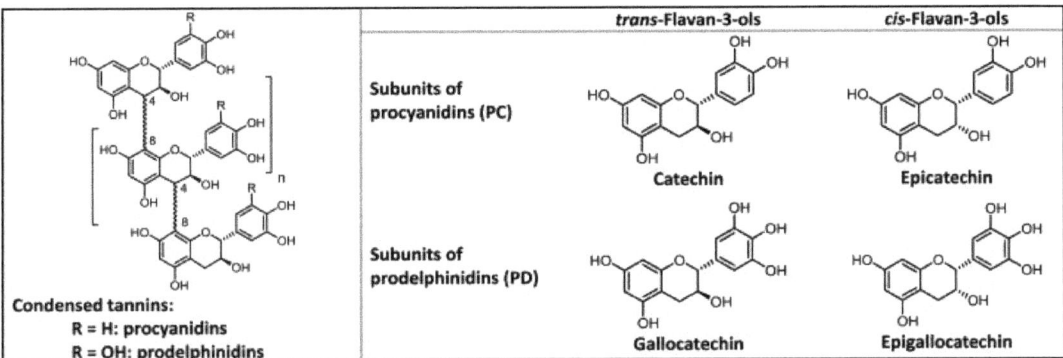

Figure 1. Structure of condensed tannin subunits [44].

Catechin and epicatechin have two hydroxyl groups present adjacent to each other on the B ring of flavon-3-ol subunits, and are categorized as procyanidin (PCs) units when found in CT structures. Gallocatechin and epigallocatechin have three hydroxyl groups adjacent to each other on the B ring and are categorized as prodelphinidin (PDs) units in CTs [33,45–47]. Additionally, both PCs and PDs can differ in their relative orientation of the C-2/C-3 carbon substituents of the C-ring, where catechin and gallocatechin have a trans-configuration, whereas epicatechin and epigallocatechin have a *cis*-configuration [48,49]. These subunits are connected through interflavan linkages, the most common of which are B type linkages. In B type linkages, the bonds between the subunits are formed either between the C-4 carbon of the C ring and the C-8 carbon of the subsequent flavan-3-ol subunit (4 → 8) or between the C-4 carbon and the C-6 carbon (4 → 6) [50,51]. When the covalent bond is formed between two flavon-3-ol subunits via a C-2 oxygen atom and a C-7 carbon in addition to the 4 → 8 B linkage, the linkage is known as A type (Figure 2) [49].

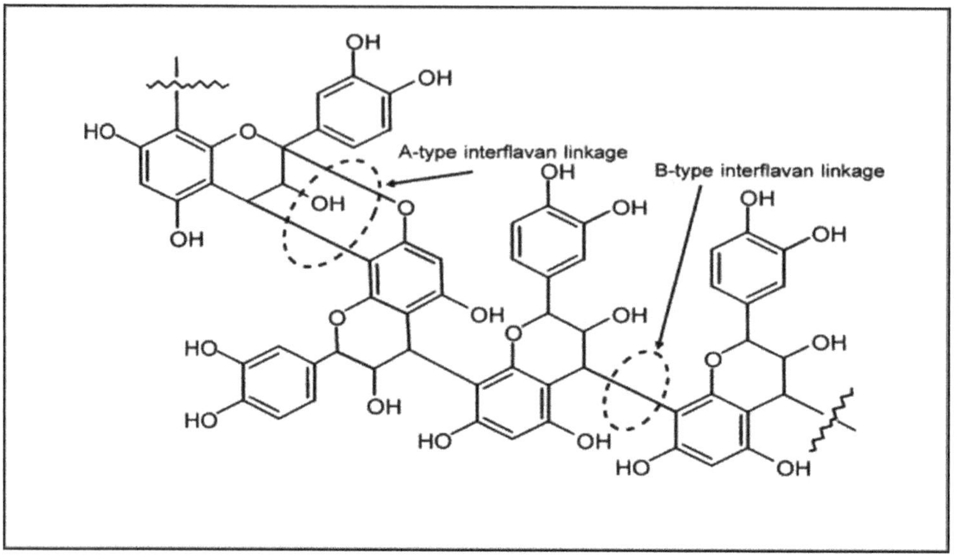

Figure 2. A- and B-type interflavan linkages in condensed tannin oligomers and polymers [52].

The proportions of PC and PD subunits, and also the type of linkages within CTs, vary substantially both across and within plant species [53,54]. These variations combined with the varying degrees of polymerization can lead to a multitude of combinations in structures and hence, a wide range of bioactive properties of CTs [55]

2.2. Functional Attributes of Tannins

The bioactive properties of tannins are either a result of their protein precipitation capacity (PPC), or their anti- or pro-oxidant behaviour. The effect of tannins on biological systems is found to be dependent on pH, with protein precipitation capacity being generally efficient in slightly to moderately acidic environments, whereas the oxidative activity is expressed in alkaline environments or by oxidative plant enzymes, such as polyphenol oxidases [56].

The fate of ingested tannins in herbivores is dependent on the physiological conditions of their gut. Tannins when consumed by herbivores with high gut pH, such as in caterpillars, undergo auto oxidation to produce semiquinone radicals and quinones. These oxidation products can bind to the nutrients in the gut lumen of the caterpillar and cause damage to the surrounding gut tissues [57]. In contrast, the effect of tannins on mammalian herbivores is dependent mainly on its PPC, as the mammalian gut has acidic to neutral gut conditions which provide an ideal environment for tannin–protein interactions [34]. When supplied in moderate quantities, the protein binding ability of tannins can improve nutrient utilisation in ruminants; however, in insects, both CTs and HTs had no impact on protein utilization [58,59]. Additionally, the efficacy of these effects is dependent on the structure of tannins. Ellagitannin-rich plants were found to be more potent in terms of their oxidative behaviour compared to plants rich in galloyl glucoses or CTs [57], and CTs are found to precipitate proteins more actively than ETs [33].

The anthelmintic and antimethanogenic bioactivity of CTs in ruminants is linked to their precipitation capacity [60,61] and their antioxidative behaviour [34,49,62]. Condensed tannins are known to form insoluble complexes with proteins by binding to the protein's surface and forming a coat and this leads to its precipitation [63,64]. These complexes are generally based on non-covalent interactions such as hydrophobic interactions and hydrogen bonding [65]. However, there have been reports on ionic interactions and covalent bonds with amino acids or sulphur on proteins [66]. Additionally, under low pH and oxidative conditions, tannins can form covalent bonds with proteins [38,65,67,68].

Independent of the bond type, within CTs, a higher PD percentage has been associated with a higher PPC, which is likely a result of the additional hydroxyl groups at carbon 5 of the B ring [49,69,70]. In addition to the PD/PC ratio, the *cis/trans* ratio, polymer size, and co-presence of galloyl groups have been identified as having effects on the PPC [71]. However, these results have been inconsistent, which is likely a result of multiple structural features being responsible for the tannins' astringency concomitantly and potentially imparting contrasting effects [59,72,73]. The polyphenolic polarity, as defined by the octanol-water partition coefficient (K_{OW}), can also influence the PPC of tannins [68]. Tannins with high K_{OW} values (e.g., acacia (*Acacia mearnsii*) leaves, K_{OW} = 13.92) are fat soluble and bind non-specifically to the proteins. They have the tendency to be adsorbed by animal tissues and exert toxic effects. Tannins with low K_{OW} values such as chestnut (*Castanea sativa*) extracts (K_{OW} = 1) bind more efficiently with proteins and lead to better nutrient utilization in animals [59]. However, the nature of these interactions is also dependent on the proteins. For example, the PCs were found to have a stronger affinity for larger proteins with open structures such as BSA (66 kD) compared to lysozyme (14.4 kD), which has a compact structure and is smaller in size [50,59,66,68]. Additionally, the isoelectric point (pI) of proteins has generally been identified to affect the tannins' protein precipitation behaviour [38], and proteins aggregate faster when the pH is close to their pI [68]. The reaction conditions also play a significant role in the strength of tannin–protein complexes. The variability in dietary composition with differences in protein chemistry (for example: proline content), amino acids, and CT composition, makes it exceptionally difficult to predict the response of CT–protein interactions. Finally, it should be mentioned that there appears to be at least a partial specificity, with plant tannins showing a higher precipitation of plant proteins, compared to animal protein. Accordingly, in a study by Zeller [51], tannins from birdsfoot trefoil (*Lotus corniculatus*) were better at precipitating proteins from lucerne (*Medicago sativa*), compared to BSA. Hence, the protein source should also be accounted for in the estimation of the PPC [74].

The link between PPC, oxidative properties, and the observed bioactivity of tannins is, however, still not clear because of the inadequate tests in many reported studies. Therefore, complementing their protein precipitation assays with the analysis of their anti-/pro-oxidative behaviour can provide a better overview and improve understanding of CT–animal interactions.

2.3. Potential of Incorporating Tannin Rich Forages in Ruminant Nutrition

As explained previously, tannins have long been considered to be non-specific anti-nutritive factors and potentially toxic, as they protect dietary protein from degradation, and because of their pro-oxidant properties [65]. These characteristics are undoubtedly true, as tannins have, indeed, been found to form strong, yet pH-dependent and reversible bonds to proline rich proteins and affect protein digestibility [75]. Some browsing herbivores have developed the ability to produce proline-rich salivary mucoproteins as an evolutionary adaption to overcome the deleterious effects of tannins [66]. Herbivore palatability of TRFs is determined on the basis of astringency resulting from the interactions between CTs and the herbivore's salivary proteins. Tanniferous forages are often considered to be less palatable and therefore, less acceptable. At a CT concentration above 5% of the herbage dry matter (DM), intake and palatability of TRFs may be depressed and feed intake is reduced. However, reported results shows this is highly variable [49]. Despite its high CT concentration, sulla (*Hedysarum coronarium*) has been found to be highly acceptable by sheep [76]. Similarly, the acceptability (and assumed palatability) of sainfoin was found to be comparable to conventional temperate forages such as alfalfa and ryegrass/clover mixtures [77]. Sainfoin has also been reported to be more palatable than birdsfoot trefoil despite its higher tannin concentration [77,78].

2.3.1. Impact of Tannins on Enteric Fermentation

Feed constituents such as carbohydrates, proteins, and other organic polymers are degraded to their monomer components in the presence of rumen microbes under anaerobic conditions [7,79,80]. Tannin-rich forages have been reported to cause alterations in rumen microflora, increase nutrient utilization efficiency, improve animal health, and consequently, influence their environmental effect [15,46,81,82]. The presence of tannins in the feed has been found to slow down the degradation of the dietary proteins by forming tannin–protein complexes in the rumen [83]. These complexes are then transported from the rumen (pH = 6–7) to the small intestine (pH > 7), where they are partially dissociated under alkaline conditions. Through this process, the excess protein is initially protected from inefficient degradation in the rumen, so it reaches the small intestine as rumen bypass protein. As a result, there is an increased amino acids absorption throughout the entire digestive tract for tannin-containing feeds compared with non-tannin-containing feeds [32,59,84]. The decrease in excess protein degradation in the rumen also results in a decrease in methanogenesis and consequently, lower CH_4 emissions. Concomitantly, the non-ammonia N transported to the small intestine leads to a higher production of milk, meat, and wool. This deviation further decreases the urinary N and slightly increases faecal N [10,59,78,85,86]. The decrease in urinary N can lead to lower indirect N losses to the environment from the urine patches, as these spatially concentrated excretions have a high risk of volatilization, nitrification, and denitrification (Figure 3) [15,87].

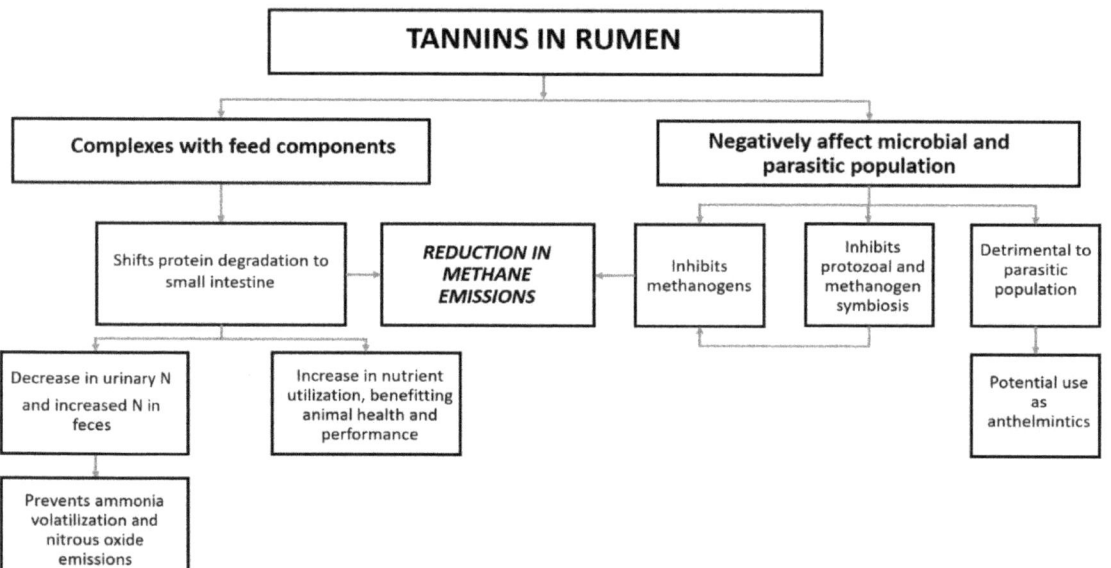

Figure 3. Beneficial effects of tannins on ruminant nutrition.

In temperate forage systems, the forage-protein concentrations in are generally higher than in tropical forages. Hence, the N use efficiency in temperate forages is often low, and in some instances, as low as around 10–20% [78,88]. Accordingly, reductions in available protein can be achieved without adversely affecting milk yields by increasing the N use efficiency, thereby concomitantly reducing the nitrogen emissions to the environment. However, the effects of tannins in the gastrointestinal tract of ruminants are complex. For example, CTs in birdsfoot trefoil have a strong effect on the proteolytic bacteria in the rumen of sheep. As a result, plant protein degradation in the rumen is decreased and non-ammonia N flow to the small intestine is increased, resulting in higher utilizable crude protein (uCP) in the small intestine [89,90]. However, even within the *Lotus* genus, big trefoil (*Lotus pedunculatus*) and birdsfoot trefoil have different modes of action in their effect on nitrogen flows. A direct comparison of these species shows that CTs from big trefoil were more effective in the degradation of Rubisco compared to those of birdsfoot trefoil. Similarly, CTs from big trefoil were able to inhibit the degradation of protein in the rumen by forming strong tannin–protein complexes, whereas birdsfoot trefoil tannins reduced degradation of proteins by directly inhibiting the proteases [91]. Additionally, big trefoil was found to have a stronger potential to reduce CH_4 emissions than birdsfoot trefoil [59].

As a result of this complexity, in vivo experimentation has not yet been able to successfully show both a reduction in CH_4 emissions and incremental improvement in N use efficiency simultaneously from tannin-containing forages. To illustrate the existing research gaps, it is important to understand how tannins influence rumen microbiota, as well as the interactions between hydrogen producers (bacteria, protozoa, fungi) and consumers (methanogens) [92].

2.3.2. Mode of Action to Lower Methane Emissions

Several mechanisms have been hypothesized by which tannins might decrease CH_4 emissions in ruminants. Efficient nutrient utilization is considered to be one of the most likely explanations, and this might increase animal productivity and reduce CH_4 production per unit of animal product. The inclusion of tannins in feed has been found to improve nutrient utilization in the ruminants, thereby reducing metabolic energy losses that would otherwise occur through CH_4 emissions [79,93,94].

Another factor which could be linked to CT's potential in reducing CH_4 emissions is its affinity to form complexes with lignocellulose and preventing fibre degradation, thereby leading to lower microbial fermentation [95]. Microbial fermentation leads to the formation of volatile fatty acids (VFA) such as acetate, propionate, and butyrate, with CO_2 and H_2. These metabolic byproducts are either absorbed by the rumen wall and used as a source of energy for animals or used as substrates by microorganisms [7,79,94,96]. Tannins have been known to reduce the CH_4 emissions of ruminants either by directly inhibiting the ruminal methanogenic population [92,97], or by hindering the methanogen-protozoa symbiosis [49]. Approximately 37% of the CH_4 from the ruminants is produced by protozoa-associated methanogens. In the methanogen-protozoa symbiosis, hydrogen (H_2) required by methanogens to produce CH_4 is provided by rumen protozoal population via transfer of H_2 produced in their hydrogenosomes. The subsequent utilization of H_2 by methanogens benefits the protozoal population as H_2 hinders their metabolism [10,98]. As the accumulation of H_2 in the rumen can impede fermentation, methanogens play an important role in feed digestibility by utilizing the rumen borne H_2. Hence, before adapting feeding strategies to achieve defaunation of the rumen, it is important to provide alternative H_2 sinks to maintain the animal's productivity and improve the utilization of metabolizable energy from the feed [99,100].

Here, tannins might be part of the solution as well, as some studies have hypothesized that tannins influence the VFA profile in rumen. They promote the shift towards the production of more propionate compared to acetate, which acts as a hydrogen sink. The reduced availability of H_2, which is the main substrate for CH_4 production, results in a reduction in methanogenesis [86,100,101]. The shift in acetate and propionate production could be attributed to changes in the composition of microbial communities and their activity [95]. However, the mechanism by which tannins influence methanogenesis and shift the VFA profile is still not well understood.

3. Current Findings on the Antimethanogenic Potential of TRFs

Recent studies have shown that the effect of tannins on ruminant nutrition is highly dependent on the tannin type, structural characteristics, dosage supplied, rumen morphology, and rumen physiology [102,103]. Numerous plant species containing tannins have been studied to determine their efficacy in ruminant nutrition, either as forages or feed additives. These species include acacia, quebracho (*Schinopsis balansae*), chestnut, valonea (*Quercus Aegilops*), leucaena (*Leucaena leucocephala*), desmodium (*Desmodium ovalifolium*), sainfoin, birdsfoot trefoil, big trefoil, Chinese bushclover (*Lespedeza cuneata*), Japanese clover (*Lespedeza striata*), white clover (*Trifolium repens*), and sulla [59,86,101].

Antimethanogenic potential was found to vary across the species. Promising temperate forage species include sainfoin, birdsfoot trefoil, big trefoil, and sulla, and among tropical forages are leucaena, desmodium, and Chinese bushclover [24,62,104,105]. A study of Friesian dairy cows found that cows that grazed on birdsfoot trefoil produced not only 17.5% less CH_4 emissions (per unit DMI) but also 32% less CH_4 emissions/kg milk solids when compared with cows grazing on perennial ryegrass (*Lolium perenne*) [106]. Similarly, leucaena, a tropical leguminous shrub, has been found to reduce CH_4 emissions in sheep and heifers without affecting DMI or organic matter intake in the animals [107,108]. In another study, when supplied with 80% leucaena in the diet compared with a basal diet of *Pennisetum purpureum*, CH_4 emissions were reduced by 61% in heifers without negatively affecting DMI and VFA production [109]. The overall performance of the lambs (approx. 6 months age) was improved when CTs were included in their basal diet (wheat straw, oat hay, and concentrate mixture). Condensed tannins in the diet were supplied as leaf meal mixture of *Ficus infectoria* and *Psidium guajava* (70:30). The diet with 2% CTs was able to suppress CH_4 emissions by approx. 26%. Additionally, improved N metabolism, wool yield, and growth performance of lambs was reported. Inclusion of CTs in the feed did not affect the intake or apparent palatability of the feed [110].

Similarly, in a study conducted on adult sheep, hazel (*Corylus avellana*) leaves when supplemented at 50% of the total diet were able to reduce CH_4 emissions by 35% (per unit OM intake) compared to the control (ryegrass hay and lucerne pellets). Concomitantly, a substantial decrease in urinary N proportion of total N intake was observed without any negative effects on forage intake, apparent palatability, or body weight of the sheep [111]. However, despite the promising findings indicated by these studies, the antimethanogenic potential of the forages is not clearly linked to the tannin concentration, as evidenced by the high variability in results from different studies (Table 1). As summarized in Table 1, the variation in CH_4 abatement by forages also depends on the phenological stage at which they are harvested and by the method of forage preservation. In addition to the changes in forage chemical composition, phenological stage also affects the CT composition and structural features. The bioactivity of sainfoin CTs was found to decrease with maturity, as shown by the increase in phenological stage. This could be attributed to the lower proportion of extractable CTs (ECTs) resulting from increase in CT polymerization with maturity [112]. Similarly, when TRFs are ensiled, the process can rupture plant cells, allowing the CTs to release and bind to other molecules. This decreases the proportion of free CTs (ECTs) and hence, there is reduced bioactivity of CTs in conserved forages compared to fresh forages in terms of their ability to reduce CH_4 emissions [21].

Furthermore, the mode of action by which these forages reduce CH_4 emissions remains largely unclear. Tannins from chestnut, quebracho [113], and leucaena [114,115] have been found to reduce CH_4 by reducing different methanogenic populations in the rumen. There was a significant effect of high molecular weight (M_W) CT fractions from *Leucaena* on richness and species diversity of rumen methanogenic and bacterial population in rumen. The study showed that CTs with high M_W had a pronounced inhibitory effect on proteolytic bacteria, *Prevotella* spp., and *Methanobrevibacter* population [97,116].

Table 1. A short overview of methane production potential of tropical and temperate forages.

Plant Species	Age	Fraction	Preservation	ECT * (%)	Animal (Rumen Fluid)	Methane (g/kg DM)	Study (Duration)	Reference
Acacia angustissima var. hirta (STX)	Mature	Leaves	Fresh	4.9	Steers	0.6	In vitro (48 h)	[90,117]
Acacia angustissima var. hirta (STP5)	Mature	Leaves	Fresh	4.4	Steers	0.8	In vitro (48 h)	[90,117]
Desmanthus illinoensis (Michx.) MacMill	Mature	Leaves	Fresh	5.1	Steers	24.9	In vitro (48 h)	[90,117]
Desmodium paniculatum var. paniculatum	Mature	Leaves	Fresh	10.3	Steers	7.9	In vitro (48 h)	[90,117]
Lespedeza cuneata	Mature	Leaves	Fresh	4.7	Steers	15.1	In vitro (48 h)	[90,117]
Lespedeza stuevei	Mature	Leaves	Fresh	9.9	Steers	4.9	In vitro (48 h)	[90,117]
Leucaena retusa	Mature	Leaves	Fresh	2.4	Steers	40.7	In vitro (48 h)	[90,117]
Mimosa strigillosa	Mature	Leaves	Fresh	9.9	Steers	7.6	In vitro (48 h)	[90,117]
Neptunia lutea	Mature	Leaves	Fresh	7.0	Steers	19.7	In vitro (48 h)	[90,117]
Onobrychis viciifolia acc LRC 3519	Early stage	Herbage	Fresh	2.5	Cross bred heifers	28.2	In vivo (24 h)	[14]
Onobrychis viciifolia acc LRC 3519	Late stage	Herbage	Fresh	0.7	Cross bred heifers	24	In vivo (24 h)	[14]
Onobrychis viciifolia acc LRC 3519	Mature	Herbage	Hay	0.6	Cross bred heifers	22.5	In vivo (24 h)	[14]
Medicago sativa	Early stage	Herbage	Fresh	0	Cross bred heifers	26.6	In vivo (24 h)	[14]
Onobrychis viciifolia cv. Perly	Mature	Herbage	Silage	3.7	Brown Swiss cows	18.75	In vitro (24 h)	[17]
Onobrychis viciifolia cv. Shoshone [1]	Early Flowering	Herbage	Hay	3.9	Holstein dairy cows	12.9	In vitro (24 h)	[118]
Lotus corniculatus cv. Norcen [1]	Early Flowering	Herbage	Hay	0.4	Holstein dairy cows	11.7	In vitro (24 h)	[118]
Lotus corniculatus cv. Oberhaunstadter [1]	Early Flowering	Herbage	Hay	0.7	Holstein dairy cows	11.8	In vitro (24 h)	[118]
Lotus corniculatus cv. Bull	Mature	Herbage	Silage	2.2	Brown Swiss cows	17.64	In vivo (24 h)	[17]
Lotus corniculatus cv. Polom	Mature	Herbage	Silage	0.8	Brown Swiss cows	18.75	In vivo (24 h)	[17]
Lotus corniculatus	Vegetative	Herbage	Silage	2.5	Friesian dairy cows	26.9	In vivo (24 h)	[119]
Lolium perenne	Mature	Herbage	Fresh	0	Friesian dairy cows	24.15	In vivo (24 h)	[106]
Lotus corniculatus	Mature	Herbage	Fresh	0.2	Friesian dairy cows	19.9	In vivo (24 h)	[106]
Lotus pedunculatus	Mature	Herbage	Fresh	8	Sheep	14.5	In vivo (24 h)	[119]
Hedysarum coronarium	Mature	Herbage	Fresh	2.8	Friesian and Jersey dairy cows	19.5	In vivo (24 h)	[120]

[1] Refers to feed supplied in total mixed ration, * Extractable condensed tannins.

Similarly, birdsfoot trefoil and sainfoin were found to inhibit the proteolytic bacterial population [49]. A study analysing the effect of different tannin sources on CH_4 emissions found that CT-rich (acacia and quebracho tannins) and HT-rich (chestnut and valonea tannins) affect rumen fermentation differently. At concentrations above 5% DM, in addition to a significant decrease in CH_4 emissions, there was also a negative effect on total VFA production. CT-rich extracts reduced the acetate/propionate ratio significantly at a concentration higher than 5%. However, the ratio was not affected by HT extracts, indicating that they had a stronger impact on methanogen population in comparison with substrate fermentation. Only valonea extracts (5% w/w) were able to reduce CH_4 emissions without any negative impact on fermentation and VFA profile. This indicates that classification based solely on tannin concentration or the type of tannins (HTs and CTs) present in the feed is not sufficient to determine their potential to reduce CH_4 emissions [86]. Similarly, a study was performed on CT-rich forages from Texas to determine the effect of different functional features (PPC and antioxidative activity) on CH_4 emissions. No effect of PPC was found on CH_4 abatement, whereas the correlation between antioxidative property of tannins and decrease in CH_4 emissions was significant. In contrast to previous studies, the decrease in acetate/propionate in this study was not correlated with a decrease in CH_4 emissions [90]. The results from these studies further reinforce the need for CT structural characterization in addition to concentration, in order to make an accurate assessment of their impact on ruminant nutrition.

4. Existing Research Gaps and Future Directions

As discussed in the previous sections, several studies have tried to explore the properties that affect tannin astringency. However, the variations in the results obtained, and their lack of reproducibility, hinder their field-scale applicability. Furthermore, their structural complexity and the varied forage chemical composition among different species present difficulties for understanding the implications that CTs have for ruminant nutrition and particularly their antimethanogenic potential. Although several studies have identified a large variability in both the concentration and structure of CTs across species and their cultivars [22,32,74], few studies have analysed the implications of this variability on the observed bioactivity. In the following sections, we present a brief overview about the factors responsible for current situation, with an apparent incongruity regarding the influence of tannins on ruminant nutrition. We also discuss the frequently used analytical techniques for qualitative and quantitative analysis of tannins and the underlying problems associated with them. Our aim here is to illustrate the importance of optimized tannin analyses and inclusion of tannin structural features in animal studies to overcome inconsistent animal responses. By avoiding these factors, which cause substantial variation in the reported studies, we can focus more precisely on CT–animal interactions.

4.1. Experimental and Analytical Incongruities

Tannin concentration and composition in plant has been reported to be substantially influenced by changes in environmental conditions, as well as by plant species and its phenological stage [40]. The preparation and handling of tannin extracts can also cause alterations in quantification of tannins [121]. In order to ensure accurate determination of tannin concentration and composition in a plant tissue, handling and storage protocols should be followed, as CT concentration is highly influenced by the environmental factors. Quantitative analyses are essential for determining CT bioavailability in the samples and spectrophotometric assays are routinely utilized due to their rapid and low-cost analysis. Due to their structural complexity, the number of derivatisation and analytical techniques are few and they have certain limitations. Substantial information about the activity of CTs can be obtained by analysing specific structural traits of tannins as it is difficult to isolate individual large polymeric CT units compared to dimers or trimers [122]. The complexity of CT structures means that they are frequently analysed by a method where multiple techniques with different functions are integrated together.

4.1.1. Growth Conditions of Experimental Plants

Tannin concentrations in plants can be up to 20% of their total dry weight [38]. Although, CTs are found in different parts of the plant and they are predominantly concentrated in young leaves and flowers [105,123]. The concentration of tannins in tropical plants is, on average, higher than in temperate plants, yet there is substantial variation across seasons and environmental factors [62]. Drought, nutrient availability, and other conditions during plant growth have also been shown to affect CT concentration and composition. Although the effect of these abiotic stresses on the CT composition has not been well researched [27,28,124], they have been shown to produce incremental effects in CT concentrations [84,125]. Accordingly, it was observed that the concentration of CTs in sulla was higher in the summer than in spring [54]. Similarly, *Quercus rubica* had higher tannin concentration and less polymerized tannins when grown in dry conditions compared with wet conditions [62]. Thus, it is important to account for and report the precise experimental conditions because of their potential to affect the observed tannin concentration, composition, and bioactivity [126].

4.1.2. Sample Preservation and Storage

After harvest, sample preservation plays an important role in the quantification of CTs, as their extraction and quantification are heavily influenced by biotic factors. For precise tannin concentration and composition analysis, samples should be freeze-dried, rather than air or oven dried. In a direct comparison, hay-drying of samples from purple prairie clover (*Dalea purpurea* Vent.) resulted in a slight decrease in ECTs from 70.2 to 64.1 g kg DM^{-1}, while the protein-bound CTs increased from 9.0 to 12.4 g kg DM^{-1}. With ensiled samples, the differences were even more pronounced, and in these samples, ECTs decreased to 27.4 g kg DM^{-1}, while protein-bound CTs increased their concentration to 44.3 g kg DM^{-1} [127]. Thus, while the total CT concentration did not differ, without freeze drying at least a part of the CTs, it can change from the available form to the protein-bound form, which is often not accounted for in the studies analysing bioactivity of tannins. Similarly, when comparing different drying methods for the concentration of HTs in white birch (*Betula pubescens*), oven drying reduced the ET concentration significantly from 10.9 to 8.4 g kg DM^{-1}, while simultaneously increasing the concentration of insoluble ET from 0.8 to 2.4 g kg DM^{-1}. In this study, neither storing the samples at $-20\ °C$ for 3 months prior to drying nor vacuum or air drying resulted in a decrease in ellagitannins or total HTs despite a minor but non-significant decrease in air dried samples [121]. This indicates that the adverse effect from air and oven drying increases with the air temperature, yet the short-term effect at room temperature appears to be negligible. The effect of temperature during post drying storage is less clear, and storage at $25\ °C$ for three weeks reduced tannins in walnuts by 20–40% (dependent on the subsequent extraction technique), with large parts of the reduction having occurred in the first week [128]. However, with regard to the HTs from birch, a one-year storage period at room temperature ($22\ °C$) yielded lower tannin concentrations compared to samples that were stored in a freezer, with 17.4 and 19.1 g kg DM^{-1}, respectively. This value was still higher than the HT concentration in the samples stored in a refrigerator at $4\ °C$, which yielded 15.7 g kg DM^{-1} [121]. Contrasting results were obtained by Kardel [129], where samples stored in a refrigerator for one year had on average 3.5% higher CT concentrations than samples stored at room temperature. In this study, also storing the samples in an oven at $60\ °C$ for 5 days had no measurable impact on CT concentrations.

4.1.3. Tannin Extraction

While the vast majority of studies use aqueous acetone to extract tannins of any kind, with either 70 or 80% acetone, there is no clear indication about the superiority of any extraction method yet. Some studies have also used methanol or hot water extraction as well. Accordingly, the extraction yields with hot water were 6% higher on average, compared to a water/methanol (1:1) extraction solvent, and 13% higher compared to an acetone/water/formic acid (70:29.5:0.5) solvent [129]. This study only determined the tannin concentration and did not evaluate potential changes in the structure due to the high temperatures. Contrary to these findings, Salminen [121] found that aqueous acetone extraction yielded on average 41% more extractable HT, compared to aqueous methanol extraction, although the study did not test hot water extraction. Pure acetone, however, yielded the lowest HT concentrations, with a reduction of almost 75% compared to acetone/water (70:30). The extraction yield of the 70% acetone was increased even further by 29%, if ascorbic acid was added to the acetone water mixture, presumably because it prevented oxidation of the HTs [121]. This is in accordance with the findings of Chavan [130] and Hagerman [131] for CTs, where 70% acetone also provided the highest extraction yields compared to all methanol mixtures and acetone mixtures with higher acetone concentrations. Acidification of aqueous acetone extraction solvent with 1mL of concentrated HCl further increased the extraction yields by around 10%.

Some recent studies have, however, indicated in general much lower performance from maceration-based techniques compared to techniques such as ultrasonic baths, microwave-assisted extraction (MAE), and Soxhlet extraction. Aspé [132] identified much larger cell wall destruction from these last three techniques, which resulted in generally much higher extraction yields compared to maceration techniques, where only minor cell wall damage has occurred. According to Chupin, et al. [133], MAE effectiveness depends on the particle size and increases with small particles. However, they generally did not identify an effect of MAE on the structure of the tannins.

In addition to the solvent, the extraction conditions can also affect the efficiency of extraction. Extraction at 4 °C in darkness led to around 14–17% higher recovery of CTs compared to extraction at room temperature, also in dark conditions [134].

4.1.4. Major Analytical Techniques I: Spectrophotometric Assays

Spectrophotometric assays are routinely used in studies due to their rapid and low-cost analysis. Vanillin and HCl-butanol assays are conventionally employed to quantify CTs. Both these assays have been widely used in the majority of studies due to their accessibility and low cost for quantifying CTs. In principle, results from the HCl-butanol assay are more reproducible than the vanillin assay. Vanillin assay with methanol has been found to lack specificity as in addition to CTs, it complexes with flavan-3-ols and dihydrochalcones [84,135]. In the HCl-butanol assay, CTs are depolymerized oxidatively to form bright coloured anthocyanidins in the presence of mineral acids [135]. The HCl- butanol assay in some instances lacks specificity and leads to over- or underestimation of results as the colour is dependent on the interflavonoid linkages and 5-OH groups [126]. For accurate quantification by the HCl-butanol assay, freeze-dried samples should be preferred as under heat treatment, tannins can bind to other macromolecules which could undergo oxidation. Macromolecules containing phenolic groups can be oxidised to form quinones which can lead to condensation reactions between tannins and other macromolecules, thereby preventing the release of anthocyanidins from the modified tannins [84]. Furthermore, the addition of iron increases the specificity of this assay [135]. The addition of acetone to the assay also leads to the complete dissociation of CTs from the plant material and this inclusion has further refined this method [136]. A wide range of studies use tannic acid, catechin, or leucocyanidin as standards for these assays. This is also one of the sources for which there is a huge variation in the results, as seen in Table 2 [84,129].

Table 2. Variability of condensed tannins concentration in frequently studied species across different studies.

Species	Condensed Tannins (% DM)	Coefficient of Variation (%)	References
Acacia angustissima	7.4–8.9	9.3	[117,137]
Acacia nilotica	0.46–8	67.4	[137–140]
Acacia senegal	0.07–7.8	138.9	[137,139]
Acacia tortilis	4.7–5.4	9.8	[137,139]
Lespedeza cuneata	0.83–5.1	36.2	[117]
Leucaena leucocephala	0.52–18	112.7	[109]
Mimosa caesalpinifolia	1.8–12.4	105.5	[117,141]
Hedysarium coronarium	0.4–3.8	68.1	[76,142,143]
Onobrychis viciifolia	2.4–14.1	113.1	[14,18,78,144]
Lotus corniculatus	1.4–7.6	45.3	[78,106,145–148]
Lotus pedunculatus	0.25–0.8	50.9	[119,147,149]
Onobrychis viciifolia (Silage)	2.6–3.7	17.4	[17,21]
Lotus corniculatus (Silage)	2.2–3.4	22.3	[17,119]

4.1.5. Major Analytical Techniques II: Liquid Chromatography Coupled with Mass Spectrometry

High-Performance Liquid Chromatography (HPLC) has been proven to be a competent and rapid method for the analysis of polyphenols apart from the highly polymerized oligomers [48,150]. Reverse phase LC (RPLC) is a commonly used chromatographical technique to analyse CTs, ranging from monomers to tetramers and in some cases, their isomers distinctly [151]. With increased CT polymerization, the intelligibility of the chromatogram in RPLC decreases due to the presence of unresolved peaks. The combination of fluorescence detection with RPLC leads to increases in selectivity and sensitivity of the method [152]. A UV-DAD detector is most frequently complemented with LC for the determination of CTs. It also helps in the direct classification of polyphenols into different subgroups such as flavonoids, ellagitannins, gallic acid derivatives, and caffeic acid derivatives, etc. [153]. Recently, to increase the specificity and resolution of the LC analysis, separation techniques have been coupled with ESI-MS or matrix-assisted laser desorption/ionization time-of-flight (MALDI-TOF) MS [154]. These are used extensively for the analysis of CTs in plant and food material such as cocoa, grapes, wine, and birch species, etc. [155,156]. Soft ionization methods such as ESI or MALDI are used to ionize non-volatile analytes such as biopolymers and detect highly polymerized CTs. MALDI-MS has identified procyanidins of degree of polymerization (DP) of 15 in unripe apples, for seed coats of soyabeans until DP of 30 [157]. Similarly, the combination of Hydrophilic Interaction Chromatography (HILIC) x RPLC with fluorescence detection and electrospray full scan mass spectrometry (ESI-MS) resulted in high resolution analysis. This method was able to detect procyanidins with DP value of 16 and gallolylation degree of 8 in the grape seed extracts [158]. These methods are constantly evolving and are now able to provide rapid quantitative and qualitative results. One such method is the Engstrom method which utilizes ultra-high performance liquid chromatography (UPLC) separation coupled with DAD and negative ion ESI-MS to generate a polyphenolic profile directly from plant extracts. In addition to the quantification of different polyphenolic groups, it provides an insight into the composition of flavonols, CTs, HTs, and structural features of condensed tannins [159,160]. These methods provide a great deal of information on tannin structural diversity, but due to high operational costs, their use is not yet widespread.

4.2. Influence of CT Structural Features on Ruminant Nutrition Is Still Ambiguous

The additive influence of CT structural features on the antimethanogenic potential of CT forages remains largely unexplored. Although research on CT structure and functional features has progressed immensely [64,158,160], only a small number of studies have assessed their impact on CH_4 abatement. These studies have shown that CT composition, and the concentration of CTs present in forages, are both significant determinants of their antimethanogenic potential [24,90,95,161]. One of the major reasons is that CTs exist as highly polymerized structures, so large polymers cannot be easily purified as individual compounds and are studied in terms of certain structural features such as molecular weight, polymer size, and prodelphinidin proportion [33]. High structural variability across and within the species adds to the difficulty for assessing their structure–activity relationship. This has been shown in studies on sainfoin cultivars where antimethanogenic potential was found to be highly variable. Hatew [24] studied the intraspecies variability by analysing 46 different accessions of sainfoin with CT concentrations ranging from 0.6 to 2.8% of DM for their antimethanogenic potential. Emissions were analysed based on CT structural properties, i.e., mDP (12 to 84), percentage of trans isomers (12 to 34%), and PD (52.7 to 94.8%) in CT. These properties have been associated with the astringency of CTs. It was observed that PD percentage was a primary CT structural characteristic responsible for reducing CH_4 emissions in this in vitro study [24]. Weight-average M_w of CTs was shown to have little impact on reduction in CH_4 production (R^2 = 0.0009) from North American native forage plants [117]. Nevertheless, it is important to note that the variation in CH_4 reduction potential of different species also arises from plant morphology, CT interaction with feed components, and the presence of other plant secondary metabolites [162,163]. The impact of CT structural features was found to be more pronounced in the studies conducted on CT extracts from plants. The additive effect of other secondary metabolites and forage quality parameters could be voided by the addition of purified CT extracts in the feed. Studies have shown that inclusion of CT extracts (40 mg/g DM) from leucaena (hybrid-Bahru) and mangosteen (*Garcinia mangostana* L) peel could reduce CH_4 emissions by 45 and 35 percent, respectively. In both studies, *Panicum maximum* substrate was used as control. The inhibitory effect of mangosteen peel extracts (M_w = 2081) was milder than leucaena (M_w = 2737) owing to its lower M_w but it was associated with fewer negative effects on in vitro DM degradability and lower protein binding affinity [164,165]. When the antimethanogenic potential of leucaena extracts with differing average molecular weights was tested, extracts with the highest M_W were able engender CH_4 to the maximum [161,166]. Table 3 summarizes the data from two different in vitro studies where at the same concentration, the effect of the molecular weight was more pronounced and it had a strong negative correlation with CH_4 emissions.

Table 3. Influence of molecular weight on methane emissions from two in vitro studies using Kedah-Kelantan cattle rumen fluid.

Extracts	CT [1] (%)	M_W [2] (Da)	Total Gas (mL g^{-1} DM)	Methane (mL g^{-1} DM)	Reference
Leucaena leucocephala hybrid-Rendang	3	1265.8	57	8.07	[161]
Leucaena leucocephala hybrid-Rendang	3	1028.6	61.6	9.2	[161]
Leucaena leucocephala hybrid-Rendang	3	652.2	67	9.35	[161]
Leucaena leucocephala hybrid-Rendang	3	562.2	67.3	10.27	[161]
Leucaena leucocephala hybrid-Rendang	3	469.6	69.7	11.06	[161]
Leucaena leucocephala hybrid-Bahru	3	1348	49.8	4.6	[166]
Leucaena leucocephala hybrid-Bahru	3	857	51.8	5.6	[166]
Leucaena leucocephala hybrid-Bahru	3	730	56	7.8	[166]
Leucaena leucocephala hybrid-Bahru	3	726	55.5	9.7	[166]
Leucaena leucocephala hybrid-Bahru	3	494	57.5	9.5	[166]
Correlation between molecular weight and methane production					−0.72

[1] CT: Condensed tannins, [2] M_W: Molecular weight of condensed tannins.

Some studies have also analysed the impact of multiple structural features simultaneously from CT extracts. Extracts from multiple sainfoin cultivars and diverse CT sources were analysed for CT structural features such as PD percentage, *cis* flavan-3-ols percentage, and average polymer size (mean degree of polymerization). PD percentage and average polymer size were found play an important role in determining antimethanogenic potential of CTs, in addition to the actual CT concentration [25,95]. This shows that high reproducibility of the results can be attained by incorporating the structural features of CTs in ruminant nutrition studies.

4.3. A Roadmap to Close the Missing Links and Possible Future Directions

As discussed above, there is an apparent incongruity between measured tannin concentrations and their bioactive effects. This may be explained by a combination of four factors: (a) the variability of tannins and their composition is large both within and amongst species, and it is affected, at least partially, by the environment; (b) most studies have used too few plants and have been conducted under non-controlled environmental conditions or not comparable conditions, to capture the variability of the tannins; (c) many studies have used inadequate or unsuitable analytical techniques (often due to lack of alternatives or resources), which do not capture the structural characteristics; and (d) the studies that investigated the antimethanogenic potential of CTs while accounting for structural attributes are still limited.

To overcome these inconsistencies, in future studies, every aspect that might affect the results, from the growing conditions to growth stage of the plant at harvest, and from sampling to extraction should be carefully considered in future studies. In the absence of laboratory infrastructure for the structural characterisation of CTs, assays to determine their astringency could be employed. Protein precipitation and radial diffusion assays are frequently used to measure the protein binding ability of tannins [167–169]. Furthermore, assays that determine their antioxidative and oxidative (at high pH) behaviour could also be utilized. They have been associated with the negative impact of tannins on the rumen microbial population [90] and their antiherbivore effect [33], respectively. Additional treatments with polyethylene and polyvinylpolypyrrolidone in vitro studies could be employed to elucidate the tannin effect on CH_4 emissions, as they bind specifically to tannins. These studies could be instrumental in distinguishing between the effect of tannins and forage chemical composition on CH_4 emissions. In vitro fermentation/CH_4 production techniques could be useful for screening these forages and determining their adequate supplementation. Using CT extracts of tanniferous forages in in vitro studies can illustrate the structure-activity relationship of CTs with methane emissions more distinctly. Condensed tannin supplementation has been found to impact the diversity and composition of the rumen microbial community [170]. Understanding the dynamics of microbial populations in rumen, and how CT-containing forages influence their abundance and diversity, can provide significant insights into their mode of action. Employing novel techniques such as metagenome and metatranscriptome analysis of the rumen microbiome under CT treatment can help in identifying the microbial population and the functional shifts in rumen microbiome which lead to CH_4 abatement [171]. As we gather more information about the relationship between the structure of CTs and their bioactivity, there are prospects for breeding plants with desired concentrations and composition of CTs [105,172]. Molecular approaches have already made it possible for white clover to reach moderate levels of CT in its leaves [173] and efforts are also being made in directions to improve the persistence of TRFs, as may be seen for birdsfoot trefoil [174]. Several questions still remain unanswered, and these are critical for ensuring a comprehensive understanding of the fate of CTs in biological systems.

- How do different environmental conditions influence the structural features of CTs?
- To what extent are the structural features responsible for the functional attributes of tannins (PPC and oxidative property) and whether these assays could be utilized as an indicator of their antimethanogenic activity?
- How does the presence of other secondary plant metabolites affect the influence of tannins on CH_4 emissions?
- How does tannin supplementation affect mineral and vitamin bioavailability in ruminants? Which properties are primarily responsible for these interactions?

- How do forage conservation methods (ensiling vs. hay drying vs. fresh material) influence the palatability/acceptability and DMI by livestock, and anthelmintic and antimethanogenic potential of TRFs?
- To what extent are the anthelmintic effects of tannins sustained during long-term trials? Is it possible for gastrointestinal parasites to develop resistance to tannins?
- How do the PPC and oxidative capacity of tannins influence their antimethanogenic potential? What is the magnitude of their effect on antimethanogenic potential?
- How do different tannin sources influence rumen microbiome diversity and abundance and whether these effects are short or long term?
- How do CTs interact with feed constituents and how do structural characteristics play a role in this?

5. Conclusions

In recent years, there have been remarkable new insights into CT structural diversity and functions with more sensitive analytical methods. However, CT bioactivity is a complex process which results from a multitude of variations occurring simultaneously in plants as well as in their effects in animals. The variability in the results from different studies focuses our attention on the need for developing and adapting a course of action for the investigation of potential of CTs to reduce CH_4 emissions. The comparison of CT fingerprints of different species could help us understand not only the factors which define their antimethanogenic potential but also provide a vital framework to assess their interactions with plant constituents and rumen microflora, benefitting overall ruminant health.

Author Contributions: S.V. and C.S.M. conceived and conceptualized the manuscript. S.V. took the lead for writing the manuscript, while C.S.M. and F.T. edited for clarity and completeness. All authors provided critical feedback and helped shape the research, analysis and manuscript. Additionally, all authors have read and agreed to the published version of the manuscript. All authors have read and agreed to the published version of the manuscript.

Funding: This research was funded by Deutsche Forschungsgemeinschaft (DFG)—Project number 406534244, grant number MA 8199'/1-1. We acknowledge financial support by DFG within the funding programme Open Access Publizieren.

Institutional Review Board Statement: Not applicable.

Informed Consent Statement: Not applicable.

Data Availability Statement: Not applicable.

Acknowledgments: The authors would like to thank Juha-Pekka Salminen for his contributions to improve the quality of this manuscript.

Conflicts of Interest: The authors declare no conflict of interest.

References

1. Haque, M.N. Dietary manipulation: A sustainable way to mitigate methane emissions from ruminants. *J. Anim. Sci. Technol.* **2018**, *60*, 1–10. [CrossRef] [PubMed]
2. Gerber, P.J.; Steinfeld, H.; Henderson, B.; Mottet, A.; Opio, C.; Dijkman, J.; Falcucci, A.; Tempio, G. *Tackling Climate Change Through Livestock: A Global Assessment of Emissions and Mitigation Opportunities*; Food and Agriculture Organization of the United Nations (FAO): Rome, Italy, 2013.
3. Ehhalt, D.; Prather, M.; Dentener, F.; Derwent, R.; Dlugokencky, E.J.; Holland, E.; Isaksen, I.; Katima, J.; Kirchhoff, V.; Matson, P. *Atmospheric Chemistry and Greenhouse Gases*; Pacific Northwest National Lab. (PNNL): Richland, WA, USA, 2001.
4. Allen, M.R.; Shine, K.P.; Fuglestvedt, J.S.; Millar, R.J.; Cain, M.; Frame, D.J.; Macey, A.H. A solution to the misrepresentations of CO_2-equivalent emissions of short-lived climate pollutants under ambitious mitigation. *NPJ Clim. Atmos.* **2018**, *1*, 1–8. [CrossRef]
5. Broucek, J. Production of methane emissions from ruminant husbandry: A review. *J. Environ. Prot.* **2014**, *5*, 1482–1493. [CrossRef]
6. Lauder, A.R.; Enting, I.G.; Carter, J.O.; Clisby, N.; Cowie, A.L.; Henry, B.K.; Raupach, M.R. Offsetting methane emissions—An alternative to emission equivalence metrics. *Int. J. Greenh. Gas. Control.* **2013**, *12*, 419–429. [CrossRef]
7. Ellis, J.L.; Dijkstra, J.; Kebreab, E.; Bannink, A.; Odongo, N.E.; McBride, B.W.; France, J. Aspects of rumen microbiology central to mechanistic modelling of methane production in cattle. *J. Agric. Sci.* **2008**, *146*, 213–233. [CrossRef]

8. Flachowsky, G.; Lebzien, P. Effects of phytogenic substances on rumen fermentation and methane emissions: A proposal for a research process. *Anim. Feed Sci. Technol.* **2012**, *176*, 70–77. [CrossRef]
9. Jayanegara, A.; Wina, E.; Takahashi, J. Meta-analysis on methane mitigating properties of saponin-rich sources in the rumen: Influence of addition levels and plant sources. *Asian Australas. J. Anim. Sci.* **2014**, *27*, 1426–1435. [CrossRef]
10. Jayanegara, A.; Goel, G.; Makkar, H.P.S.; Becker, K. *Reduction in Methane Emissions from Ruminants by Plant Secondary Metabolites: Effects of Polyphenols and Saponins*; Food and Agriculture Organization of the United Nations (FAO): Rome, Italy, 2010; pp. 151–157. ISBN 978-92-5-106697-3.
11. Truong, A.H.; Kim, M.; Nguyen, T.; Nguyen, N.; Quang Trung, N. Methane, nitrous oxide and ammonia emissions from livestock farming in the red river delta, Vietnam: An inventory and projection for 2000–2030. *Sustainability* **2018**, *10*, 3826. [CrossRef]
12. Baert, N.; Pellikaan, W.F.; Karonen, M.; Salminen, J.-P. A study of the structure-activity relationship of oligomeric ellagitannins on ruminal fermentation in vitro. *J. Dairy Sci.* **2016**, *99*, 8041–8052. [CrossRef] [PubMed]
13. Zucker, W.V. Tannins: Does structure determine function? An ecological perspective. *Am. Nat.* **1983**, *121*, 335–365. [CrossRef]
14. Chung, Y.H.; Mc Geough, E.J.; Acharya, S.; McAllister, T.A.; McGinn, S.M.; Harstad, O.M.; Beauchemin, K.A. Enteric methane emission, diet digestibility, and nitrogen excretion from beef heifers fed sainfoin or alfalfa. *J. Anim. Sci.* **2013**, *91*, 4861–4874. [CrossRef]
15. Huyen, N.T.; Desrues, O.; Alferink, S.J.J.; Zandstra, T.; Verstegen, M.W.A.; Hendriks, W.H.; Pellikaan, W.F. Inclusion of sainfoin (*Onobrychis viciifolia*) silage in dairy cow rations affects nutrient digestibility, nitrogen utilization, energy balance, and methane emissions. *J. Dairy Sci.* **2016**, *99*, 3566–3577. [CrossRef]
16. Guglielmelli, A.; Calabrò, S.; Primi, R.; Carone, F.; Cutrignelli, M.; Raffaella, T.; Piccolo, G.; Ronchi, B.; Danieli, P.P. In vitro fermentation patterns and methane production of sainfoin (*Onobrychis viciifolia* Scop.) hay with different condensed tannin contents. *Grass Forage Sci.* **2011**, *66*, 488–500. [CrossRef]
17. Grosse Brinkhaus, A.; Wyss, U.; Arrigo, Y.; Girard, M.; Bee, G.; Zeitz, J.O.; Kreuzer, M.; Dohme-Meier, F. In vitro ruminal fermentation characteristics and utilisable CP supply of sainfoin and birdsfoot trefoil silages and their mixtures with other legumes. *Animal* **2017**, *11*, 580–590. [CrossRef] [PubMed]
18. McMahon, L.R.; Majak, W.; McAllister, T.A.; Hall, J.W.; Jones, G.A.; Popp, J.D.; Cheng, K.J. Effect of sainfoin on in vitro digestion of fresh alfalfa and bloat in steers. *Can. J. Anim. Sci.* **1999**, *79*, 203–212. [CrossRef]
19. Wang, Y.; Berg, B.P.; Barbieri, L.R.; Veira, D.M.; McAllister, T.A. Comparison of alfalfa and mixed alfalfa-sainfoin pastures for grazing cattle: Effects on incidence of bloat, ruminal fermentation, and feed intake. *Can. J. Anim. Sci.* **2006**, *86*, 383–392. [CrossRef]
20. Aufrere, J.; Dudilieu, M.; Andueza, D.; Poncet, C.; Baumont, R. Mixing sainfoin and lucerne to improve the feed value of legumes fed to sheep by the effect of condensed tannins. *Animal* **2013**, *7*, 82–92. [CrossRef] [PubMed]
21. Theodoridou, K.; Aufrere, J.; Andueza, D.; Le Morvan, A.; Picard, F.; Pourrat, J.; Baumont, R. Effects of condensed tannins in wrapped silage bales of sainfoin (*Onobrychis viciifolia*) on in vivo and in situ digestion in sheep. *Animal* **2012**, *6*, 245–253. [CrossRef] [PubMed]
22. Malisch, C.S.; Lüscher, A.; Baert, N.; Engström, M.T.; Studer, B.; Fryganas, C.; Suter, D.; Mueller-Harvey, I.; Salminen, J.-P. Large variability of proanthocyanidin content and composition in sainfoin (*Onobrychis viciifolia*). *J. Agric. Food Chem.* **2015**, *63*, 10234–10242. [CrossRef]
23. Stringano, E.; Hayot Carbonero, C.; Smith, L.M.; Brown, R.H.; Mueller-Harvey, I. Proanthocyanidin diversity in the EU 'HealthyHay' sainfoin (*Onobrychis viciifolia*) germplasm collection. *Phytochemistry* **2012**, *77*, 197–208. [CrossRef]
24. Hatew, B.; Hayot Carbonero, C.; Stringano, E.; Sales, L.F.; Smith, L.M.J.; Mueller-Harvey, I.; Hendriks, W.H.; Pellikaan, W.F. Diversity of condensed tannin structures affects rumen in vitro methane production in sainfoin (*Onobrychis viciifolia*) accessions. *Grass Forage Sci.* **2015**, *70*, 474–490. [CrossRef]
25. Hatew, B.; Stringano, E.; Mueller-Harvey, I.; Hendriks, W.H.; Carbonero, C.H.; Smith, L.M.; Pellikaan, W.F. Impact of variation in structure of condensed tannins from sainfoin (*Onobrychis viciifolia*) on in vitro ruminal methane production and fermentation characteristics. *J. Anim. Physiol. Anim. Nutr.* **2016**, *100*, 348–360. [CrossRef] [PubMed]
26. Jayanegara, A.; Marquardt, S.; Wina, E.; Kreuzer, M.; Leiber, F. In vitro indications for favourable non-additive effects on ruminal methane mitigation between high-phenolic and high-quality forages. *Br. J. Nutr.* **2013**, *109*, 615–622. [CrossRef] [PubMed]
27. Malisch, C.; Salminen, J.-P.; Kölliker, R.; Engström, M.; Suter, D.; Studer, B.; Lüscher, A. Drought effects on proanthocyanidins in sainfoin (*Onobrychis viciifolia* Scop.) are dependent on the plant's ontogenetic stage. *J. Agric. Food Chem.* **2016**, *64*, 9307–9316. [CrossRef] [PubMed]
28. Selmar, D.; Kleinwaechter, M. stress enhances the synthesis of secondary plant products: The impact of stress-related over-reduction on the accumulation of natural products. *Plant. Cell Physiol.* **2013**, *54*, 817–826. [CrossRef] [PubMed]
29. Zhang, L.H.; Shao, H.B.; Ye, G.F.; Lin, Y.M. Effects of fertilization and drought stress on tannin biosynthesis of *Casuarina equisetifolia* seedlings branchlets. *Acta. Physiol. Plant.* **2012**, *34*, 1639–1649. [CrossRef]
30. Top, S.M.; Preston, C.M.; Dukes, J.S.; Tharayil, N. Climate influences the content and chemical composition of foliar tannins in green and senesced tissues of *Quercus rubra*. *Front. Plant. Sci.* **2017**, *8*, 423. [CrossRef] [PubMed]
31. Haslam, E. Plant polyphenols (syn. vegetable tannins) and chemical defense—A reappraisal. *J. Chem. Ecol.* **1988**, *14*, 1789–1805. [CrossRef] [PubMed]

32. Mueller-Harvey, I.; Bee, G.; Dohme-Meier, F.; Hoste, H.; Karonen, M.; Kölliker, R.; Luscher, A.; Nidekorn, V.; Pellikaan, W.F.; Salminen, J.-P.; et al. Benefits of condensed tannins in forage legumes fed to ruminants: Importance of structure, concentration and diet composition. *Crop Sci.* **2017**, *59*, 861–885. [CrossRef]
33. Salminen, J.-P.; Karonen, M. Chemical ecology of tannins and other phenolics: We need a change in approach. *Funct. Ecol.* **2011**, *25*, 325–338. [CrossRef]
34. Salminen, J.P.; Karonen, M.; Sinkkonen, J. Chemical ecology of tannins: Recent developments in tannin chemistry reveal new structures and structure-activity patterns. *Chem. Eur. J.* **2011**, *17*, 2806–2816. [CrossRef] [PubMed]
35. Sharma, P.; Jha, A.; Dubey, R.; Pessarakli, M. Reactive oxygen species, oxidative damage, and antioxidative defense mechanism in plants under stressful conditions. *J. Bot.* **2012**, *2012*, 217037. [CrossRef]
36. Chomel, M.; Guittonny-Larchevêque, M.; Fernandez, C.; Gallet, C.; DesRochers, A.; Paré, D.; Jackson, B.G.; Baldy, V. Plant secondary metabolites: A key driver of litter decomposition and soil nutrient cycling. *J. Ecol.* **2016**, *104*, 1527–1541. [CrossRef]
37. Adamczyk, B.; Karonen, M.; Adamczyk, S.; Engström, M.T.; Laakso, T.; Saranpää, P.; Kitunen, V.; Smolander, A.; Simon, J. Tannins can slow-down but also speed-up soil enzymatic activity in boreal forest. *Soil Biol. Biochem.* **2017**, *107*, 60–67. [CrossRef]
38. Adamczyk, B.; Simon, J.; Kitunen, V.; Adamczyk, S.; Smolander, A. Tannins and their complex interaction with different organic nitrogen compounds and enzymes: Old paradigms versus recent advances. *Chem. Open* **2017**, *6*, 610–614. [CrossRef] [PubMed]
39. Adamczyk, S.; Kitunen, V.; Lindroos, A.-J.; Adamczyk, B.; Smolander, A. Soil carbon and nitrogen cycling processes and composition of terpenes five years after clear-cutting a Norway spruce stand: Effects of logging residues. *For. Ecol. Manag.* **2016**, *381*, 318–326. [CrossRef]
40. Kagiya, N.; Reinsch, T.; Taube, F.; Salminen, J.-P.; Kluß, C.; Hasler, M.; Malisch, C.S. Turnover rates of roots vary considerably across temperate forage species. *Soil Biol. Biochem.* **2019**, *139*, 107614. [CrossRef]
41. Salami, S.A.; Valenti, B.; Bella, M.; O'Grady, M.N.; Luciano, G.; Kerry, J.P.; Jones, E.; Priolo, A.; Newbold, C.J. Characterisation of the ruminal fermentation and microbiome in lambs supplemented with hydrolysable and condensed tannins. *FEMS Microbiol. Ecol.* **2018**, 94. [CrossRef]
42. Ekambaram, S.P.; Perumal, S.S.; Balakrishnan, A. Scope of hydrolysable tannins as possible antimicrobial agent. *Phytother. Res.* **2016**, *30*, 1035–1045. [CrossRef] [PubMed]
43. Koleckar, V.; Kubikova, K.; Rehakova, Z.; Kuca, K.; Jun, D.; Jahodar, L.; Opletal, L. Condensed and hydrolysable tannins as antioxidants influencing the health. *Mini Rev. Med. Chem.* **2008**, *8*, 436–447. [CrossRef]
44. Klongsiriwet, C.; Quijada, J.; Williams, A.R.; Mueller-Harvey, I.; Williamson, E.M.; Hoste, H. Synergistic inhibition of *Haemonchus contortus* exsheathment by flavonoid monomers and condensed tannins. *Int. J. Parasitol. Drugs Drug Resist.* **2015**, *5*, 127–134. [CrossRef] [PubMed]
45. Desrues, O.; Fryganas, C.; Ropiak, H.M.; Mueller-Harvey, I.; Enemark, H.L.; Thamsborg, S.M. Impact of chemical structure of flavanol monomers and condensed tannins on in vitro anthelmintic activity against bovine nematodes. *Parasitology* **2016**, *143*, 444–454. [CrossRef]
46. Williams, A.R.; Fryganas, C.; Ramsay, A.; Mueller-Harvey, I.; Thamsborg, S.M. Direct anthelmintic effects of condensed tannins from diverse plant sources against *Ascaris Suum*. *PLoS ONE* **2014**, *9*, e97053. [CrossRef] [PubMed]
47. Rasmussen, S.E.; Frederiksen, H.; Krogholm, K.S.; Poulsen, L. Dietary proanthocyanidins: Occurrence, dietary intake, bioavailability, and protection against cardiovascular disease. *Mol. Nutr. Food Res.* **2005**, *49*, 159–174. [CrossRef] [PubMed]
48. Zeller, W.E.; Ramsay, A.; Ropiak, H.M.; Fryganas, C.; Mueller-Harvey, I.; Brown, R.H.; Drake, C.; Grabber, J.H. 1H-13C HSQC NMR spectroscopy for estimating procyanidin/prodelphinidin and cis/trans-flavan-3-ol ratios of condensed tannin samples: Correlation with thiolysis. *J. Agric. Food Chem.* **2015**, *63*, 1967–1973. [CrossRef]
49. Naumann, H.D.; Tedeschi, L.O.; Zeller, W.E.; Huntley, N.F. The role of condensed tannins in ruminant animal production: Advances, limitations and future directions. *Rev. Bras. Zootec.* **2017**, *46*, 929–949. [CrossRef]
50. Barry, T.N.; McNabb, W.C. The implications of condensed tannins on the nutritive value of temperate forages fed to ruminants. *Br. J. Nutr.* **1999**, *81*, 263–272. [CrossRef]
51. Zeller, W.E.; Sullivan, M.L.; Mueller-Harvey, I.; Grabber, J.H.; Ramsay, A.; Drake, C.; Brown, R.H. Protein precipitation behavior of condensed tannins from *Lotus pedunculatus* and *Trifolium repens* with different mean degrees of polymerization. *J. Agric. Food Chem.* **2015**, *63*, 1160–1168. [CrossRef]
52. Kimura, H.; Ogawa, S.; Akihiro, T.; Yokota, K. Structural analysis of A-type or B-type highly polymeric proanthocyanidins by thiolytic degradation and the implication in their inhibitory effects on pancreatic lipase. *J. Chromatogr. A* **2011**, *1218*, 7704–7712. [CrossRef]
53. Girard, M.; Dohme-Meier, F.; Silacci, P.; Ampuero Kragten, S.; Kreuzer, M.; Bee, G. Forage legumes rich in condensed tannins may increase n-3 fatty acid levels and sensory quality of lamb meat. *J. Sci. Food Agric.* **2016**, *96*, 1923–1933. [CrossRef]
54. Tibe, O.; Meagher, L.P.; Fraser, K.; Harding, D.R.K. Condensed tannins and flavonoids from the forage legume sulla (*Hedysarum coronarium*). *J. Agric. Food Chem.* **2011**, *59*, 9402–9409. [CrossRef] [PubMed]
55. Aerts, R.J.; Barry, T.N.; McNabb, W.C. Polyphenols and agriculture: Beneficial effects of proanthocyanidins in forages. *Agric. Ecosyst. Environ.* **1999**, *75*, 1–12. [CrossRef]
56. Kim, J.; Päljärvi, M.; Karonen, M.; Salminen, J.-P. Distribution of enzymatic and alkaline oxidative activities of phenolic compounds in plants. *Phytochemistry* **2020**, *179*, 112501. [CrossRef]

57. Barbehenn, R.V.; Jones, C.P.; Hagerman, A.E.; Karonen, M.; Salminen, J.P. Ellagitannins have greater oxidative activities than condensed tannins and galloyl glucoses at high pH: Potential impact on caterpillars. *J. Chem. Ecol.* **2006**, *32*, 2253–2267. [CrossRef] [PubMed]
58. Barbehenn, R.V.; Peter Constabel, C. Tannins in plant–herbivore interactions. *Phytochemistry* **2011**, *72*, 1551–1565. [CrossRef] [PubMed]
59. Mueller-Harvey, I. Unravelling the conundrum of tannins in animal nutrition and health. *J. Sci. Food Agric.* **2006**, *86*, 2010–2037. [CrossRef]
60. Desrues, O.; Mueller-Harvey, I.; Pellikaan, W.F.; Enemark, H.L.; Thamsborg, S.M. Condensed tannins in the gastrointestinal tract of cattle after sainfoin (*Onobrychis viciifolia*) intake and their possible relationship with anthelmintic effects. *J. Agric. Food Chem.* **2017**, *65*, 1420–1427. [CrossRef]
61. Zhang, L.L.; Lin, Y.M. HPLC, NMR and MALDI-TOF MS analysis of condensed tannins from *Lithocarpus glaber* leaves with potent free radical scavenging activity. *Molecules* **2008**, *13*, 2986–2997. [CrossRef] [PubMed]
62. Aboagye, I.A.; Beauchemin, K.A. Potential of molecular weight and structure of tannins to reduce methane emissions from ruminants: A review. *Animals* **2019**, *9*, 856. [CrossRef] [PubMed]
63. Adamczyk, B.; Salminen, J.-P.; Smolander, A.; Kitunen, V. Precipitation of proteins by tannins: Effects of concentration, protein/tannin ratio and pH. *Int. J. Food Sci. Technol.* **2011**, *47*, 875–878. [CrossRef]
64. Zeller, W.E. Activity, purification, and analysis of condensed tannins: Current state of affairs and future endeavors. *Crop Sci.* **2019**, *59*, 886–904. [CrossRef]
65. Prigent, S.V.E.; Voragen, A.G.J.; van Koningsveld, G.A.; Baron, A.; Renard, C.M.G.C.; Gruppen, H. Interactions between globular proteins and procyanidins of different degrees of polymerization. *J. Dairy Sci.* **2009**, *92*, 5843–5853. [CrossRef] [PubMed]
66. McMahon, L.R.; McAllister, T.A.; Berg, B.P.; Majak, W.; Acharya, S.N.; Popp, J.D.; Coulman, B.E.; Wang, Y.; Cheng, K.J. A review of the effects of forage condensed tannins on ruminal fermentation and bloat in grazing cattle. *Can. J. Plant. Sci.* **2000**, *80*, 469–485. [CrossRef]
67. Engström, M.T.; Sun, X.; Suber, M.P.; Li, M.; Salminen, J.-P.; Hagerman, A.E. The oxidative activity of ellagitannins dictates their tendency to form highly stabilized complexes with bovine serum albumin at increased pH. *J. Agric. Food Chem.* **2016**, *64*, 8994–9003. [CrossRef] [PubMed]
68. Hagerman, A.E. Fifty years of polyphenol–protein complexes. In *Recent Advances in Polyphenol Research*; Cheynier, V., Sarni-Manchado, P., Quideau, S., Eds.; Wiley: Hoboken, NJ, USA, 2012; Volume 3, pp. 71–97.
69. Leppä, M.M.; Laitila, J.E.; Salminen, J.-P. Distribution of protein precipitation capacity within variable proanthocyanidin fingerprints. *Molecules* **2020**, *25*, 5002. [CrossRef]
70. Schofield, P.; Mbugua, D.M.; Pell, A.N. Analysis of condensed tannins: A review. *Anim. Feed Sci. Technol.* **2001**, *91*, 21–40. [CrossRef]
71. Ropiak, H.M.; Desrues, O.; Williams, A.R.; Ramsay, A.; Mueller-Harvey, I.; Thamsborg, S.M. Structure-activity relationship of condensed tannins and synergism with *trans*-cinnamaldehyde against *Caenorhabditis elegans*. *J. Agric. Food Chem.* **2016**, *64*, 8795–8805. [CrossRef] [PubMed]
72. Naumann, H.D.; Armstrong, S.A.; Lambert, B.D.; Muir, J.P.; Tedeschi, L.O.; Kothmann, M.M. Effect of molecular weight and concentration of legume condensed tannins on in vitro larval migration inhibition of *Haemonchus contortus*. *Vet. Parasitol.* **2014**, *199*, 93–98. [CrossRef]
73. Naumann, H.D.; Hagerman, A.E.; Lambert, B.D.; M-uir, J.P.; Tedeschi, L.O.; Kothmann, M.M. Molecular weight and protein-precipitating ability of condensed tannins from warm-season perennial legumes. *J. Plant Interact.* **2014**, *9*, 212–219. [CrossRef]
74. McAllister, T.A.; Martinez, T.; Bae, H.D.; Muir, A.D.; Yanke, L.J.; Jones, G.A. Characterization of condensed tannins purified from legume forages: Chromophore production, protein precipitation, and inhibitory effects on cellulose digestion. *J. Chem. Ecol.* **2005**, *31*, 2049–2068. [CrossRef] [PubMed]
75. O'Donovan, L.; Brooker, J.D. Effect of hydrolysable and condensed tannins on growth, morphology and metabolism of *Streptococcus gallolyticus* (*S. caprinus*) and *Streptococcus bovis*. *Microbiology* **2001**, *147*, 1025–1033. [CrossRef]
76. Stienezen, M.; Waghorn, G.C.; Douglas, G.B. Digestibility and effects of condensed tannins on digestion of sulla (*Hedysarum coronarium*) when fed to sheep. *N. Zeal. J. Agric. Res.* **1996**, *39*, 215–221. [CrossRef]
77. Häring, D.A.; Scharenberg, A.; Heckendorn, F.; Dohme, F.; Lüscher, A.; Maurer, V.; Suter, D.; Hertzberg, H. Tanniferous forage plants: Agronomic performance, palatability and efficacy against parasitic nematodes in sheep. *Renew. Agric. Food Syst.* **2008**, *23*, 19–29. [CrossRef]
78. Grosse Brinkhaus, A.; Bee, G.; Silacci, P.; Kreuzer, M.; Dohme-Meier, F. Effect of exchanging *Onobrychis viciifolia* and *Lotus corniculatus* for *Medicago sativa* on ruminal fermentation and nitrogen turnover in dairy cows. *J. Dairy Sci.* **2016**, *99*, 4384–4397. [CrossRef]
79. Johnson, K.A.; Johnson, D.E. Methane emissions from cattle. *J. Anim. Sci.* **1995**, *73*, 2483–2492. [CrossRef]
80. Morgavi, D.P.; Forano, E.; Martin, C.; Newbold, C.J. Microbial ecosystem and methanogenesis in ruminants. *Animal* **2010**, *4*, 1024–1036. [CrossRef]
81. Peng, K.; Jin, L.; Niu, Y.D.; Huang, Q.; McAllister, T.A.; Yang, H.E.; Denise, H.; Xu, Z.; Acharya, S.; Wang, S.; et al. Condensed tannins affect bacterial and fungal microbiomes and mycotoxin production during ensiling and upon aerobic exposure. *Appl. Environ. Microbiol.* **2018**, *84*, e02274-17. [CrossRef] [PubMed]

82. Waghorn, G.C.; Ulyatt, M.J.; John, A.; Fisher, M.T. The effect of condensed tannins on the site of digestion of amino acids and other nutrients in sheep fed on *Lotus corniculatus* L. *Br. J. Nutr.* **1987**, *57*, 115–126. [CrossRef]
83. Athanasiadou, S.; Kyriazakis, I.; Jackson, F.; Coop, R.L. Consequences of long-term feeding with condensed tannins on sheep parasitised with *Trichostrongylus colubriformis*. *Int. J. Parasitol.* **2000**, *30*, 1025–1033. [CrossRef]
84. Makkar, H. Chemical, protein precipitation and bioassays for tannins, tannin levels and activity in unconventional feeds, and effects and fate of tannins. In *Quantification of Tannins in Tree and Shrub Foliage: A Laboratory Manual*; Makkar, H.P.S., Ed.; Springer: Dordrecht, The Netherlands, 2003; pp. 1–42.
85. Gunun, P.; Wanapat, M.; Gunun, N.; Cherdthong, A.; Sirilaophaisan, S.; Kaewwongsa, W. Effects of condensed tannins in Mao (*Antidesma thwaitesianum* Muell. Arg.) seed meal on rumen fermentation characteristics and nitrogen utilization in Goats. *Asian-Australas. J. Anim. Sci.* **2016**, *29*, 1111–1119. [CrossRef] [PubMed]
86. Hassanat, F.; Benchaar, C. Assessment of the effect of condensed (acacia and quebracho) and hydrolysable (chestnut and valonea) tannins on rumen fermentation and methane production in vitro. *J. Sci. Food Agric.* **2013**, *93*, 332–339. [CrossRef] [PubMed]
87. Sordi, A.; Dieckow, J.; Bayer, C.; Alburquerque, M.A.; Piva, J.T.; Zanatta, J.A.; Tomazi, M.; da Rosa, C.M.; de Moraes, A. Nitrous oxide emission factors for urine and dung patches in a subtropical Brazilian pastureland. *Agric. Ecosyst. Environ.* **2014**, *190*, 94–103. [CrossRef]
88. Azuhnwi, B.N.; Hertzberg, H.; Arrigo, Y.; Gutzwiller, A.; Hess, H.D.; Mueller-Harvey, I.; Torgerson, P.R.; Kreuzer, M.; Dohme-Meier, F. Investigation of sainfoin (*Onobrychis viciifolia*) cultivar differences on nitrogen balance and fecal egg count in artificially infected lambs. *J. Anim. Sci.* **2013**, *91*, 2343–2354. [CrossRef]
89. Min, B.R.; Attwood, G.T.; Reilly, K.; Sun, W.; Peters, J.S.; Barry, T.N.; McNabb, W.C. *Lotus corniculatus* condensed tannins decrease in vivo populations of proteolytic bacteria and affect nitrogen metabolism in the rumen of sheep. *Can. J. Microbiol.* **2002**, *48*, 911–921. [CrossRef] [PubMed]
90. Naumann, H.; Sepela, R.; Rezaire, A.; Masih, S.E.; Zeller, W.E.; Reinhardt, L.A.; Robe, J.T.; Sullivan, M.L.; Hagerman, A.E. Relationships between structures of condensed tannins from texas legumes and methane production during in vitro rumen digestion. *Molecules* **2018**, *23*, 2123. [CrossRef] [PubMed]
91. Aerts, R.J.; McNabb, W.C.; Molan, A.; Brand, A.; Barry, T.N.; Peters, J.S. Condensed tannins from *Lotus corniculatus* and *Lotus pedunculatus* exert different effects on the in vitro rumen degradation of ribulose-1,5-bisphosphate carboxylase/oxygenase (Rubisco) protein. *J. Sci. Food Agric.* **1999**, *79*, 79–85. [CrossRef]
92. Cieslak, A.; Szumacher-Strabel, M.; Stochmal, A.; Oleszek, W. Plant components with specific activities against rumen methanogens. *Animal* **2013**, *7*, 253–265. [CrossRef] [PubMed]
93. Hristov, A.N.; Oh, J.; Firkins, J.L.; Dijkstra, J.; Kebreab, E.; Waghorn, G.; Makkar, H.P.S.; Adesogan, A.T.; Yang, W.; Lee, C.; et al. Special topics—Mitigation of methane and nitrous oxide emissions from animal operations: I. A review of enteric methane mitigation options. *J. Anim. Sci.* **2013**, *91*, 5045–5069. [CrossRef] [PubMed]
94. Pedreira, M.d.S.; Oliveira, S.G.d.; Primavesi, O.; Lima, M.A.d.; Frighetto, R.T.S.; Berchielli, T.T. Methane emissions and estimates of ruminal fermentation parameters in beef cattle fed different dietary concentrate levels. *Rev. Bras. Zootec.* **2013**, *42*, 592–598. [CrossRef]
95. Huyen, N.T.; Fryganas, C.; Uittenbogaard, G.; Mueller-Harvey, I.; Verstegen, M.W.A.; Hendriks, W.H.; Pellikaan, W.F. Structural features of condensed tannins affect in vitro ruminal methane production and fermentation characteristics. *J. Agric. Sci.* **2016**, *154*, 1474–1487. [CrossRef]
96. Hristov, A.N.; Oh, J.; Giallongo, F.; Frederick, T.W.; Harper, M.T.; Weeks, H.L.; Branco, A.F.; Moate, P.J.; Deighton, M.H.; Williams, S.R.O.; et al. An inhibitor persistently decreased enteric methane emission from dairy cows with no negative effect on milk production. *Proc. Natl. Acad. Sci. USA* **2015**, *112*, 10663–10668. [CrossRef] [PubMed]
97. Saminathan, M.; Sieo, C.C.; Gan, H.M.; Abdullah, N.; Wong, C.M.V.L.; Ho, Y.W. Effects of condensed tannin fractions of different molecular weights on population and diversity of bovine rumen methanogenic archaea in vitro, as determined by high-throughput sequencing. *Anim. Feed Sci. Technol.* **2016**, *216*, 146–160. [CrossRef]
98. Belanche, A.; de la Fuente, G.; Newbold, C.J. Study of methanogen communities associated with different rumen protozoal populations. *FEMS Microbiol. Ecol.* **2014**, *90*, 663–677. [CrossRef]
99. Beauchemin, K.; McAllister, T.; McGinn, S. Dietary mitigation of enteric methane from cattle. *CAB Rev.* **2009**, *4*, 1–18. [CrossRef]
100. Tavendale, M.H.; Meagher, L.P.; Pacheco, D.; Walker, N.; Attwood, G.T.; Sivakumaran, S. Methane production from in vitro rumen incubations with *Lotus pedunculatus* and *Medicago sativa*, and effects of extractable condensed tannin fractions on methanogenesis. *Anim. Feed Sci. Technol.* **2005**, *123*, 403–419. [CrossRef]
101. Jayanegara, A.; Leiber, F.; Kreuzer, M. Meta-analysis of the relationship between dietary tannin level and methane formation in ruminants from in vivo and in vitro experiments. *J. Anim. Physiol. Anim. Nutr.* **2012**, *96*, 365–375. [CrossRef] [PubMed]
102. Van Gastelen, S.; Dijkstra, J.; Bannink, A. Are dietary strategies to mitigate enteric methane emission equally effective across dairy cattle, beef cattle, and sheep? *J. Dairy Sci.* **2019**, *102*, 6109–6130. [CrossRef] [PubMed]
103. Min, B.R.; Solaiman, S.; Waldrip, H.M.; Parker, D.; Todd, R.W.; Brauer, D. Dietary mitigation of enteric methane emissions from ruminants: A review of plant tannin mitigation options. *Anim. Nutr.* **2020**, *6*, 231–236. [CrossRef]
104. Fagundes, G.M.; Benetel, G.; Santos, K.C.; Welter, K.C.; Melo, F.A.; Muir, J.P.; Bueno, I.C.S. Tannin-rich plants as natural manipulators of rumen fermentation in the livestock industry. *Molecules* **2020**, *25*, 2943. [CrossRef]

105. Waghorn, G. Beneficial and detrimental effects of dietary condensed tannins for sustainable sheep and goat production-Progress and challenges. *Anim. Feed Sci. Technol.* **2008**, *147*, 116–139. [CrossRef]
106. Woodward, S.L.; Waghorn, G.C.; Laboyrie, P.G. Condensed tannins in birdsfoot trefoil (*Lotus corniculatus*) reduce methane emissions from dairy cows. *Proc. N. Z. Soc. Anim. Prod.* **2004**, *64*, 160–164.
107. Soltan, Y.A.; Morsy, A.S.; Lucas, R.C.; Abdalla, A.L. Potential of mimosine of *Leucaena leucocephala* for modulating ruminal nutrient degradability and methanogenesis. *Anim. Feed Sci. Technol.* **2017**, *223*, 30–41. [CrossRef]
108. Soltan, Y.A.; Morsy, A.S.; Sallam, S.M.A.; Lucas, R.C.; Louvandini, H.; Kreuzer, M.; Abdalla, A.L. Contribution of condensed tannins and mimosine to the methane mitigation caused by feeding *Leucaena leucocephala*. *Arch. Anim. Nutr.* **2013**, *67*, 169–184. [CrossRef] [PubMed]
109. Pineiro-Vazquez, A.T.; Canul-Solis, J.R.; Jimenez-Ferrer, G.O.; Alayon-Gamboa, J.A.; Chay-Canul, A.J.; Ayala-Burgos, A.J.; Aguilar-Perez, C.F.; Ku-Vera, J.C. Effect of condensed tannins from *Leucaena leucocephala* on rumen fermentation, methane production and population of rumen protozoa in heifers fed low-quality forage. *Asian-Australas. J. Anim. Sci.* **2018**, *31*, 1738–1746. [CrossRef] [PubMed]
110. Pathak, A.K.; Dutta, N.; Pattanaik, A.K.; Chaturvedi, V.B.; Sharma, K. Effect of condensed tannins from *Ficus infectoria* and *Psidium guajava* leaf meal mixture on nutrient metabolism, methane emission and performance of lambs. *Asian-Australas. J. Anim. Sci.* **2017**, *30*, 1702–1710. [CrossRef] [PubMed]
111. Wang, S.; Terranova, M.; Kreuzer, M.; Marquardt, S.; Eggerschwiler, L.; Schwarm, A. Supplementation of pelleted hazel (*Corylus avellana*) leaves decreases methane and urinary nitrogen emissions by sheep at unchanged forage intake. *Sci. Rep.* **2018**, *8*, 5427. [CrossRef]
112. Theodoridou, K.; Aufrère, J.; Andueza, D.; Le Morvan, A.; Picard, F.; Stringano, E.; Pourrat, J.; Mueller-Harvey, I.; Baumont, R. Effect of plant development during first and second growth cycle on chemical composition, condensed tannins and nutritive value of three sainfoin (*Onobrychis viciifolia*) varieties and lucerne. *Grass Forage Sci.* **2011**, *66*, 402–414. [CrossRef]
113. Diaz Carrasco, J.; Cabral, C.; Redondo, L.; Daniela Pin Viso, N.; Colombatto, D.; Diana Farber, M.; Fernandez Miyakawa, M. Impact of chestnut and quebracho tannins on rumen microbiota of bovines. *Biomed. Res. Int.* **2017**, *2017*, 9610810. [CrossRef] [PubMed]
114. Phesatcha, K.; Wanapat, M. Tropical legume supplementation influences microbial protein synthesis and rumen ecology. *J. Anim. Physiol. Anim. Nutr.* **2017**, *101*, 552–562. [CrossRef]
115. Tan, H.Y.; Sieo, C.C.; Abdullah, N.; Liang, J.B.; Huang, X.D.; Ho, Y.W. Effects of condensed tannins from *Leucaena* on methane production, rumen fermentation and populations of methanogens and protozoa in vitro. *Anim. Feed Sci. Technol.* **2011**, *169*, 185–193. [CrossRef]
116. Saminathan, M.; Sieo, C.C.; Gan, H.M.; Ravi, S.; Venkatachalam, K.; Abdullah, N.; Wong, C.M.V.L.; Ho, Y.W. Modulatory effects of condensed tannin fractions of different molecular weights from a *Leucaena leucocephala* hybrid on the bovine rumen bacterial community in vitro. *J. Sci. Food Agric.* **2016**, *96*, 4565–4574. [CrossRef]
117. Naumann, H.D.; Tedeschi, L.O.; Muir, J.P.; Lambert, B.D.; Kothmann, M.M. Effect of molecular weight of condensed tannins from warm-season perennial legumes on ruminal methane production in vitro. *Biochem. Syst. Ecol.* **2013**, *50*, 154–162. [CrossRef]
118. Williams, C.M.; Eun, J.S.; MacAdam, J.W.; Young, A.J.; Fellner, V.; Min, B.R. Effects of forage legumes containing condensed tannins on methane and ammonia production in continuous cultures of mixed ruminal microorganisms. *Anim. Feed Sci. Technol.* **2011**, *166*, 364–372. [CrossRef]
119. Woodward, S.L.; Waghorn, G.C.; Lassey, K. Early indications that feeding *Lotus* will reduce methane emissions from ruminants. *Proc. N. Z. Soc. Anim. Prod.* **2001**, *61*, 23–26.
120. Woodward, S.L.; Waghorn, G.; Lassey, K.; Laboyrie, P. Does feeding sulla (*Hedysarum coronarium*) reduce methane emissions from dairy cows? *Proc. N. Z. Soc. Anim. Prod.* **2002**, *62*, 227–230.
121. Salminen, J.-P. Effects of sample drying and storage, and choice of extraction solvent and analysis method on the yield of birch leaf hydrolyzable tannins. *J. Chem. Ecol.* **2003**, *29*, 1289–1305. [CrossRef] [PubMed]
122. Neilson, A.P.; O'Keefe, S.F.; Bolling, B.W. High-molecular-weight proanthocyanidins in foods: Overcoming analytical challenges in pursuit of novel dietary bioactive components. *Annu. Rev. Food Sci. Technol.* **2016**, *7*, 43–64. [CrossRef] [PubMed]
123. Lees, G.L. Condensed tannins in some forage legumes: Their role in the prevention of ruminant pasture bloat. In *Plant Polyphenols: Synthesis, Properties, Significance*, 1992/01/01 ed.; Hemingway, R.W., Laks, P.E., Eds.; Springer: Boston, MA, USA, 1992; pp. 915–934.
124. Häring, D.A.; Suter, D.; Amrhein, N.; Lüscher, A. Biomass allocation is an important determinant of the tannin concentration in growing plants. *Ann. Bot.* **2007**, *99*, 111–120. [CrossRef] [PubMed]
125. Frutos, P.; Hervás, G.; Giráldez, F.; Mantecón, A. Review. Tannins and ruminant nutrition. *Span. J. Agric. Res.* **2004**, *2*, 191–202. [CrossRef]
126. Hummer, W.; Schreier, P. Analysis of proanthocyanidins. *Mol. Nutr. Food Res.* **2008**, *52*, 1381–1398. [CrossRef] [PubMed]
127. Peng, K.; Huang, Q.; Xu, Z.; McAllister, T.A.; Acharya, S.; Mueller-Harvey, I.; Drake, C.; Cao, J.; Huang, Y.; Sun, Y.; et al. Characterization of condensed tannins from purple prairie clover (*Dalea purpurea* Vent.) conserved as either freeze-dried forage, sun-cured hay or silage. *Molecules* **2018**, *23*, 586. [CrossRef] [PubMed]
128. Sze-Tao, K.W.C.; Schrimpf, J.E.; Teuber, S.S.; Roux, K.H.; Sathe, S.K. Effects of processing and storage on walnut (*Juglans regia* L) tannins. *J. Sci. Food Agric.* **2001**, *81*, 1215–1222. [CrossRef]

129. Kardel, M.; Taube, F.; Schulz, H.; Schütze, W.; Gierus, M. Different approaches to evaluate tannin content and structure of selected plant extracts—Review and new aspects. *J. Appl. Bot. Food Qual.* **2013**, *86*, 154–166. [CrossRef]
130. Chavan, U.; Shahidi, F.; Naczk, M. Extraction of condensed tannins from beach pea (*Lathyrus maritimus* L.) as affected by different solvents. *Food Chem.* **2001**, *75*, 509–512. [CrossRef]
131. Hagerman, A.E. Extraction of tannin from fresh and preserved leaves. *J. Chem. Ecol.* **1988**, *14*, 453–461. [CrossRef]
132. Aspé, E.; Fernández, K. The effect of different extraction techniques on extraction yield, total phenolic, and anti-radical capacity of extracts from *Pinus radiata* Bark. *Ind. Crops Prod.* **2011**, *34*, 838–844. [CrossRef]
133. Chupin, L.; Maunu, S.; Reynaud, S.; Pizzi, A.P.; Charrier, B.; Charrier—El Bouhtoury, F. Microwave assisted extraction of maritime pine (*Pinus pinaster*) bark: Impact of particle size and characterization. *Ind. Crops Prod.* **2015**, *65*, 142–149. [CrossRef]
134. Cork, S.J.; Krockenberger, A.K. Methods and pitfalls of extracting condensed tannins and other phenolics from plants: Insights from investigations on *Eucalyptus* leaves. *J. Chem. Ecol.* **1991**, *17*, 123–134. [CrossRef] [PubMed]
135. Grabber, J.H.; Zeller, W.E.; Mueller-Harvey, I. Acetone enhances the direct analysis of procyanidin- and prodelphinidin-based condensed tannins in *Lotus* species by the butanol-HCl-iron assay. *J. Agric. Food Chem.* **2013**, *61*, 2669–2678. [CrossRef]
136. Hixson, J.L.; Bindon, K.A.; Smith, P.A. Evaluation of direct phloroglucinolysis and colorimetric depolymerization assays and their applicability for determining condensed tannins in grape marc. *J. Agric. Food Chem.* **2015**, *63*, 9954–9962. [CrossRef] [PubMed]
137. Rubanza, C.D.K.; Shem, M.N.; Otsyina, R.; Bakengesa, S.S.; Ichinohe, T.; Fujihara, T. Polyphenolics and tannins effect on in vitro digestibility of selected *Acacia* species leaves. *Anim. Feed Sci. Technol.* **2005**, *119*, 129–142. [CrossRef]
138. Rira, M.; Morgavi, D.P.; Genestoux, L.; Djibiri, S.; Sekhri, I.; Doreau, M. Methanogenic potential of tropical feeds rich in hydrolyzable tannins. *J. Anim. Sci.* **2019**, *97*, 2700–2710. [CrossRef]
139. Pal, K.; Patra, A.; Sahoo, A.; Kumawat, P. Evaluation of several tropical tree leaves for methane production potential, degradability and rumen fermentation in vitro. *Livest. Sci.* **2015**, *180*, 98–105. [CrossRef]
140. Muir, J.P.; Terrill, T.H.; Mosjidis, J.A.; Luginbuhl, J.-M.; Miller, J.E.; Burke, J.M.; Coleman, S.W. Harvest regimen changes *sericea lespedeza* condensed tannin, fiber and protein concentrations. *Grassl. Sci.* **2018**, *64*, 137–144. [CrossRef]
141. Guimarães-Beelen, P.; Berchielli, T.; Beelen, R.; Filho, J.; Oliveira, S. Characterization of condensed tannins from native legumes of the Brazilian Northeastern semi-arid. *Sci. Agric.* **2006**, *63*, 522–528. [CrossRef]
142. Piluzza, G.; Sulas, L.; Bullitta, S. Tannins in forage plants and their role in animal husbandry and environmental sustainability: A review. *Grass Forage Sci.* **2014**, *69*, 32–48. [CrossRef]
143. Tibe, O.; Sutherland, I.A.; Lesperance, L.; Harding, D.R. The effect of purified condensed tannins of forage plants from Botswana on the free-living stages of gastrointestinal nematode parasites of livestock. *Vet. Parasitol.* **2013**, *197*, 160–167. [CrossRef]
144. Rufino-Moya, P.J.; Blanco, M.; Bertolin, J.R.; Joy, M. Methane production of fresh sainfoin, with or without PEG, and fresh alfalfa at different stages of maturity is similar but the fermentation end products vary. *Animals* **2019**, *9*, 197. [CrossRef] [PubMed]
145. Grabber, J.; Coblentz, W.; Riday, H.; Griggs, T.; Min, D.-H.; MacAdam, J.; Cassida, K. Protein and dry-matter degradability of european- and mediterranean-derived birdsfoot trefoil cultivars grown in the colder continental USA. *Crop. Sci.* **2015**, *55*, 1356. [CrossRef]
146. Grabber, J.H.; Riday, H.; Cassida, K.A.; Griggs, T.C.; Min, D.H.; MacAdam, J.W. Yield, morphological characteristics, and chemical composition of european-and mediterranean-derived birdsfoot trefoil cultivars grown in the colder continental United States. *Crop. Sci.* **2014**, *54*, 1893–1901. [CrossRef]
147. Kelman, W.; Tanner, G. Foliar Condensed Tannin Levels in Lotus Species Growing on Limed and Unlimed Soils in South-Eastern Australia. *Proc. N. Z. Grassl. Assoc.* **1990**, *52*, 51–54.
148. Wang, Y.; Waghorn, G.C.; Barry, T.N.; Shelton, I.D. The effect of condensed tannins in *Lotus corniculatus* on plasma metabolism of methionine, cystine and inorganic sulphate by sheep. *Br. J. Nutr.* **1994**, *72*, 923–935. [CrossRef] [PubMed]
149. Barry, T.N. The role of condensed tannins in the nutritional value of *Lotus pedunculatus* for sheep. Rates of body and wool growth. *Br. J. Nutr.* **1985**, *54*, 211–217. [CrossRef]
150. Diez, M.T.; Garcia del Moral, P.; Resines, J.A.; Arin, M.J. Determination of phenolic compounds derived from hydrolysable tannins in biological matrices by RP-HPLC. *J. Sep. Sci.* **2008**, *31*, 2797–2803. [CrossRef]
151. Tuominen, A.; Karonen, M. Variability between organs of proanthocyanidins in *Geranium sylvaticum* analyzed by off-line 2-dimensional HPLC-MS. *Phytochemistry* **2018**, *150*, 106–117. [CrossRef] [PubMed]
152. Kelm, M.A.; Hammerstone, J.F.; Schmitz, H.H. Identification and quantitation of flavanols and proanthocyanidins in foods: How good are the datas? *Clin. Dev. Immunol.* **2005**, *12*, 35–41. [CrossRef] [PubMed]
153. Moilanen, J.; Sinkkonen, J.; Salminen, J.-P. Characterization of bioactive plant ellagitannins by chromatographic, spectroscopic and mass spectrometric methods. *Chemoecology* **2013**, *23*, 165–179. [CrossRef]
154. Yanagida, A.; Shoji, T.; Shibusawa, Y. Separation of proanthocyanidins by degree of polymerization by means of size-exclusion chromatography and related techniques. *J. Biochem. Biophy. Meth.* **2003**, *56*, 311–322. [CrossRef]
155. Karonen, M.; Ossipov, V.; Sinkkonen, J.; Loponen, J.; Haukioja, E.; Pihlaja, K. Quantitative analysis of polymeric proanthocyanidins in birch leaves with normal-phase HPLC. *Phytochem Anal.* **2006**, *17*, 149–156. [CrossRef] [PubMed]
156. Kelm, M.A.; Johnson, J.C.; Robbins, R.J.; Hammerstone, J.F.; Schmitz, H.H. High-performance liquid chromatography separation and purification of cacao (*Theobroma cacao* L.) procyanidins according to degree of polymerization using a diol stationary phase. *J. Agric. Food Chem.* **2006**, *54*, 1571–1576. [CrossRef] [PubMed]

157. Mouls, L.; Mazauric, J.P.; Sommerer, N.; Fulcrand, H.; Mazerolles, G. Comprehensive study of condensed tannins by ESI mass spectrometry: Average degree of polymerisation and polymer distribution determination from mass spectra. *Anal. Bioanal. Chem.* **2011**, *400*, 613–623. [CrossRef]
158. Kalili, K.M.; Vestner, J.; Stander, M.A.; de Villiers, A. Toward unraveling grape tannin composition: Application of online hydrophilic interaction chromatography × reversed-phase liquid chromatography–time-of-flight mass spectrometry for grape seed analysis. *Anal. Chem.* **2013**, *85*, 9107–9115. [CrossRef] [PubMed]
159. Salminen, J.-P. Two-dimensional tannin fingerprints by liquid chromatography tandem mass spectrometry offer a new dimension to plant tannin analyses and help to visualize the tannin diversity in plants. *J. Agric. Food Chem.* **2018**, *66*, 9162–9171. [CrossRef] [PubMed]
160. Engström, M.T.; Pälijärvi, M.; Fryganas, C.; Grabber, J.H.; Mueller-Harvey, I.; Salminen, J.-P. Rapid qualitative and quantitative analyses of proanthocyanidin oligomers and polymers by UPLC-MS/MS. *J. Agric. Food Chem.* **2014**, *62*, 3390–3399. [CrossRef]
161. Saminathan, M.; Sieo, C.C.; Abdullah, N.; Wong, C.M.V.L.; Ho, Y.W. Effects of condensed tannin fractions of different molecular weights from a *Leucaena leucocephala* hybrid on in vitro methane production and rumen fermentation. *J. Sci. Food Agric.* **2015**, *95*, 2742–2749. [CrossRef] [PubMed]
162. MacAdam, J.W.; Villalba, J.J. Beneficial effects of temperate forage legumes that contain condensed tannins. *Agriculture* **2015**, *5*, 475–491. [CrossRef]
163. Waghorn, G.C.; Woodward, S.L.; Tavendale, M.; Clark, D.A. Inconsistencies in rumen methane production—Effects of forage composition and animal genotype. *Int. Congr. Ser.* **2006**, *1293*, 115–118. [CrossRef]
164. Huang, X.; Liang, J.; Tan, H.; Yahya, R.; Bodee, K.; Ho, Y. Molecular weight and protein binding affinity of *Leucaena* condensed tannins and their effects on in vitro fermentation parameters. *Anim. Feed Sci. Technol.* **2010**, *159*, 81–87. [CrossRef]
165. Paengkoum, P.; Phonmun, T.; Liang, J.B.; Huang, X.D.; Tan, H.Y.; Jahromi, M.F. Molecular weight, protein binding affinity and methane mitigation of condensed tannins from mangosteen-peel (*Garcinia mangostana* L). *Asian-Australas. J. Anim. Sci.* **2015**, *28*, 1442–1448. [CrossRef] [PubMed]
166. Huang, X.D.; Liang, J.B.; Tan, H.Y.; Yahya, R.; Ho, Y.W. Effects of *Leucaena* condensed tannins of differing molecular weights on in vitro CH_4 production. *Anim. Feed Sci. Technol.* **2011**, *166*, 373–376. [CrossRef]
167. Hagerman, A.E.; Butler, L.G. Protein precipitation method for the quantitative determination of tannins. *J. Agric. Food Chem.* **1978**, *26*, 809–812. [CrossRef]
168. Makkar, H.P.; Dawra, R.K.; Singh, B. Protein precipitation assay for quantitation of tannins: Determination of protein in tannin-protein complex. *Anal. Biochem.* **1987**, *166*, 435–439. [CrossRef]
169. Ropiak, H.M.; Lachmann, P.; Ramsay, A.; Green, R.J.; Mueller-Harvey, I. Identification of structural features of condensed tannins that affect protein aggregation. *PLoS ONE* **2017**, *12*, e0170768. [CrossRef]
170. Tan, H.Y.; Sieo, C.C.; Lee, C.M.; Abdullah, N.; Liang, J.B.; Ho, Y.W. Diversity of bovine rumen methanogens in vitro in the presence of condensed tannins, as determined by sequence analysis of 16S rRNA gene library. *Int. J. Microbiol.* **2011**, *49*, 492–498. [CrossRef]
171. Denman, S.E.; Martinez Fernandez, G.; Shinkai, T.; Mitsumori, M.; McSweeney, C.S. Metagenomic analysis of the rumen microbial community following inhibition of methane formation by a halogenated methane analog. *Front. Microbiol.* **2015**, *6*, 1087. [CrossRef] [PubMed]
172. Fay, M.F.; Dale, P.J. Condensed tannins in *Trifolium* species and their significance for taxonomy and plant breeding. *Genet. Resour. Crop. Evol.* **1993**, *40*, 7–13. [CrossRef]
173. Roldan, M.B.; Cousins, G.; Fraser, K.; Hancock, K.R.; Collette, V.; Demmer, J.; Woodfield, D.R.; Caradus, J.R.; Jones, C.; Voisey, C.R. Elevation of condensed tannins in the leaves of Ta-MYB14-1 white clover (*Trifolium repens* L.) outcrossed with high anthocyanin lines. *J. Agric. Food Chem.* **2020**, *68*, 2927–2939. [CrossRef] [PubMed]
174. Real, D.; Sandral, G.; Rebuffo, M.; Hughes, S.; Kelman, W.; Mieres, J.; Dods, K.; Crossa, J. Breeding of an early-flowering and drought-tolerant *Lotus corniculatus* L. variety for the high-rainfall zone of southern Australia. *Crop. Pasture Sci.* **2012**, *63*, 848–857. [CrossRef]

Article

Technical Efficiency in the European Dairy Industry: Can We Observe Systematic Failures in the Efficiency of Input Use?

Lukáš Čechura and Zdeňka Žáková Kroupová *

Department of Economics, Faculty of Economics and Management of the Czech University of Life Sciences in Prague, 16500 Prague, Czech Republic; cechura@pef.czu.cz
* Correspondence: kroupovaz@pef.czu.cz

Abstract: The paper provides findings on the technical efficiency of the European dairy processing industry, which is one of the most important subsectors of the food processing industry in the European Union (EU). The ability to efficiently use inputs in the production of outputs is a prerequisite for the sustainability and competitiveness of the agri-food sector as well as for food security. Thus, the aim of this paper is to provide a robust estimate of technical efficiency by employing new advances in productivity and efficiency analysis, and to investigate the efficiency of input use in 10 selected European countries. The analysis is based on two-stage stochastic frontier modelling incorporating country-specific input distance function (IDF) estimates and a meta-frontier input distance function estimate, both in specification of the four-component model, which currently represents the most advanced approach to technical efficiency analysis. To provide a robust estimate of these models, the paper employs methods that control for the potential endogeneity of netputs in the multi-step estimation procedure. The results, based on the Amadeus dataset, reveal that companies manufacturing dairy products greatly exploited their production possibilities in 2006–2018. The dairy processing industry in the analysed countries cannot generally be characterized by a considerable waste of resources. The potential cost reduction is estimated at 4–8%, evaluated on the country samples mean. The overall technical inefficiency (OTE) is mainly a result of short-term shocks and unsystematic failures. However, the meta-frontier estimates also reveal a certain degree of systematic failure, e.g., permanent managerial failures and structural problems in European dairy processing industry.

Keywords: technical efficiency; four-component model; endogeneity; input distance function; meta-frontier; stochastic frontier analysis; dairy processing industry; European Union

Citation: Čechura, L.; Žáková Kroupová, Z. Technical Efficiency in the European Dairy Industry: Can We Observe Systematic Failures in the Efficiency of Input Use? *Sustainability* **2021**, *13*, 1830. https://doi.org/10.3390/su13041830

Academic Editor: Rajeev Bhat
Received: 19 January 2021
Accepted: 5 February 2021
Published: 8 February 2021

Publisher's Note: MDPI stays neutral with regard to jurisdictional claims in published maps and institutional affiliations.

Copyright: © 2021 by the authors. Licensee MDPI, Basel, Switzerland. This article is an open access article distributed under the terms and conditions of the Creative Commons Attribution (CC BY) license (https://creativecommons.org/licenses/by/4.0/).

1. Introduction

A functional, sustainable, competitive, and structurally balanced agri-food sector has an irreplaceable position in the European economy. The importance of this sector in modern society is emphasized by global trends, among which population explosions, migration waves, and climate change can be highlighted. Additionally, the COVID-19 pandemic has strengthened the role of national food security.

The food processing industry is one of the most important sectors of the European economy. According to Eurostat [1], it is the largest manufacturing sector in the European Union (EU), representing 14% of total manufacturing employment, 12% of total manufacturing turnover, and 10% of value added in 2018. The manufacture of dairy products, with a 17% share of total food industry turnover, 12% share of value added, and 10% share of total food industry employment, is one of the most important subsectors of the European food processing industry.

The development of the European food processing industry in recent decades has been affected by several economic, social, and technological trends and challenges, especially the globalization and liberalization of food markets, the global financial crisis, the change in consumer preferences towards healthy foods, socially responsible consumption and organic foods, and the implementation of EU regulations focusing on food safety and environmental issues [2,3]. Additionally, the COVID-19 pandemic crisis has created a new era in which the food processing industry is facing various challenges, including the change of consumer purchasing behaviour, transportation network disturbances, labour absenteeism, and the closure of various food manufacturing industries [4]. These trends, connected with a high level of market saturation and concentration of the food retailing sector [5], contribute to a highly competitive environment. To enhance the sustainability and competitiveness of companies/sectors, managers and policymakers have dealt with factors determining productivity, namely, the technology in use, the quality of inputs, the ability to efficiently use inputs in the production of outputs, and the exploitation of economies of scale [6].

These factors have also received special attention in the research of the food processing industry over the last decade. For example, Náglová and Šimpachová Pechrová [7]; Čechura and Hockmann [8]; Rudinskaya [9]; Rezitis and Kalandzi [10]; Špička [11], Popović and Panić [12], and Setiawan et al. [13] investigated the technical efficiency of the food processing industry. Soboh et al. [14] and Dimara et al. [15] dealt with technical and scale efficiency, that was investigated also by Baran [16]. Allendorf and Hirsch [5] analysed technical change and technical efficiency. Čechura et al. [6] assessed the exploitation of economies of scale and production possibilities, along with the impact of technical change. Kapelko et al. [3,17,18], Rudinskaya and Kuzmenko [19], and Čechura [20] investigated productivity dynamics based on technical change, technical inefficiency change, and scale inefficiency change. The majority of these studies are oriented only to one selected country. The exceptions are Kapelko [3], who investigated meat manufacturing firms, fruit and vegetable processing firms, dairy manufacturing firms, and bakery and farinaceous products manufacturing firms in 18 European member states; Allendorf and Hirsch [5], who analysed the dairy and meat processing sectors in eight European countries; Čechura et al. [6], who investigated the milling, fruit and vegetable, dairy, and meat processing sectors in 24 EU member countries; and Soboh et al. [14], who focused on dairy processing firms in six European countries.

With regard to the dairy processing industry, several studies that investigate technical efficiency of European dairy processing industry can be highlighted. Table 1 summarizes these studies. These studies revealed that evaluated on the sample means dairy processing companies in majority of analysed countries greatly exploited their production possibilities in the short-term. However, substantial differences were found between the best and worst producers as well as among countries representing the potential for costs reduction. The studies in Table 1 are in agreement on lower technical efficiency of the Eastern European dairy processing industries compare to the Western European ones. The technical efficiency development was predominantly positive in recent decades and contributed to productivity growth. The need of technical efficiency improvements was prompted by an increase in the price of the main input—milk. Significant short-term changes in technical efficiency were found in the period of the financial crisis. The shortcoming of these studies is that they take into account only transient (short-term) technical efficiency (TTE), which varies across companies because of the shocks associated with new production technologies, human capital, and learning-by-doing. Persistent (long-term) inefficiency (PTE), which could arise due to the presence of rigidity within a firm's organization and production process, is unrecognized in these studies despite the fact that it can lead to the underestimation of technical inefficiency. To the best of our knowledge, there are no empirical studies examining both parts of overall technical efficiency in the European dairy processing industry.

Table 1. Recent studies of technical efficiency of European dairy processing industry. Source: authors.

Study	Countries	Years	Method
Čechura and Hockmann [8]	CZ	2003–2012	Stochastic frontier analysis (SFA)
Čechura et al. [6]	AT, BE, BG, CZ, DE, DK, EE, ES, FI, FR, UK, GR, HU, IR, IT, LT, LV, NL, PL, PT, RO, SW, SL, SI, SR	2003–2012	SFA
Kapelko and Oude Lansink [21]	AT, BE, FI, FR, DE, LU, NL, NO, CH, BIH, BG, HR, CZ, EE, HU, PL, RO, SR, SL, SI, IT, PT, ES	2005–2012	Data envelopment analysis (DEA)
Kapelko and Oude Lankink [22]	ES	2001–2009	DEA
Rezitis and Kalandzi [10]	GR	1984–2007	DEA
Rudinskaya [9]	CZ	2005–2012	SFA
Soboh et al. [14]	BE, DK, FR, DE, IR, NL	1995–2005	SFA
Špička [11]	CZ, PL, SK	2008–2013	DEA

Note: AT denotes Austria, BE denotes Belgium, BG denotes Bulgaria, BIH denotes Bosnia and Herzegovina, CH denotes Switzerland, CZ denotes Czechia, DE denotes Germany, DE denotes Denmark, EE denotes Estonia, ES denotes Spain, FI denotes Finland, FR denotes France, GR denotes Greece, HR denotes Croatia, HU denotes Hungary, IT denotes Italy, LV denotes Latvia, LT denotes Lithuania, LU denotes Luxembourg, PL denotes Poland, PT denotes Portugal, RO denotes Romania, NL denotes the Netherlands, NO denotes Norway, SK denotes Slovakia, SI denotes Slovenia, SR denotes Serbia, SW denotes Sweden, UK denotes the United Kingdom.

The aim of this study is to fill the gap in the empirical literature on the efficiency of the European food processing industry by providing technical efficiency estimates in selected European countries, along with a meta-frontier analysis for cross-country efficiency comparison in the dairy processing industry. In particular, we aim to provide a robust estimate of technical efficiency by employing new advances in productivity and efficiency analysis, and to investigate the efficiency of input use in the analysed countries. The study provides an insight into the decomposition of overall technical efficiency into transient technical efficiency, representing short-term deviations from the frontier, and persistent technical efficiency, which captures systematic deviations from frontier technology.

The study addresses the following research questions: (i) The first relates to the impacts and dynamics of the overall technical efficiency. The aim is to assess whether there is indication that the countries follows a sustainable development path characterized by reduced waste of resource due to inefficient input use. (ii) The second question deals with the sources of overall technical efficiency. The study evaluates whether we can observe systematic failures in the efficiency of input use or whether the deviations from the frontier technologies are due to the transient reasons. Answering these research questions provides important information for policy makers. The structure of the paper is as follows: Section 2 presents the theoretical background of our approach; Section 3 then introduces the model, empirical strategy, and dataset; Section 4 presents the results; and Section 5 discusses our findings and compares them with other studies.

2. Theoretical Background

Technical efficiency was originally defined by Koopmans [23] as the situation where an increase in any output is impossible without a reduction in at least one other output or an increase in at least one input and where a reduction in any input requires an increase in at least one other input or a reduction in at least one output. Debreu [24] and Farrell [25] introduced radial measures of technical inefficiency. Fried et al. [26] suggested an input-conserving orientation. Their measure is defined as the maximum equi-proportionate reduction in all inputs that is feasible with a given technology and outputs. With an output-augmenting orientation, the measure is defined as the maximum equi-proportionate expansion in all outputs that is feasible with a given technology and inputs. A value of unity indicates technical efficiency because no radial adjustment is feasible, and a value different from unity indicates the severity of technical inefficiency.

According to Coelli et al. [27], different methods have been considered for the estimation of the technical inefficiency of the production plan. Two widely used approaches are data envelopment analysis (DEA), which is non-parametric and deterministic, and stochastic frontier analysis (SFA), which is, on the contrary, parametric and stochastic. There are advantages and disadvantages to each approach (e.g., [28]). The DEA approach is computationally simple, but it is sensitive to outliers since the model ignores measurement error and other sources of statistical noise—all deviations from the frontier are interpreted as the results of technical inefficiency [27].

The stochastic modelling of technical efficiency introduced independently by Meeusen and van den Broeck [29] and Aigner et al. [30] is based on a decomposition of the error term into a symmetric random error and one-sided technical inefficiency. Since the introduction of this frontier modelling within the context of panel-data models, there has been considerable research done in order to extend it to generate consistent and unbiased estimates, see Figure 1. As mentioned by Alem [31], several models have been developed that are based on assumptions about the temporal behaviour of the inefficiency and account for heterogeneity, which can be divided into two categories: observed, which is reflected in the measured variables, and unobserved, which is typically interpreted as the effect of unobservable factors on the outcome of interest [26,32].

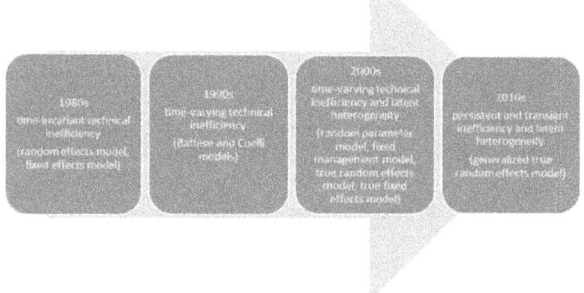

Figure 1. Development of stochastic frontier models. Source: authors.

Initial studies considering heterogeneity assumed that the time-invariant part of the model represents inefficiency, and the time-varying part represents firm-specific heterogeneity. Later, the firm-specific heterogeneity was assumed to be time-invariant, while the time-variant part was considered as inefficiency [28]. The latest approach emphasises the importance of considering latent heterogeneity in generating an unbiased estimate of time-invariant technical inefficiency, as well as the possibility of efficiency improvement [33,34]. In this approach, the overall technical inefficiency of a producer can be decomposed into transient and persistent parts. Transient technical inefficiency arises as a result of non-systematic managerial failures that can be resolved in the short term. According to Pisulewski and Marzec [35] as well as Njuki and Bravo-Ureta [36], the transient part of inefficiency relates to shocks associated with new production technologies and human capital. Kumbhakar and Lien [37] add that transient technical efficiency can represent the managerial ability to learn by doing. Persistent technical inefficiency represents structural problems in the organization of the production process or a systematic lack of managerial skills [38]. Moreover, it can be an indicator of a non-competitive market condition. Badunenko and Kumbhakar [34] state that persistent inefficiency could not exist in a competitive market, i.e., persistently inefficient firms would not survive in the business. The distinction between persistent and transient technical inefficiency has significant political implications because the persistent part of technical inefficiency is unchangeable without a new policy or fundamental change in the ownership and management of companies. Transient technical inefficiency can be adjusted over time without a major policy change [28].

The properties of the above-mentioned models are described by Kumbhakar et al. [28] and others. A comparison of the majority of SFA panel-data models, proving the sensitivity of technical efficiency estimates to the model specification, is presented by Alem [31] and Rashidghalam et al. [39]. Moreover, Badunenko and Kumbhakar [34] provide a judgement on the reliability of transient and persistent technical efficiency estimates.

3. Materials and Methods

3.1. Methodology Used in the Study

Two-step stochastic frontier modelling is applied to get a robust comparison of the efficiency of input use in the dairy processing industry among the analysed countries. The first step includes the estimation of the country-specific input distance functions, technical efficiency scores, and efficient output levels of dairy processing companies in 10 European member states (Austria (AT), Belgium (BE), Czechia (CZ), Germany (DE), Finland (FI), France (FR), Italy (IT), Spain (ES), Sweden (SW), and the United Kingdom (UK)) using the SFA approach. The second step includes the meta-frontier estimation and a comparison of the technical efficiency among the countries.

3.1.1. Input Distance Function

The analysis is based on an assumption that the transformation process is well approximated by an input distance function (IDF) that measures the largest factor of proportionality ρ by which the input vector x can be scaled down in order to produce a given output vector y with the technology existing at a particular time t [40]. The IDF is formally defined as

$$D^I(y, x, t) = max\left\{\rho : \frac{x}{\rho} \in L(y)\right\} \quad (1)$$

where x denotes the input vector, y denotes the output vector, t is a time variable, and $L(y)$ represents the input requirement set. For any input-output combination (x,y) belonging to the technology set, the input distance function takes a value no smaller than unity. According to Karagiannis et al. [41], a value of unity simply indicates that the input-output combination (x,y) belongs to the input isoquant, which represents the minimum input quantities that are necessary to produce a given output vector y.

In other words, if $D^I(y, x, t) = 1$, the given output vector y is produced with the minimum amount of inputs at a given time and with the given technology, and the firm is technically efficient [40]. That is, the IDF provides a measure of technical efficiency since it is reciprocal to the Farrell [25,41] input-based technical efficiency:

$$TE^I = 1/(D^I(y, x, t)) \quad (2)$$

According to Greene [42], the IDF exhibits the following properties: symmetry, monotonicity, linear homogeneity, and concavity in inputs and quasi-concavity in outputs. For the interpretation of the empirical estimates, the duality between the cost and input distance functions is another important property:

$$C(w, y, t) = \min_x\left\{wx : D^I(y, x, t) \geq 1\right\} \quad (3)$$

where w denotes a vector of input prices. The minimisation problem provides the relation between the derivatives of the IDF and the cost function [43]. In particular, the derivative with respect to the jth input gives:

$$\frac{\delta D^I(x^*(w, y, t), y, t)}{\delta x_j} = \frac{w_j}{C(w, y, t)} = r_j^*(x, y, t) \quad (4)$$

That is, the derivative of the input distance function with respect to a particular input is equal to the cost-deflated shadow price of that input. More conveniently, in terms of the log derivative of the distance function, we can rewrite this expression to

$$\frac{\delta \ln D^I(x^*(w,y,t),y,t)}{\delta \ln x_j} = \frac{w_j x_j^*(w,y,t)}{C(w,y,t)} = S_{j,t} \quad (5)$$

where $S_{j,t}$ is a cost-share of the particular input.

With respect to the output vector y, application of the envelope theorem to the minimisation problem $C(w,y,t) = \min_x \{wx : D^I(y,x,t) \geq 1\}$ leads to

$$\frac{\delta \ln D^I(x^*(w,y,t),y,t)}{\delta \ln y_m} = -\frac{\delta \ln C(w,y,t)}{\delta \ln y_m} = e_{mi,t} \quad (6)$$

Hence, the elasticity of the IDF with respect to the mth output is therefore equal to the negative of the cost elasticity of that output, and as such it indicates the importance of output in terms of cost [44].

In this study, we assume that the transformation process can be well approximated by the IDF in a translog functional form. This second-order local approximation of any twice-differentiable function satisfies Diewert's minimum flexibility requirement for flexible form [35]. The translog input distance function for output (y), J-inputs (x), and time (t) is defined as

$$\ln D_{it}^I = \alpha_0 + \alpha_m ln y_{it} + 1/2\alpha_{mm}(ln y_{it})^2 + \sum_{j=1}^J \gamma_{mj} \ln y_{it} \ln x_{j,it} + \sum_{j=1}^J \beta_j \ln x_{j,it} + \\ +1/2\sum_{j=1}^J \sum_{k=1}^K \beta_{jk} \ln x_{j,it} \ln x_{k,it} + \delta_t t + 1/2\delta_{tt}t^2 + \delta_{mt} ln y_{m,it} t + \sum_{j=1}^J \delta_{jt} \ln x_{j,it} t \quad (7)$$

where subscripts i, with $i = 1,2, \ldots ,N$, and t, with $t = 1, \ldots ,T$, refer to a certain company and time (year), respectively. α, β, γ, and δ are vectors of the parameters to be estimated. The symmetry restrictions imply that $\beta_{jk} = \beta_{kj}$. The time trend included in the IDF allows for capturing the joint effects of embedded knowledge, technology improvements and learning-by-doing in input quality improvements [45]. Here, δ_t and δ_{tt} capture the global effect of technical change on the IDF, while δ_{mt} and δ_{jt} measure the bias of technical change.

The IDF is homogenous of degree 1 in inputs. According to Sipiläinen [46], it requires

$$\sum_{j=1}^J \beta_j = 1; \sum_{j=1}^J \beta_{jk} = 0; \sum_{j=1}^J \gamma_{mj} = 0; \sum_{j=1}^J \delta_{jt} = 0. \quad (8)$$

Implying the homogeneity property of the IDF [47], which is imposed by normalising all the inputs by one input, we can rewrite the IDF as

$$\ln D_{it}^I - \ln x_{1,it} = \alpha_0 + \alpha_m ln y_{it} + 1/2\alpha_{mm}(ln y_{it})^2 + \sum_{j=2}^J \gamma_{mj} \ln y_{it} \ln \tilde{x}_{j,it} + \sum_{j=2}^J \beta_j \ln \tilde{x}_{j,it} + \\ +1/2\sum_{j=2}^J \sum_{k=2}^K \beta_{jk} \ln \tilde{x}_{j,it} \ln \tilde{x}_{k,it} + \delta_t t + 1/2\delta_{tt}t^2 + \delta_{mt} ln y_{m,it} t + \sum_{j=2}^J \delta_{jt} \ln \tilde{x}_{j,it} t \quad (9)$$

where $\ln \tilde{x}_{j,it} = \ln x_{j,it} - \ln x_{1,it}$

Moreover, all variables in logarithm are normalized by their sample mean. In this case, the first-order parameters can be interpreted as output elasticity and input cost shares, evaluated on the sample mean, respectively.

After introducing a statistical error term (v_{it}) and latent heterogeneity (μ_i), and replacing $\ln D_{it}^I$ with inefficiency terms, persistent technical inefficiency (η_i) and transient technical inefficiency (u_{it}), that is $\eta_i + u_{it} = \ln D_{it}^I$, the IDF takes the form of a generalized true random effects model (GTRE, [33]):

$$- \ln x_{1,it} = \alpha_0 + \alpha_m ln y_{it} + 1/2\alpha_{mm}(ln y_{it})^2 + \sum_{j=2}^J \gamma_{mj} \ln y_{it} \ln \tilde{x}_{j,it} + \\ +\sum_{j=2}^J \beta_j \ln \tilde{x}_{j,it} + 1/2\sum_{j=2}^J \sum_{k=2}^K \beta_{jk} \ln \tilde{x}_{j,it} \ln \tilde{x}_{k,it} + \delta_t t + 1/2\delta_{tt}t^2 + \\ +\delta_{mt} ln y_{m,it} t + \sum_{j=2}^J \delta_{jt} \ln \tilde{x}_{j,it} t - \eta_i - u_{it} + \mu_i + v_{it} \quad (10)$$

where $v_{it} \sim N(0, \sigma_v^2), u_{it} \sim N^+(0, \sigma_u^2), \eta_i \sim N^+(0, \sigma_\eta^2), \mu_i \sim N(0, \sigma_\mu^2)$.

3.1.2. Heterogeneity in Technology

The literature provides broad evidence for significant heterogeneity in dairy processing technology. Since we are estimating a joint country input distance function for the entire food processing industry (due to a data limitation, the low number of observations in an industry does not allow us to estimate country IDF for the dairy processing sector in the majority of countries), we need to consider two types of heterogeneity for processing technology, i.e., the potential existence of inter- and intra-sector heterogeneity. The intra-sector heterogeneity is captured by μ_i in (10). To capture the inter-sector heterogeneity, first-order parameters in (10) are expanded based on dummy variables for four major sectors in the food processing industry (namely the manufacture of dairy products, processing of meat, milling, and manufacture of bakery and farinaceous products):

$$\begin{aligned} \alpha_{ms} &= \alpha_m + \sum_s \alpha_s d_s, \forall m \\ \beta_{js} &= \beta_j + \sum_s \beta_s d_s, \forall j \\ \delta_{ts} &= \delta_t + \sum_s \delta_s d_s \end{aligned} \quad (11)$$

where d represents dummy variables which account for inter-sectoral differences in technology.

3.1.3. Estimation Strategy

Since the endogeneity problem usually frustrates researchers in productivity and efficiency analysis and leads to inconsistent estimates, this study uses methods which control for the potential endogeneity of netputs and thereby allow us to obtain consistent estimates of technology as well as efficiency measures. The study addresses two potential sources of endogeneity (due to the heterogeneity and due to simultaneity of input with technical efficiency) by using the system generalized method of moments (GMM) estimator. In particular, the GMM estimator allows us to deal with the endogeneity bias that arises when one or more explanatory variables are correlated with the error term. This correlation can have different origins, e.g., measurement errors, omitted variables bias, and simultaneity [48].

Moreover, the study compares the GMM estimates with the generalized true random effects model estimates. Analogously to the random effect model, the GTRE assumes that μ_i are independent of explanatory variables. The violation of this assumption can originate from the heterogeneity bias as a kind of omitted variable bias, which is a typical problem of hierarchical data. To deal with this heterogeneity bias, the study applies Mundlak's [49] formulation and adds group-means for each time-varying explanatory variable in the first-order level.

Since the GMM model deals with both sources of endogeneity, it represents our model of first choice. If the GMM model does not provide consistent estimates, the GTRE model with Mundlak's adjustment is our second choice. The standard GTRE model allows for an overall model comparison.

The estimation of the GTRE model and the GTRE model with Mundlak's adjustment is undertaken as a multistep procedure. We follow Kumbhakar et al. [50] and rewrite the model in (10) as

$$\begin{aligned} -\ln x_{1,it} = &\alpha_0^* + \alpha_m ln y_{it} + 1/2\alpha_{mm}(ln y_{it})^2 + \sum_{j=2}^J \gamma_{mj} \ln y_{it} \ln \widetilde{x}_{j,it} + \\ &+ \sum_{j=2}^J \beta_j \ln \widetilde{x}_{j,it} + 1/2 \sum_{j=2}^J \sum_{k=2}^K \beta_{jk} \ln \widetilde{x}_{j,it} \ln \widetilde{x}_{k,it} + \delta_t t + 1/2\delta_{tt} t^2 + \\ &+ \delta_{mt} ln y_{m,it} t + \sum_{j=2}^J \delta_{jt} \ln \widetilde{x}_{j,it} t + \alpha_i + \varepsilon_{it} \end{aligned} \quad (12)$$

where $\alpha_0^* = \alpha_0 - E(\eta_i) - E(u_{it}), \alpha_i = \mu_i - (\eta_i) - E(\eta_i))$ and $\varepsilon_{it} = v_{it} - (u_{ii} - E(u_{it}))$.

This specification ensures that α_i and ε_{it} have zero mean and constant variance. The multistep procedure consists of three steps. In step 1, standard random effect panel regression is used to estimate β, γ, δ, α_m, and theoretical values of α_i and ε_{it}, denoted by $\hat{\alpha}_i$ and $\hat{\varepsilon}_{it}$. In step 2, the transient technical inefficiency, u_{it}, is estimated using $\hat{\varepsilon}_{it}$ and the standard stochastic frontier technique with assumptions: $v_{it} \sim N(0, \sigma_v^2)$, $u_{it} \sim N^+(0, \sigma_u^2)$. In step 3, the persistent technical inefficiency, η_i, is estimated using and the stochastic frontier model with the following assumptions: $\eta_i \sim N^+(0, \sigma_\eta^2)$, $\mu_i \sim N(0, \sigma_\mu^2)$. These steps are done in the SW STATA 14.0.

Furthermore, the total technical efficiency (OTE) is quantified based on Kumbhakar et al. [50]:

$$OTE_{it} = \exp(-\hat{\eta}_i) \times \exp(-\hat{u}_{it}). \qquad (13)$$

The GMM model extends this estimation procedure. The study follows the four-step procedure of Bokusheva and Čechura [51]. In step 1, the two-step system generalized method of moments estimator [52,53] is used to obtain consistent estimates of the IDF parameters. The Arellano and Bover [52]/Blundell and Bond [53] approach builds a system of two equations, the original equation (in levels) and the transformed one (in differences), and employs two types of instruments: level instruments for the differenced equations and lagged differences for the equations in levels [54]. The validity of these instruments is tested by the Hansen J-test [55], which analyses the joint validity of the instruments, and the Arellano-Bond test [56], which analyses the autocorrelation in the idiosyncratic disturbance term (v_{it}) that could render some lags invalid as instruments. In step 2, residuals are used from the system GMM level equations to estimate a random effects panel model employing the generalized least squares (GLS) estimator. The transient and persistent technical inefficiency is estimated in steps 3 and 4 based on the same procedure as described above. These estimates are also done in the SW STATA 14.0.

3.2. Data Used in the Study

The study uses an unbalanced panel data set drawn from the Amadeus database collected by Bureau van Dijk—Moody's Analytics company. The database contains information on around 21 million European companies and provides detailed information about company financials in a standard format, financial strength indicators, sectoral activities, and corporate structures. Moreover, the database is unified between different countries to guarantee the comparability of data. The scope of the database and the comparability of data are the reasons why this database is widely used in economic research. These are examples of studies of technical efficiency of dairy processing industry based on Amadeus database: Kapelko and Oude Lansink [21], Čechura et al. [6], Soboh et al. [14], and Špička [11].

This study uses information from the final accounts of companies whose main activity was food processing (Division C10: Manufacture of food products according to the Statistical classification of economic activities in the European Community, abbreviated as NACE) in the period from 2006 till 2018. Moreover, the study is concentrated on an analysis of C10.5: Manufacture of dairy products (including subgroups: 10.51, 10.52). The study focuses on 10 countries which together represents 76% of European food processing turnover according to Eurostat [1]. Specifically, the analysis uses the following data: output (y), labour (xL), capital (xC), and material (xM). The output is represented by operating revenue (turnover) deflated by the sectoral index of food processing prices (EU level or country level if it was disposable; 2010 = 100) and changes in a company's stock. Labour is represented by the cost of employees deflated by the index of producer prices in the industry (country level; 2010 = 100). Capital is the book value of fixed assets deflated by the index of producer prices in the industry (country level; 2010 = 100). Finally, material is the total cost of materials and energy deflated by the index of producer prices in the industry (country level; 2010 = 100). The source of price indexes is the EUROSTAT database.

Since not all information can be found in the database, only those companies with non-zero and positive values are used for the variable of interest. Moreover, companies with less than three consecutive observations are rejected from the dataset. This procedure decreases the problem associated with the entry and exit of producers from the database and allows the use of the GMM estimator with a sufficient number of lagged instruments. The study is constrained by that to the use of an unbalanced panel data set containing 11,605 food processing companies with 95,003 observations in the first step: country-specific IDF estimation. The second step: meta-frontier IDF estimation uses only data of dairy processing companies, and the structure of this data sub-set is presented in Table 2.

Table 2. Structure of the data set. Source: Amadeus database and Eurostat.

Country	AT	BE	CZ	DE	ES	FI	FR	IT	SW	UK	Total
I	15	40	58	80	114	36	211	97	22	62	735
N	131	336	498	658	1108	285	1940	915	150	496	6517
RS1	48	40	43	32	41	84	33	58	54	60	48
RS2	44	74	79	59	82	97	48	95	70	90	63

Note: I is the number of dairy processing firms (10.5), N is the total number of observations (10.5) RS1 is the revenue share (in %) of Amadeus sample food processing firms on the total revenue of food processing firms, RS2 is the revenue share (in %) of Amadeus dairy processing firms on the total revenue of dairy processing industry; AT denotes Austria, BE denotes Belgium, CZ denotes Czechia, DE denotes Germany, ES denotes Spain, FI denotes Finland, FR denotes France, IT denotes Italy, SW denotes Sweden, UK denotes the United Kingdom.

4. Results

The country-specific IDF estimates (country-specific IDF estimates are available on request from the authors or in [57]) reveal that the technology of the food processing industry exhibits common patterns in the analysed European countries. Common features are found in the case of cost elasticity, as well as in the cost structure. As expected, and in line with the information provided by the dataset, the highest cost shares have material inputs between 0.50 and 0.64. The labour cost share is approximately 30%, with the lowest cost share in Germany (25.8%) and the highest in France (38.3%). The rest is represented by the share of capital. Moreover, food processors in general cannot benefit greatly from the exploitation of economies of scale since the size seems to be optimal in the majority of cases. The elasticities of output, which reveal whether the technology exhibits increasing, constant, or decreasing returns to scale (i.e., whether a proportionate increase in all inputs results in a larger, equal, or less than proportionate increase in the aggregate output, respectively [58]), are close to one in the majority of countries, evaluated on the sample mean. The deflection towards a higher absolute value than one is a characteristic of the Nordic countries (Finland and Sweden).

As far as heterogeneity is concerned, the results do not reveal any significant heterogeneity effect on output elasticities for the majority of countries. In other words, inter-sectoral heterogeneity in cost elasticity is not a common feature of European food processing sectors. When the focus is primarily on the manufacture of dairy products, an exception can be found in Spain, where an absolute value of output elasticity slightly higher than one reveals that dairy processors exhibit moderate economies of scale, evaluated on the country sample mean. Inter-sectoral heterogeneity is not a statistically significant feature of food processing industries, even in the case of cost shares. This fact holds true especially for labour. The heterogeneity parameters of labour are statistically significant (at the 10% level) only in the Spanish and Swedish dairy industries (without a common pattern). More frequent occurrence of significant inter-sectoral heterogeneity is revealed in the case of material cost shares. The dairy processing sector is characterized by higher material cost shares in the majority of countries compared to the rest of the food processing sectors.

The common patterns of the food processing industry in the analysed European countries are also revealed in the term of technical change. The technology exhibits positive technical change in most of the analysed countries, and this is accelerating over time. With a focus on the dairy processing industry and an evaluation on the country sample means, a technological regression is revealed only in Belgium and Finland. This may indicate a low level of innovation (especially process one) in these countries and may imply significant structural weakness [59]. In general, the lack of innovation activities may result in lower competitiveness. On the contrary, a successful innovation process resulting in cost diminution is an important source of competitive advantage. These results have important political implications. Strengthening the innovative activity of companies is unlikely to be possible without investment incentives and continued support for the building of an innovation-friendly business environment. Supporting the innovation activity of food processors is an important source not only of profitability and competitiveness but also of food security, safety, and sustainability.

Table 3 summarises parameter estimates of the stochastic meta-frontier input distance function models in the three alternative specifications for the dairy processing industry in 10 European countries. Considering theoretical consistency, the IDF estimates should satisfy the properties of monotonicity and concavity in inputs. The violation of these assumptions may result in misleading values of computed elasticities and technical efficiency scores. The monotonicity assumption is met if the IDF is non-increasing in output and non-decreasing in inputs. In order to verify this property, it is sufficient that the first-order parameters are negative for the output ($\beta_m \leq 0$) and positive for the inputs ($\beta_j \geq 0$ for $j = L, K$) and $\beta_L + \beta_M < 1$, where L stands for labour and M represents material [60]. Concavity in inputs requires that the Hessian matrix of second-order derivatives of the IDF of the function with respect to the inputs is negatively semidefinite, according to Diewert and Wales [61]. This is fulfilled on the sample mean, if $\beta_{jj} + \beta^2_j - \beta_j \leq 0\ j = L, M$. Table 3 proves that these conditions hold for all model specifications, evaluated on the sample mean.

Table 3. Meta-frontier model estimates. Source: authors' own calculation.

10.5	GTRE			GTRE with Mundlak			GMM		
Variable	Coef.	St.Er.	P > |z|	Coef.	St.Er.	P > |z|	Coef.	St.Er.	P > |t|
ln_y	−0.9601	0.0064	0.0000	−0.8886	0.0207	0.0000	−0.9801	0.0039	0.0000
ln_xL	0.2976	0.0161	0.0000	0.3039	0.0230	0.0000	0.2620	0.0088	0.0000
ln_xM	0.6483	0.0146	0.0000	0.6375	0.0222	0.0000	0.6864	0.0084	0.0000
t	−0.0074	0.0008	0.0000	−0.0090	0.0009	0.0000	−0.0069	0.0010	0.0000
ln_y_2	−0.0032	0.0057	0.5740	−0.0007	0.0053	0.8890	0.0045	0.0052	0.3920
ln_xL_2	0.0538	0.0072	0.0000	0.0582	0.0077	0.0000	0.0656	0.0127	0.0000
ln_xM_2	0.1322	0.0098	0.0000	0.1326	0.0090	0.0000	0.1472	0.0080	0.0000
ln_xLxM	−0.0847	0.0079	0.0000	−0.0863	0.0077	0.0000	−0.0949	0.0103	0.0000
t_2	0.0005	0.0004	0.1400	0.0000	0.0003	0.9240	0.0009	0.0005	0.0780
ln_yt	0.0005	0.0005	0.2800	0.0008	0.0005	0.0950	0.0004	0.0010	0.6780
ln_xLt	−0.0001	0.0011	0.9380	0.0003	0.0011	0.7700	−0.0014	0.0030	0.6390
ln_xMt	−0.0001	0.0012	0.9650	−0.0004	0.0012	0.7760	0.0033	0.0026	0.1920
ln_yxL	−0.0055	0.0073	0.4570	−0.0016	0.0069	0.8200	0.0008	0.0079	0.9180
ln_yxM	−0.0080	0.0059	0.1760	−0.0101	0.0057	0.0750	−0.0246	0.0069	0.0000
_cons	−0.0714	0.0136	0.0000	−0.0772	0.0135	0.0000	−0.0915	0.0095	0.0000
ln_y_gmean				−0.0867	0.0200	0.0000			
ln_xL_gmean				−0.0049	0.0201	0.8080			
ln_xM_gmean				−0.0013	0.0232	0.9550			
t_gmean				0.0007	0.0066	0.9110			
	Mean	Std.D.		Mean	Std.D.		Mean	Std.D.	
Overall TE	0.9553	0.0139		0.9624	0.0099		0.9200	0.0179	
Transient TE	0.9561	0.0139		0.9629	0.0099		0.9500	0.0177	
Persistent TE	0.9992	0.0000		0.9994	0.0000		0.9684	0.0044	

Note: GMM model–AR (2) = −0.06 and Hanset test = 347.25 (No. of instruments–226).

The first-order parameters of the IDFs are highly significant even at the 1% significance level and have a similar pattern, indicating the high material intensity of the dairy industry. The share of material in the total input is about 60%, the share of labour is about 30%, and the output elasticity varies between (−0.98) and (−0.89). The time parameter (δ_t) has a low negative value representing the cost decrease with a decelerating rate ($\delta_{tt} > 0$) over the analysed period. However, the second-order time parameter (δ_{tt}) is statistically significant only at the 10% significance level in the GMM model. The parameters of biased technical change are not statistically significant even at the 10% significance level. The rest of the second-order parameters have a similar pattern as well.

The results reveal high overall technical efficiency. Evaluated at the sample averages, the overall technical efficiency is between 92% and 96%, that is, dairy processors can reduce their costs from 4% up to 8%, subject to the model specification, to produce the same volumes of outputs. As far as the standard deviations are concerned, we may observe that the dairy producers operate very close to the estimated mean value of technical efficiency. Figure 2 shows that the models provide a similar overall technical efficiency distribution.

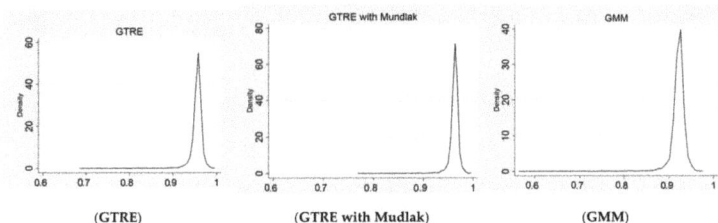

Figure 2. Kernel density comparison (overall technical efficiency). Source: authors' own calculation.

Persistent technical inefficiency is estimated only in the GMM model and is quite low. That is, with similar estimates of transient technical efficiency, we do not see significant differences in overall technical efficiency among the model specifications. These results suggest that the efficiency estimates seem to be robust.

Table 4 provides the statistical characteristics of the country-specific overall technical efficiency and its decomposition into transient and persistent parts. The overall technical efficiency estimates indicate that, evaluated at the country sample means, dairy processing companies can reduce their costs from 5% up to 24% while producing the same volumes of output.

Table 4. Country-specific overall technical efficiency and its decomposition (GMM). Source: authors' own calculation.

Country	Overall Technical Efficiency				Transient Technical Efficiency				Persistent Technical Efficiency			
	Mean	Std.Dev.	Min.	Max.	Mean	Std.Dev.	Min.	Max.	Mean	Std.Dev.	Min.	Max.
Austria	0.7548	0.0541	0.5414	0.8577	0.8319	0.0438	0.6454	0.9153	0.9066	0.0300	0.8234	0.9371
Belgium	0.9297	0.0158	0.794	0.9684	0.9309	0.0159	0.7949	0.9695	0.9988	0.0000	0.9988	0.9988
Czechia	0.9121	0.0185	0.723	0.9686	0.9123	0.0185	0.7232	0.9688	0.9997	0.0000	0.9997	0.9997
Finland	0.9028	0.0219	0.784	0.9635	0.903	0.0219	0.7842	0.9637	0.9997	0.0000	0.9997	0.9997
France	0.9320	0.0243	0.3968	0.9947	0.9328	0.0243	0.3971	0.9954	0.9992	0.0000	0.9992	0.9992
Germany	0.9303	0.0271	0.4859	0.9846	0.9307	0.0271	0.4862	0.9851	0.9995	0.0000	0.9995	0.9995
Italy	0.8393	0.0345	0.5913	0.9398	0.9227	0.0279	0.6312	0.9783	0.9095	0.0240	0.8168	0.9620
Spain	0.9267	0.0249	0.6573	0.9806	0.9269	0.0249	0.6574	0.9808	0.9998	0.0000	0.9998	0.9998
Sweden	0.9466	0.0192	0.8194	0.9892	0.9483	0.0192	0.8209	0.991	0.9982	0.0000	0.9981	0.9982
United Kingdom	0.9427	0.0266	0.7332	0.9847	0.9434	0.0266	0.7337	0.9854	0.9993	0.0000	0.9993	0.9993

As far as intra-sectoral differences are concerned, the overall technical efficiency is relatively dense around the mean and the distribution is skewed to higher values. In the majority of the analysed countries, the interquartile ranges are quite narrow, see Figure 3. The outliers in boxplots indicate high polarization between the most technically efficient producer and the least technically efficient one. The highest polarization is evident in Austria. On the other hand, the lowest polarization can be seen in Czechia and Finland. As the overall technical efficiency is associated with cost savings, it can be assumed that Austria, with the highest polarization, will face structural changes in the future. That is, we can expect changes in business activities or even the cessation of production for firms with the lowest technical efficiency and resource reallocation to more successful producers.

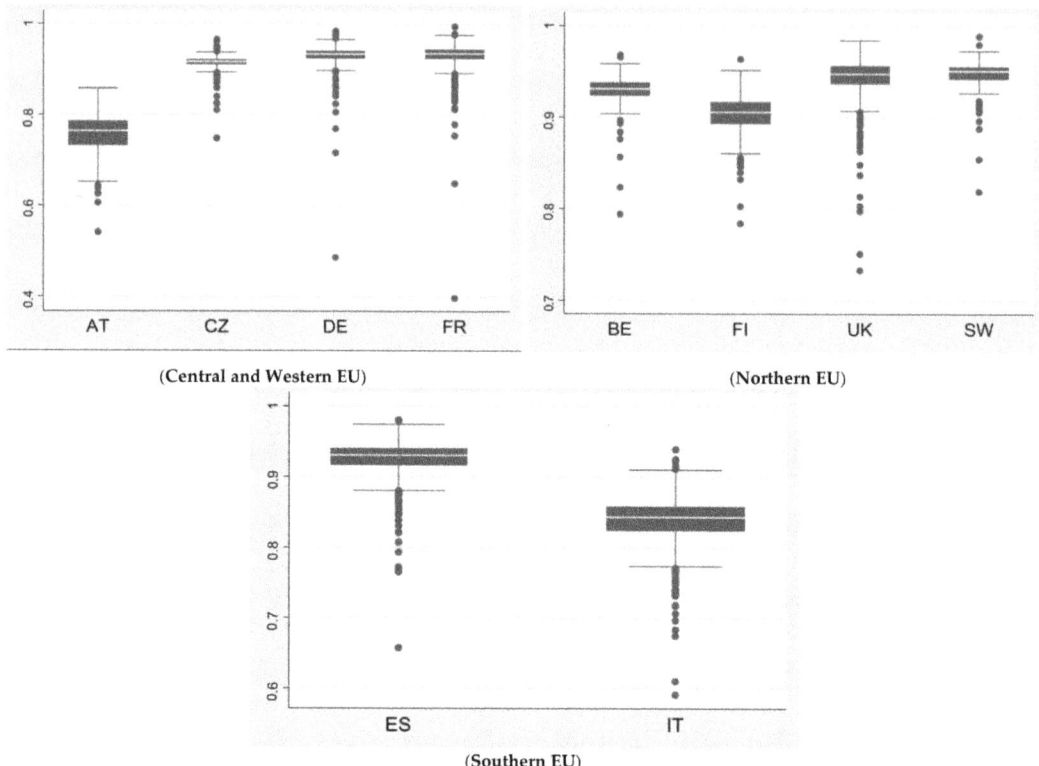

Figure 3. Country-specific overall technical efficiency (GMM). Source: authors' own calculation. Note: AT denotes Austria, BE denotes Belgium, CZ denotes Czechia, DE denotes Germany, ES denotes Spain, FI denotes Finland, FR denotes France, IT denotes Italy, SW denotes Sweden, UK denotes the United Kingdom.

The overall technical efficiency consists of the transient and persistent technical efficiency. The persistent part of overall technical efficiency indicates systematic failures or structural problems and unsuitable factor allocations, which are difficult to change over time, as well as non-competitive market conditions. However, the persistent technical efficiency is statistically significant only in the cases of Austria and Italy. Table 4 shows that the average persistent technical efficiency scores are about 91% in these two countries. A slightly higher variability for persistent technical efficiency is revealed in Austria. However, the Italian dairy sector is characterized by the highest polarization between the most efficient producer and the least efficient. This suggests that many firms systematically fail to catch up to best practices and thus show considerable resource inefficiency as compared to firms operating on the technological frontier.

Table 4 also provides country-specific transient technical efficiency scores. The intra-sectoral differences of transient technical efficiency are pronounced especially in Austria, Italy, and Germany. However, the highest polarization is revealed in France, followed by Germany, where the developments of transient technical efficiency over time have a similar pattern and the lowest values are connected with the 2009 crisis; see Figure 4. The fluctuations of technical efficiency in the short-term may also be the result of shocks associated with the introduction of new technologies or changes in human capital. Considering the development presented in Figure 4, we can conclude that the most important feature is an increasing trend in transient technical efficiency in the analysed period.

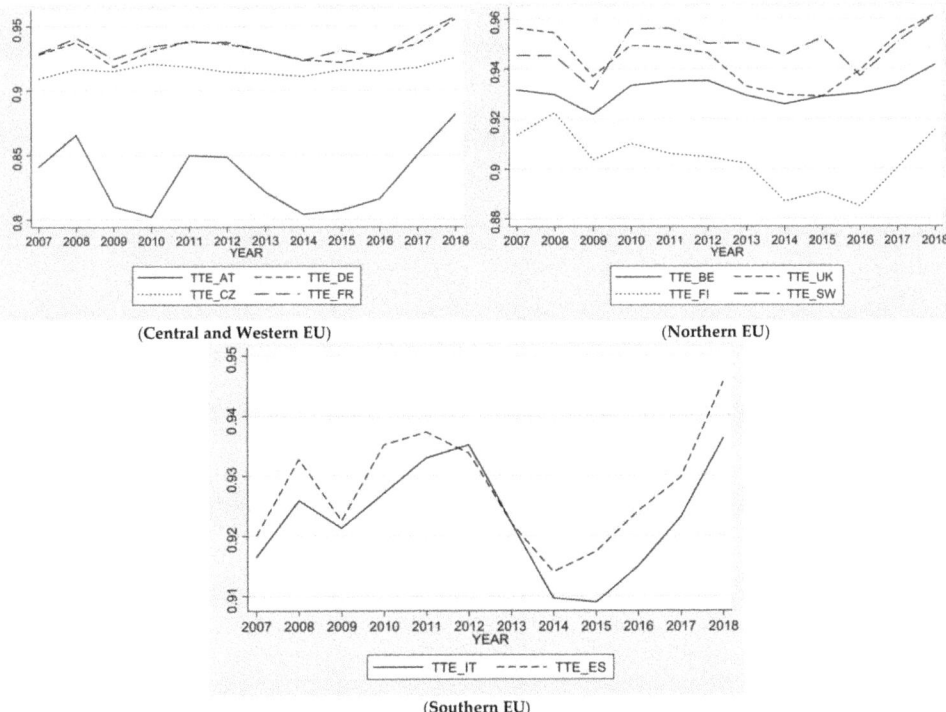

Figure 4. Country-specific transient technical efficiency development (GMM). Source: authors' own calculation. Note: AT denotes Austria, BE denotes Belgium, CZ denotes Czechia, DE denotes Germany, ES denotes Spain, FI denotes Finland, FR denotes France, IT denotes Italy, SW denotes Sweden, UK denotes the United Kingdom.

Table 5 presents meta-frontier technical efficiency estimates and indicates that the overall technical efficiency is high in all countries and there are no considerable differences among them. The lowest value of overall technical efficiency can be observed in Sweden, evaluated on mean values. Moreover, the Swedish dairy industry can be characterized by the highest variability measured by standard deviation. On the contrary, the highest average value of overall technical efficiency is found in France and Finland.

Table 5. Meta-frontier overall technical efficiency and its decomposition (GMM). Source: authors' own calculation.

Country	Overall Technical Efficiency				Transient Technical Efficiency				Persistent Technical Efficiency			
	Mean	Std.Dev.	Min.	Max.	Mean	Std.Dev.	Min.	Max.	Mean	Std.Dev.	Min.	Max.
Austria	0.9202	0.0092	0.8608	0.9392	0.9511	0.0093	0.8888	0.9682	0.9675	0.0017	0.9631	0.9701
Belgium	0.9181	0.0200	0.7490	0.9537	0.9495	0.0155	0.8379	0.9873	0.9669	0.0101	0.8696	0.9750
Czechia	0.9191	0.0217	0.6862	0.9639	0.9492	0.0220	0.7069	0.9894	0.9682	0.0033	0.9586	0.9832
Finland	0.9212	0.0183	0.8179	0.9741	0.9498	0.0181	0.8326	0.9871	0.9699	0.0051	0.9562	0.9868
France	0.9212	0.0152	0.5673	0.9715	0.9506	0.0153	0.5903	0.9963	0.9690	0.0034	0.9611	0.9844
Germany	0.9185	0.0254	0.6251	0.9673	0.9494	0.0249	0.6484	0.9893	0.9674	0.0059	0.9318	0.9831
Italy	0.9206	0.0139	0.7570	0.9576	0.9505	0.0141	0.7785	0.9832	0.9686	0.0027	0.9627	0.9757
Spain	0.9197	0.0145	0.7566	0.9578	0.9503	0.0147	0.7784	0.9820	0.9678	0.0030	0.9591	0.9754
Sweden	0.9171	0.0314	0.7384	0.9732	0.9454	0.0317	0.7705	0.9945	0.9700	0.0062	0.9520	0.9785
United Kingdom	0.9190	0.0195	0.7105	0.9613	0.9495	0.0191	0.7491	0.9825	0.9679	0.0044	0.9484	0.9806

Figure 5 illustrates the competitive position of selected countries in terms of a country-average technical efficiency comparison with the average value of the whole set. As can be seen, the Swedish, Belgian, and German dairy processing industries are failing to catch up with the best-practice technology. While in Belgium and Germany the dairy processing industry lags behind in both transient (TTE) and persistent (PTE) technical efficiency, in Sweden the transient technical inefficiency poses a greater problem for the dairy processing industry than the persistent component.

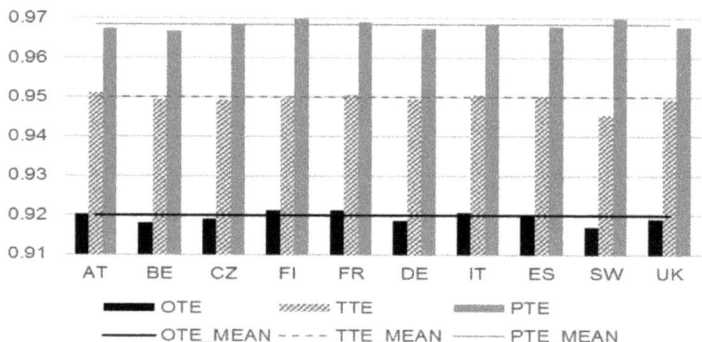

Figure 5. Meta-frontier technical efficiency comparison (GMM). Source: authors' own calculation. Note: AT denotes Austria, BE denotes Belgium, CZ denotes Czechia, DE denotes Germany, ES denotes Spain, FI denotes Finland, FR denotes France, IT denotes Italy, SW denotes Sweden, UK denotes the United Kingdom.

The loss of resources due to structural problems and permanent managerial failures in the production process is pronounced in all countries and reaches a similar level. However, it holds in all cases that the persistent inefficiencies are not large, evaluated on the sample means. On the other hand, the outliers in boxplots (Figure 6) indicate high polarization between the most long-term technically efficient producer and the least technically efficient one in Belgium, Germany, and the United Kingdom. In other words, it reveals the existence of firms that are systematically failing to catch up to best practices and thus show considerable resource inefficiency. Different paths can be followed to eliminate this waste of resources and promote the sustainability of the dairy processing industry. According to Pisulewski and Marzec [35], management should focus on changes in the organization of the production process and managerial competency. According to Dimara et al. [15], policy makers should provide access to biotech innovations and food supply networks.

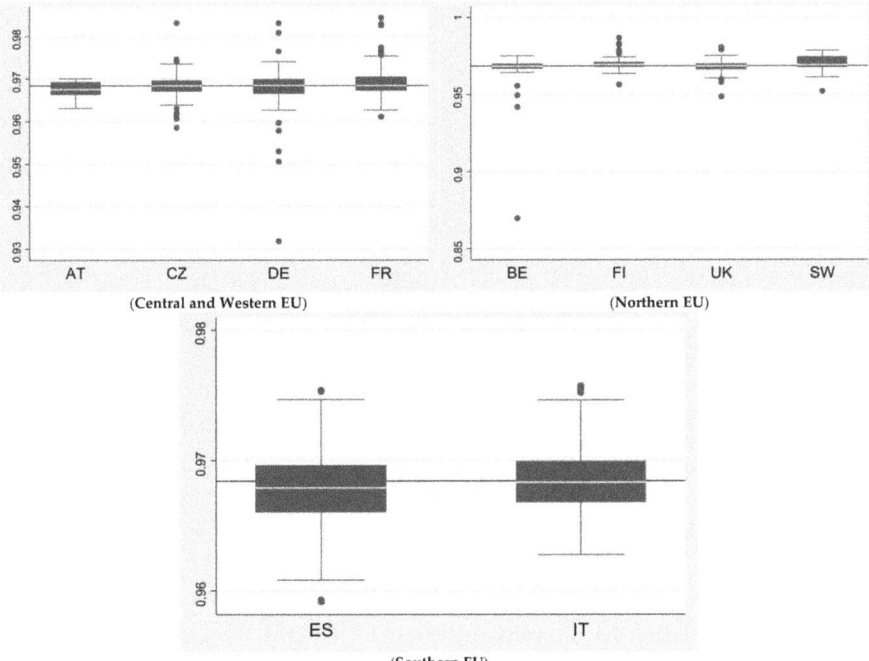

Figure 6. Meta-frontier persistent technical efficiency comparison (GMM). Source: authors' own calculation. Note: Reference line represents the sample average transient technical efficiency score; AT denotes Austria, BE denotes Belgium, CZ denotes Czechia, DE denotes Germany, ES denotes Spain, FI denotes Finland, FR denotes France, IT denotes Italy, SW denotes Sweden, UK denotes the United Kingdom.

5. Discussion

The country as well as meta-frontier technical efficiency estimates revealed high overall technical efficiency for all countries. The only exception is the country estimate for Austria. These results are consistent with the literature devoted to technical efficiency analysis in the food processing industry. For example, Čechura et al. [6] found high technical efficiency in the food processing industry in the analysis of entire EU food processing sectors. In addition, their meta-frontier analysis provides similar results as well. The authors show that the differences in average technical efficiency are not large for all EU member countries, including Serbia, even when there are huge differences between the best and worst food processors. High technical efficiency of the top 10% of food processors is a common feature for all countries in all analysed sectors. Similar findings were provided for Czechia by Čechura and Hockmann [8] and Špička [11]. On the other hand, lower technical efficiency scores were estimated for the Czech food processing industry by Náglová and Šimpachová Petrová [7]. The differences are due to the use of both a different data set and a different model specification. The authors analysed the manufacturing of food products and beverages. Moreover, they did not consider heterogeneity among sectors, which may result in higher inefficiency scores. The treatment of the unobserved heterogeneity component in the model specification is a potential source of the slightly lower technical efficiency scores obtained by Soboh et al. [14] as well.

Our technical efficiency complements the research of technical efficiency in the food processing industry by the decomposition of overall technical efficiency into transient and persistent components. Transient technical efficiency dominates in the technical efficiency country estimates. That is, the overall technical efficiency estimated for each country is largely due to the transient technical efficiency. Persistent technical efficiency is pronounced only in Austria and Italy. These findings suggest that in the majority of countries, the inefficiencies in input use do not have a systematic character. This holds true for the food processing sector as a whole, since the country estimates are based on a joint IDF function for the four processing sectors. That is, the estimated inefficiencies have non-systematic sources such as non-systematic managerial failures, market shocks, shocks associated with new production technologies, and human capital. However, the meta-frontier estimates that compare the dairy frontier technologies among the countries show a certain degree of systematic failure. In particular, the average persistent technical efficiency is high and similar for all countries. The same holds true for transient technical efficiency. Thus, the meta-frontier analysis shows high overall technical efficiency in the dairy processing sectors in all analysed countries. These findings are fully in line with the meta-frontier analysis provided by Čechura et al. [6].

6. Conclusions

The study aimed to (i) assess whether there is indication that the countries follows a sustainable development path characterized by reduced waste of resources due to inefficient input use and (ii) whether we can observe systematic failures in the efficiency of input use or whether the deviations from the frontier technologies are due to the transient reasons. The results revealed that both components of overall technical efficiency are high. That is, we cannot observe considerable systematic failures in efficiency of input used as well as inefficiencies due to the transient reason, evaluated on sample means. This suggests that the European dairy processing industry as a whole seems to be highly competitive and the companies highly efficient. On the other hand, the figures show that there are companies that are falling behind and may leave the market.

The findings of this study seem to be of high relevance for policy makers since there is still limited knowledge about the inefficiency of input use in the processing industry and their sources and thus the potential contribution the sustainability improvements in the whole food value chain.

There are some limitations of the conducted research. This is especially related to the fact that the Amadeus data allows to work only with aggregate output and thus the diversification or economies of scope cannot be considered. On the other hand, the advantages are the employment of new advances in productivity and efficiency analysis and thus the robust estimate of technical efficiency in European dairy industry.

Future research should focus on investigation of the socioeconomic and environmental factors that can explain differences in persistent as well as transient inefficiency between dairy processing companies. Understanding these factors has important political implications especially in food processing industry that face the challenge of ensuring a sufficient food supply from limited resources for a growing population. Industrial policy agents can use the knowledge of these factors to make a business environment that improves sustainability and competitiveness of food processing industry.

Author Contributions: Conceptualization, L.Č. and Z.Ž.K.; methodology, L.Č. and Z.Ž.K.; validation, L.Č.; formal analysis, L.Č. and Z.Ž.K.; investigation, L.Č. and Z.Ž.K.; resources, L.Č. and Z.Ž.K.; writing, L.Č. and Z.Ž.K.; visualization, Z.Ž.K.; supervision, L.Č.; project administration, L.Č.; funding acquisition, L.Č. Both authors have read and agreed to the published version of the manuscript.

Funding: The VALUMICS project "Understanding Food Value Chain and Network Dynamics" received funding from the European Union's Horizon 2020 research and innovation programme, under grant agreement No. 727243. https://valumics.eu/.

Institutional Review Board Statement: Not applicable.

Informed Consent Statement: Not applicable.

Data Availability Statement: Data was obtained from the CULS and are not available from the authors. The data can be bought from Bureau van Dijk—Moody's Analytics company.

Acknowledgments: The results are the part of the solution of the VALUMICS project "Understanding Food Value Chain and Network Dynamics". The project received funding from the European Union's Horizon 2020 research and innovation programme, under grant agreement No. 727243. https://valumics.eu/.

Conflicts of Interest: The authors declare no conflict of interest. The funders had no role in the design of the study; in the collection, analyses, or interpretation of data; in the writing of the manuscript; or in the decision to publish the results.

References

1. Eurostat. Annual Detailed Enterprise Statistics for Industry (NACE Rev. 2, B-E) [sbs_na_ind_r2]. Last Update: 11 October 2020. Available online: https://appsso.eurostat.ec.europa.eu/nui/show.do?dataset=sbs_na_ind_r2&lang=en (accessed on 15 November 2020).
2. European Commission. *The Competitive Position of the European Food and Drink Industry*; Publications Office of the European Union: Luxembourg, 2016.
3. Kapelko, M. Measuring productivity change accounting for adjustment costs: Evidence from the food industry in the European Union. *Ann. Oper. Res.* **2019**, *278*, 215–234. [CrossRef]
4. Nakat, Z.; Bou-Mitri, C. COVID-19 and the food industry: Readiness assessment. *Food Control* **2021**, *121*, 107661. [CrossRef] [PubMed]
5. Allendorf, J.; Hirsch, S. Dynamic productivity growth in the European food processing industry. In Proceedings of the 55th Annual Conference of German Association of Agricultural Economists (GEWISOLA), Giessen, Germany, 23–25 September 2015.
6. Čechura, L.; Hockmann, H.; Kroupová, Z. *Productivity and Efficiency of European Food Processing Industry*; COMPETE Working Paper N7; IAMO: Halle, Germany, 2014.
7. Náglová, Z.; Šimpachová Pechrová, M. Subsidies and technical efficiency of Czech food processing industry. *Agric. Econ. Czech* **2019**, *65*, 151–159. [CrossRef]
8. Čechura, L.; Hockmann, H. Heterogeneity in Production Structures and Efficiency: An Analysis of the Czech Food Processing Industry. *Pac. Econ. Rev.* **2017**, *22*, 702–719. [CrossRef]
9. Rudinskaya, T. Heterogeneity and efficiency of food processing companies in the Czech Republic. *Agric. Econ. Czech* **2017**, *63*, 411–420. [CrossRef]
10. Rezitis, A.N.; Kalantzi, M.A. Investigating Technical Efficiency and Its Determinants by Data Envelopment Analysis: An Application in the Greek Food and Beverages Manufacturing Industry. *Agribus. Int. J.* **2015**, 37. [CrossRef]
11. Špička, J. The efficiency improvement of central European corporate milk processors in 2008–2013. *AGRIS Line Pap. Econ. Inform.* **2015**, *7*, 175–188. [CrossRef]
12. Popović, R.; Panić, D. Technical Efficiency of Serbian Dairy Processing Industry. *Econ. Agric.* **2018**, *65*, 569–581. [CrossRef]
13. Setiawan, M.; Emvalomatis, G.; Oude Lansink, A. The relationship between technical efficiency and industrial concentration: Evidence from the Indonesian food and beverages industry. *J. Asian Econ.* **2012**, *23*, 466–475. [CrossRef]
14. Soboh, R.A.M.E.; Oude Lansink, A.; Van Dijk, G. Efficiency of European Dairy Processing Firm. *NJAS Wagening. J. Life Sci.* **2014**, *70–71*, 53–59. [CrossRef]
15. Dimara, E.; Skuras, D.; Tsekouras, K.; Tzelepid, D. Productive efficiency and firm exit in the food sector. *Food Policy* **2008**, *33*, 185–196. [CrossRef]
16. Baran, J. Efficiency of the production scale of Polish dairy companies based on data envelopment analysis. *Oeconomia* **2013**, *12*, 5–13.
17. Kapelko, M.; Oude Lansink, A.; Stefanou, S.H. The impact of the 2008 financial crisis on dynamic productivity growth of the Spanish food manufacturing industry. An impulse response analysis. *Agric. Econ.* **2017**, *48*, 561–571. [CrossRef]
18. Kapelko, M.; Oude Lansink, A.; Stefanou, S.H. Effect of Food Regulation on the Spanish Food Processing Industry: A Dynamic Productivity Analysis. *PLoS ONE* **2015**, *10*, e0128217. [CrossRef]
19. Rudinskaya, T.; Kuzmenko, E. Investments, Technical Change and Efficiency: Empirical Evidence from Czech Food Processing. *AGRIS Line Pap. Econ. Inform.* **2019**, *11*, 93–103. [CrossRef]
20. Čechura, L. Technological change in the Czech food processing industry: What did we experience in the last decade? In Proceedings of the 131st EAAE Seminar, 'Innovation for Agricultural Competitiveness and Sustainability of Rural Areas', Prague, Czech Republic, 18–19 September 2012.
21. Kapelko, M.; Oude Lansink, A. Dynamic multi-directional inefficiency analysis of European dairy manufacturing firms. *Eur. J. Oper. Res.* **2017**, *257*, 338–344. [CrossRef]
22. Kapelko, M.; Oude Lansink, A. Technical Efficiency of the Spanish Dairy Processing Industry: Do Size and Exporting Matter? In *Efficiency Measures in the Agricultural Sector*; Mendes, A., Soares da Silva, E.L.D.G., Azevedo Santos, J., Eds.; Springer: Dordrecht, The Netherlands, 2013; pp. 93–106. [CrossRef]
23. Koopmans, T.C. Analysis of Production as an Efficient Combination of Activities. In *Activity Analysis of Production and Allocation (Proceedings of Conference)*; Koopmans, T.C., Ed.; John Wiley & Sons, Inc.: New York, NY, USA, 1951; pp. 33–97.
24. Debreu, G. The coefficient of resource utilization. *Econometrica* **1951**, *19*, 273–292. [CrossRef]
25. Farrel, M.J. The Measurement of Productive Efficiency. *J. R. Stat. Soc.* **1957**, *120*, 253–290. [CrossRef]

26. Fried, H.O.; Knox Lovell, C.A.; Schmidt, S.S. *The Measurement of Productive Efficiency and Productivity Growth*; Oxford University Press: New York, NY, USA, 2008. [CrossRef]
27. Coelli, T.J.; Rao, D.S.P.; O'Donnell, C.J.; Battese, G.E. *An Introduction to Efficiency and Productivity Analysis*, 2nd ed.; Springer: New York, NY, USA, 2005. [CrossRef]
28. Kumbhakar, S.C.; Wang, H.-J.; Horncastle, A.P. *A Practitioner's Guide to Stochastic Frontier Analysis Using Stata*; Cambridge University Press: New York, NY, USA, 2015. [CrossRef]
29. Meeusen, W.; van den Broeck, J. Efficiency Estimation from Cobb-Douglas Production Functions with Composed Error. *Int. Econ. Rev.* **1977**, *18*, 435–444. [CrossRef]
30. Aigner, D.; Lovell, K.C.A.; Schmidt, P. Formulation and estimation of stochastic frontier production function model. *J. Econom.* **1977**, *6*, 21–37. [CrossRef]
31. Alem, H. Effects of model specification, short-run, and long-run inefficiency: An empirical analysis of stochastic frontier models. *Agric. Econ. Czech* **2018**, *64*, 508–516. [CrossRef]
32. Bartolucci, F.; Belotti, F.; Peracchi, F. Testing for time-invariant unobserved heterogeneity in generalized linear models for panel data. *J. Econom.* **2015**, *184*, 111–123. [CrossRef]
33. Tsionas, M.; Kumbhakar, S. Firm-Heterogeneity, Persistent and Transient Technical Inefficiency: A generalized true random-effects model. *J. Appl. Econom.* **2014**, *29*, 110–132. [CrossRef]
34. Badunenko, O.; Kumbhakar, S.C. When, where and how to estimate persistent and transient efficiency in stochastic frontier panel data models. *Eur. J. Oper. Res.* **2016**, *255*, 272–287. [CrossRef]
35. Pisulewski, A.; Marzec, J. Heterogeneity, transient and persistent technical efficiency of Polish crop farms. *Span. J. Agric. Res.* **2019**, *17*, e0106. [CrossRef]
36. Njuki, E.; Bravo-Ureta, B.E. The economic cost of environmental regulation in U.S. dairy farming: A directional distance function approach. *Am. J. Agric. Econ.* **2015**, *97*, 1087–1106. [CrossRef]
37. Kumbhakar, S.C.; Lien, G. Yardstick Regulation of Electricity Distribution—Disentangling Short-run and Long-run Inefficiencies. *Energy J.* **2017**, *38*, 17–37. [CrossRef]
38. Filippini, M.; Greene, W. Persistent and transient productive inefficiency: A maximum simulated likelihood approach. *J. Product. Anal.* **2016**, *45*, 187–196. [CrossRef]
39. Rashidghalam, M.; Heshmati, A.; Dashti, G.; Pishbahar, E. *A Comparison of Panel Data Models in Estimating Technical Efficiency*; IZA Discussion Paper No. 9807; IZA: Bonn, Germany, 2016.
40. Caves, D.W.; Christensen, L.R.; Diewert, W.E. The Economic Theory of Index Numbers and the Measurement of Input, Output, and Productivity. *Econometrica* **1982**, *50*, 1393–1414. [CrossRef]
41. Karagiannis, G.; Midmore, P.; Tzouvelekas, V. Parametric Decomposition of Output Growth Using a Stochastic Input Distance Function. *Am. J. Agric. Econ.* **2004**, *86*, 1044–1057. [CrossRef]
42. Greene, W. Reconsidering heterogeneity in panel data estimators of the stochastic frontier model. *J. Econom.* **2005**, *126*, 269–303. [CrossRef]
43. Irz, X.; Thirtle, C. Dual Technological Development in Botswana Agriculture: A Stochastic Input Distance Function Approach. *J. Agric. Econ.* **2005**, *55*, 455–478. [CrossRef]
44. Singbo, A.; Larue, B. Scale Economies, Technical Efficiency, and the Sources of Total Factor Productivity Growth of Quebec Dairy Farms. *Can. J. Agric. Econ.* **2016**, *64*, 339–363. [CrossRef]
45. Čechura, L.; Grau, A.; Hockmann, H.; Levkovych, I.; Kroupová, Z. Catching up or falling behind in Eastern European agriculture—The case of milk production. *J. Agric. Econ.* **2017**, *68*, 206–227. [CrossRef]
46. Sipiläinen, T. Sources of productivity growth on Finnish dairy farms—Application of input distance function. *Acta Agric. Scand. Sect. C* **2007**, *4*, 65–67. [CrossRef]
47. Lovell, K.C.A.; Richardson, S.; Travers, P.; Wood, L.L. Resources and Functionings: A New View of Inequality in Austria. In *Models and Measurement of Welfare and Inequality*; Eichhorn, W., Ed.; Springer: New York, NY, USA, 1994; pp. 787–807.
48. Ullah, S.; Akhar, P.; Yaeferian, G. Dealing with endogeneity bias: The generalized method of moments (GMM). *Ind. Mark. Manag.* **2018**, *71*, 69–78. [CrossRef]
49. Mundlak, Y. On the Pooling of Time Series and Cross Section Data. *Econometrica* **1978**, *46*, 69–85. [CrossRef]
50. Kumbhakar, S.C.; Lien, G.; Hardaker, J.B. Technical efficiency in competing panel data models: A study of Norwegian grain farming. *J. Product. Anal.* **2014**, *41*, 321–337. [CrossRef]
51. Bokusheva, R.; Čechura, L. *Evaluating Dynamics, Sources and Drivers of Productivity Growth at the Farm Level*; OECD Food, Agriculture and Fisheries Papers, No. 106; OECD Publishing: Paris, France, 2017.
52. Arellano, M.; Bover, O. Another look at the instrumental variable estimation of error-components models. *J. Econom.* **1995**, *68*, 29–51. [CrossRef]
53. Blundell, R.; Bond, S. Initial conditions and moment restrictions in dynamic panel data models. *J. Econom.* **1998**, *87*, 115–143. [CrossRef]
54. Roodman, D. How to do xtabond2: An introduction to difference and system GMM in Stata. *Stata J.* **2009**, *9*, 86–136. [CrossRef]
55. Hansen, L.P. Large Sample Properties of Generalized Method of Moments Estimator. *Econometrica* **1982**, *50*, 1029–1054. [CrossRef]
56. Arellano, M.; Bond, S. Some tests of specification for panel data: Monte Carlo evidence and an application to employment equations. *Rev. Econ. Stud.* **1991**, *58*, 277–297. [CrossRef]

57. Čechura, L.; Žáková Kroupová, Z.; Rumánková, L.; Jaghdani, T.J.; Samoggia, A.; Thakur, M. *Assessment of Economics of Scale and Technical Change along the Food Chain*; The VALUMICS project funded by EU Horizon 2020 G.A. No 727243 Deliverable, D5.6; Czech University of Life Sciences: Prague, Czech Republic, 2020.
58. Baumol, W.J.; Panzar, J.C.; Willig, R.D. *Contestable Markets and the Theory of Industry Structure*; Harcourt Brace Jovanovich: New York, NY, USA, 1982.
59. Török, Á.; Tóth, J.; Balogh, J.M. Push or Pull? The nature of innovation process in the Hungarian food SMEs. *J. Innov. Knowl.* **2019**, *4*, 234–239. [CrossRef]
60. Kumbhakar, S.C.; Lien, G.; Flaten, O.; Tveterås, R. Impacts of Norwegian Milk Quotas on Output Growth: A Modified Distance Function Approach. *J. Agric. Econ.* **2008**, *59*, 350–369. [CrossRef]
61. Diewert, W.E.; Wales, T.J. Flexible Functional Forms and Global Curvature Conditions. *Econometrica* **1987**, *55*, 43–68. [CrossRef]

Article

Environmental Impacts of Milking Cows in Latvia

Janis Brizga [1,*], Sirpa Kurppa [2] and Hannele Heusala [2]

1 Department of Environmental Governance, University of Latvia, LV-1586 Riga, Latvia
2 LUKE, Finnish Natural Resource Institute, FI-00790 Helsinki, Finland; sirpa.kurppa@outlook.com (S.K.); hannele.heusala@luke.fi (H.H.)
* Correspondence: janis.brizga@lu.lv

Abstract: Increasing pressures surrounding efficiency and sustainability are key global drivers in dairy farm management strategies. However, for numerous resource-based, social, and economic reasons sustainable intensification strategies are herd-size dependent. In this study, we investigated the environmental impacts of Latvia's dairy farms with different management practices. The herd size-dependent management groups varied from extensively managed small herds with 1–9 cows, extending to stepwise more intensively managed herds with 10–50, 51–100, 100–200, and over 200 milking cows. The aim is to compare the environmental impacts of different size-based production strategies on Latvia's dairy farms. The results show that the gross greenhouse gas emissions differ by 29%: from 1.09 kg CO_2 equivalents (CO_{2e}) per kg of raw milk for the farms with 51–100 cows, down to 0.84 kg CO_{2e}/kg milk for farms with more than 200 cows. However, the land use differs even more—the largest farms use 2.25 times less land per kg of milk than the smallest farms. Global warming potential, marine eutrophication, terrestrial acidification, and ecotoxicity were highest for the mid-sized farms. If current domestic, farm-based protein feeds were to be substituted with imported soy feed (one of the most popular high-protein feeds) the environmental impacts of Latvian dairy production would significantly increase, e.g., land use would increase by 18% and the global warming potential by 43%. Environmental policy approaches for steering the farms should consider the overall effects of operation size on environmental quality, in order to support the best practices for each farm type and steer systematic change in the country. The limitations of this study are linked to national data availability (e.g., national data on feed production, heifer breeding, differences among farms regards soil type, manure management, the proximity to marine or aquatic habitats) and methodological shortcomings (e.g., excluding emissions of carbon sequestration, the use of proxy allocation, and excluding social and biodiversity impacts in life-cycle assessment). Further research is needed to improve the data quality, the allocation method, and provide farm-size-specific information on outputs, heifer breeding, manure storage, and handling.

Keywords: LCA; dairy; environmental impacts; intensification

Citation: Brizga, J.; Kurppa, S.; Heusala, H. Environmental Impacts of Milking Cows in Latvia. *Sustainability* **2021**, *13*, 784. https://doi.org/10.3390/su13020784

Received: 26 October 2020
Accepted: 11 January 2021
Published: 15 January 2021

Publisher's Note: MDPI stays neutral with regard to jurisdictional claims in published maps and institutional affiliations.

Copyright: © 2021 by the authors. Licensee MDPI, Basel, Switzerland. This article is an open access article distributed under the terms and conditions of the Creative Commons Attribution (CC BY) license (https://creativecommons.org/licenses/by/4.0/).

1. Introduction

Dairy production systems are important and complex sources of environmental impacts including global warming, nutrient losses, water and land use [1]. The sustainable intensification of animal production is generally promoted as a way to mitigate these environmental impacts while increasing productivity [2,3]. Nevertheless, the dairy sector has shown a global trend toward intensification [4].

During the last decades following the global trend, dairy cow management systems in Latvia are intensifying rapidly [5–7], i.e., the role of grazing has decreased, and indoor feeding and high protein diets have increased. This intensification process has been in parallel with the concentration of the dairy industry, where small farms leave the market whereas the most competitive farms continue growing and increasing their herds [7]. Nevertheless, the dairy systems in Latvia are still very diverse and are largely determined by the number of cows in a herd. Eighty-four percent of the dairy farms are small—less than 10 animals per farm—producing 18% of the country's raw milk. Most of them are self-sufficient farms that do not deliver dairy products to the market. A similar situation exists in many Eastern European countries as well as in developing countries [8–10].

Data obtained from a survey in Latvia shows significant differences in milk yield depending on the size of the herd (see Table 1). Milk yield depends on diet and the cows' ability to consume dry matter (DM) [11]. While consumption of DM is affected by the quality and digestibility of feed and the optimum ratio of roughage and feed concentrate in the feed ration [12], diets, on the other hand, are linked to the size of the herd affecting the milk yield. The average milk yield of the farms that graze their dairy cows is lower than that of the farms who do not graze their dairy cows [13].

Table 1. Description of the technical management routes.

Factors	Dairy Farm (1–9)	Dairy Farm (10–50)	Dairy Farm (51–100)	Dairy Farm (101–200)	Dairy Farm (>200)	Total/ Average	Sources
Number of farms	14,463	2295	306	131	91	17,286	CSB [6]
Number of cows	33,717	45,514	20,758	17,605	36,430	154,024	CSB [6]
The average number of cows on the farm	2	20	68	134	400		Calculated
% of total (farms)	84%	13%	2%	1%	1%		Calculated
% of total (cows)	22%	30%	13%	11%	24%		Calculated
Average milk yield kg/cow/day	19.5	20.1	23.3	27.1	29.4		Calculated from ADC [14]
Average milk yield kg/cow/year	5961	6139	7116	8269	8979	7147	ADC [14]
Average fat content (%)	4.17	4.17	4.18	4.13	4.04	4.07	Calculated from ADC [14]
Average protein content (%)	3.26	3.25	3.3	3.33	3.36	3.32	Calculated from ADC [14]
Total milk yield (kg/year)	201.0	279.4	147.7	145.6	327.1	1 101	Calculated
% of total milk yield	18%	25%	13%	13%	30%		Calculated
Fat and protein corrected milk (FPCM) (kg/cow/year)	6066	6242	7272	8419	9064	7218	Calculated

Apart from the milk yields and diets of different size dairy farms, they also differ by other parameters, e.g., the milking technologies and cowsheds (heat-insulated and uninsulated) used [13], as well as the cow breeds. Two dairy cow breeds account for almost 75% of the total dairy cows in Latvia—Latvian Brown, and Black and White Holstein. However, there is no reliable data that shows how these factors create differences in a farm's output among different farm sizes.

Empirical research on the environmental effects of intensification shows that increased dairy production per ha leads to increased impacts per ha, but environmental impacts per kg of raw milk are not as clear [15]. Other studies show that increasing milk production per ha may lead to reduced global warming potential (GWP) per kg of milk [16,17] and improvements in feed efficiency, fertility, and cow longevity are important parameters to reach increased milk yield and lower climate and land footprints per kg of milk [18,19].

This study aims to compare the environmental impacts of raw milk production by different size dairy farms in Latvia. The goal was to use the life-cycle impact assessment (LCA) method to gain insights into differences among farms in terms of their environmental impacts. In addition, the effect of protein feed substitution is also assessed. These results will help better tailor environmental policies to different size farms, and thus decrease the environmental impacts of dairy farming.

2. Materials and Methods

Life-cycle assessment (LCA) is a widely acknowledged and standardized method to evaluate environmental impacts during the entire life cycle of a product (ISO, 2006). Therefore, we developed the LCA model to capture the most important interactions in complex dairy production systems and determine cradle-to-farm gate (all processes involved in milking farms and production of inputs used in farms) environmental impacts of milking cows in Latvia.

Thus, the model describes the diversity in diets and the main inputs (energy, water, and minerals) for different milking farm management strategies in Latvia. More specifically, the system boundaries of this study include the following processes and flows: emissions from manure storage and application and enteric fermentation by ruminants, direct and indirect emissions from animal feed, energy, and water use. Diet composition is an input to this module which includes the cultivation and processing emissions of all feed resources, such as grass (including grazing and hay), alfalfa and maize silage, fodder, roots, as well as molasses, rapeseed expeller and meal, salt, minerals, and vitamins. However, infrastructure and machinery are excluded from the scope of this study.

Although heifers are an important part of the dairy production system, the growing of heifers could not be included in this study as the database used does not have information on heifer rearing processes. Heifer rearing would need a specific study because there are also big differences in the existing production practices. Small farms mostly grow the heifers in the same farm, heifer farming being a side role of the farm. Big farms, however, mostly buy heifers externally from farms that specifically concentrate on heifer farming.

In LCA, the environmental impact is expressed per functional unit, which in this study was one kg of fat and protein corrected milk (FPCM) leaving the farm-gate, which allows us to compare the environmental impacts of different farming systems—the farms with different herd sizes. The following equation was used to calculate FPCM [20]:

$$\text{FPCM (kg/a)} = \text{milk (kg/a)} \times ((0.1226 \times \text{fat \%}) + (0.0776 \times \text{protein \%}) + 0.2534)) \quad (1)$$

Dairy farms generate several outputs to the technosphere. The herd at the average Latvian dairy farm consists of nine dairy cows. Dairy cows which are severely injured or old are slaughtered and hardly any male animals are kept, while most female calves are kept and raised for herd replacement. Most of the male calves and a small number of the female calves which are not needed for herd replacement are sold shortly after birth. Therefore, the environmental impacts in this study have been allocated across these co-products utilizing a proxy mass-based allocation approach—raw milk (85.95%), cows for slaughter (12.35%), and calves (1.7%) [21]. In this study these percentual allocations are applied equally across all the farms as we are missing farm-size-specific information.

Most of the data were collected from secondary sources (see Table 1). Data on herd sizes were obtained from the Central Statistics Bureau (CSB) database [6]. Information about the protein and fat content in the milk for different size farms, used to calculate the FPCM, was obtained from the national Agriculture Data Center (ADC) [14]. These are reliable data sources that are also used in the national inventory to calculate agricultural gross greenhouse gas (GHG) emissions [22].

Diet composition as kg dry matter per cow (see Table 2) and average daily nutrient provision in feed rations for cows in different size farms (see Table 3) were obtained from the survey done by Degola et al. [23], where they obtained data about feed intake values in different size herds from 24 animal feeding experts from the Latvian Rural Advisory and Training Centre. It should be mentioned that the dietary intakes in Table 2 represent average data for each of the fodder in a particular farm group. In smaller farms, there is a wide diversity in feeding strategies, and thus the daily intake in DM differs from the daily DM consumption per cow in Table 3. However, feed related impact calculations are based on the total daily DM consumption from Table 3. Emission factors of feed production were obtained from a life-cycle inventory data from the EcoInvent 3 database [21].

Table 2. Average daily feed intake per cow in different management groups (kg DM).

	Dairy Farm (1–9)	Dairy Farm (10–50)	Dairy Farm (51–100)	Dairy Farm (101–200)	Dairy Farm (>200)
Grass	8.1	3.3	2.5	1.8	2.6
Haylage/grass silage	5.3	7.1	6.7	5.2	6.6
Alfalfa silage	9.6	8.4	8.3	6.1	4.7
Straw	1.2	0.6	0.4	0.8	0.7
Corn silage	-	1.1	1.2	2.8	4.2
Fodder oat	0.9	1	1	1.1	1.2
Fodder pea	2	2.1	2.2	2.5	2.7
Mixed roots	1.8	1	-	-	-
Molasses	0.3	0.4	0.4	0.4	0.5
Expeller and meal	0.5	0.8	1	1.6	1.4

Source: Degola, Cielava, Trūpa and Aplociņa [23].

Table 3. The average daily nutrient provision in feed rations for cows in different management groups.

	Dairy Farm (1–9)	Dairy Farm (10–50)	Dairy Farm (51–100)	Dairy Farm (101–200)	Dairy Farm (>200)
Daily DM consumption, kg	16.7	20.9	23.5	22.3	24.6
Crude protein, g	2388	2743	2528	2745	3259
Crude fats, g	594	537	668	701	724
Crude fiber, kg	5.6	5.8	4.5	7.0	5.5
Nitrogen free extract, kg	10.3	10.5	9.9	11.0	13.1

Source: Degola, Cielava, Trūpa and Aplociņa [23].

We also had to use several assumptions to differentiate the farms. For heating energy use we assumed that the smallest farms (1–9 cows) were using non-heated barns, while other farms (>9 cows) were using wood biomass for the milking room heating [24]. Electricity consumption was differentiated based on the assumption that smaller farms are not using power for ventilation, milk cooling, and milking (see Table 4). However, we have to acknowledge that some of the information is missing and not covered by this study, e.g., the proximity of the farms to marine or aquatic habitats or a difference in soil quality, fertility, type, and topography among farm sizes.

Table 4. The annual energy consumption in different management groups.

	Dairy Farm (1–9)	Dairy Farm (10–50)	Dairy Farm (51–100)	Dairy Farm (101–200)	Dairy Farm (>200)
Electricity consumption, kWh/cow	246	289	421	487	524
Heat consumption, MJ/cow	-	179	417	483	520

The cumulative life-cycle inventory data were assessed at the mid-point level using the ReCiPe 2016 (H) V1.02 method. Following the most relevant environmental impact categories for dairy production proposed by the European Dairy Association [25], assessments were made of:

- Gross GWP— GHG emissions from cattle (feed production, enteric fermentation, and manure), excluding carbon sequestration, expressed in kg CO_2 equivalents (CO_{2e});
- Acidification terrestrial—NO_x, NH_3, or SO_2 related emissions expressed in kg SO_2 equivalents (SO_{2e});
- Marine eutrophication—emissions of nitrogen to water and soil expressed in kg N equivalents (N_e);

- Freshwater eutrophication—emissions of phosphorus to water and soil expressed in kg P equivalents (P_e);
- Land use—dairy products being at the top of the food pyramid play important role in the competition for arable land through feed production and grazing areas, and thus have a significant impact on terrestrial species via change of land cover and the actual use of the new land expressed in m^2 of crop equivalents ($crop_{eq}$.) [26];
- Water use—the use of water (m^3) in such a way that it is evaporated, incorporated into products, transferred to other watersheds, or disposed into the sea [26];
- Fossil resource scarcity—the dairy supply chain is still heavily reliant on fossil fuel use in feed production, transport, as well as on-farm activities. Fossil resource scarcity (kg oil_{eq}.) is defined as the ratio between the energy content of the fossil resource x and the energy content of crude oil, and is based on the higher heating value of each fossil resource (crude oil, natural gas, hard coal, brown coal, and peat) [26].

Additionally, we analyzed freshwater ecotoxicity as we believe it is one of the key issues for sustainable agriculture and that dairy farmers can perform actions to reduce the toxic impact of dairy production on ecosystems [11]. Sometimes insecticides are also used directly on the animal to protect them from insects during hot summer days. However, because of the lack of reliable data, this is not included in the study.

With regards to the GWP, in this study, we tried to cover all the major emissions, but because of the lack of reliable data, soil carbon sequestration is not included in the calculation of the carbon footprint of crop production in this study. Similarly, nitrogen fixation by plants is excluded from the study as it is a soil quality dependent process and soil data for the different farm types is not available.

Annual methane emissions from enteric fermentation per cow were calculated using the following equation, provided by the United Nations Food and Agriculture Organization (FAO) [2]:

$$\text{Annual } CH_4 \text{ emission} = ((DMa \times Ym/100)) \times (18.55/55.65) \qquad (2)$$

where:
- DMa is the annual consumption of dry matter;
- The methane conversion factor (Ym) value of 6.5 is realized at a digestibility of 65% (according to the IPCC 2006 guidelines at Tier 2 level);
- The factor 55.65—the energy content of methane (MJ/kg CH_4).
- The factor 18.45—Energy intensity of feed (MJ/kg DM).

Methane emissions from enteric fermentation were calculated to be from 85.8 in the small farms to 148 kg CH_4 cow^{-1} $year^{-1}$ in the largest farms (>200 cows). The average annual national methane emission factor was calculated to be at 118 kg CH_4 cow^{-1}, while the national methane emission factor used for the year 2016 in the National Inventory was 136 kg CH_4 cow^{-1} $year^{-1}$ [22]. Latvia uses a higher emission factor for dairy cows based on a different feeding situation that is not fully characterized as stall-fed (set for Tier 1). Besides, the digestibility used for calculations of emission coefficient is lower (65% against 70% for Tier 1).

Methane emissions depend on the composition and amount of manure produced (anaerobic conditions, found at the bottom of deep lagoons, increase methane emissions) and the type of storage used [2]. Methane productivity is also determined by the quantity of volatile solids in the manure which depends on the digestibility of feed—low digestibility feed results in a high amount of volatile solids in manure which are then converted to methane. In Latvian small farms, milk cows are still mainly stanchioned, producing farmyard manure, but with increasing intensification there is a gradual transition to the producing of liquid manure [27]. It is expected that the diets of cows (high levels of grass and silage) on the smaller farms will result in higher N excretion because the amount and form of the N excreted is affected by the concentration and ruminal degradability of crude protein in the diet. However, it should be noted that these factors are not fully considered in this study as there are no data available on manure management for the different sizes of farms in Latvia. Therefore in this study, annual methane emissions of manure management and N excretion rates for dairy cows were taken from the national inventory and are set as 17.28 kg CH_4 cow^{-1} $year^{-1}$ and 113.9 kg N cow^{-1} yr^{-1} respectively [22].

3. Results

The analysis of the environmental impacts of milking cows by farms practicing different management systems in Latvia shows that the relationship between farm management practices/size and environmental impacts is not straightforward, i.e., in some impact categories (water footprint and land use) the small farms create the most burden, but in other categories (e.g., global warming, marine eutrophication, fossil resource scarcity, terrestrial acidification) the highest impacts are caused by the mid-sized farms (see Table 5).

Table 5. Results of the characterization: comparison between environmental pressures of different size dairy farms in Latvia per kg of FPCM.

Impact Category	Unit	Dairy Farm (1–9)	Dairy Farm (10–50)	Dairy Farm (51–100)	Dairy Farm (101–200)	Dairy Farm (>200)	Coefficient of Variation
GWP	kg CO_{2e}	0.88	1.02	1.09	0.93	0.84	10%
Terrestrial acidification	kg SO_{2e}	0.015	0.016	0.016	0.014	0.013	8%
Freshwater eutrophication	kg P_e	0.0002	0.0002	0.0002	0.0002	0.0001	22%
Marine eutrophication	kg N_e	0.0018	0.0025	0.0037	0.0027	0.0018	28%
Freshwater ecotoxicity	kg 1,4 Di-chloro-benzene ($DCB_{eq.}$)	0.0061	0.0073	0.0068	0.0057	0.0042	18%
Land use	m^2 $crop_{eq.}$	1.71	1.42	1.35	1.03	0.76	26%
Fossil resource scarcity	kg $oil_{eq.}$	0.054	0.060	0.062	0.051	0.039	15%
Water use	m^3	0.099	0.049	0.040	0.030	0.028	53%

For all the farms, the largest GWP is associated with enteric fermentation and manure (see Figure 1). For smaller farms, these factors account for 61% of the GWP, but for farms with 200 or more cows these factors account for 68% of all the impact per kg of raw milk produced. The second most important process is alfalfa-grass silage, which peaks (15% of GWP) in the mid-sized farms. The next most important factor is concentrates (in our case pea and oat), which combined accounted for 9 to 15% of GWP. These feed-related GHG emissions are mostly from the energy used in field management and harvest. Emissions associated with electricity production are responsible for 2 to 3% of GWP, but in this case, smaller farms have relatively smaller electricity consumption than larger farms. Results for GWP depend mainly on different diet efficiencies and energy use. Global warming potential in our study varies from 1.1 kg CO_{2e}/kg of milk from the farms with 51–100 cows to 0.84 kg CO_{2e}/kg of milk from farms above 200 cows. Thus, the national average gross GWP is 0.9 kg CO_{2e}/kg of raw milk, but GWP per cow is between 5200 kg of CO_{2e} in the smallest farms and 7700 kg of CO_{2e} in the farms with 51 to 100 cows (see Table 6). The total cradle-to-farm gate GWP of the milking cows in Latvia is estimated to be 1 million tons of CO_{2e}.

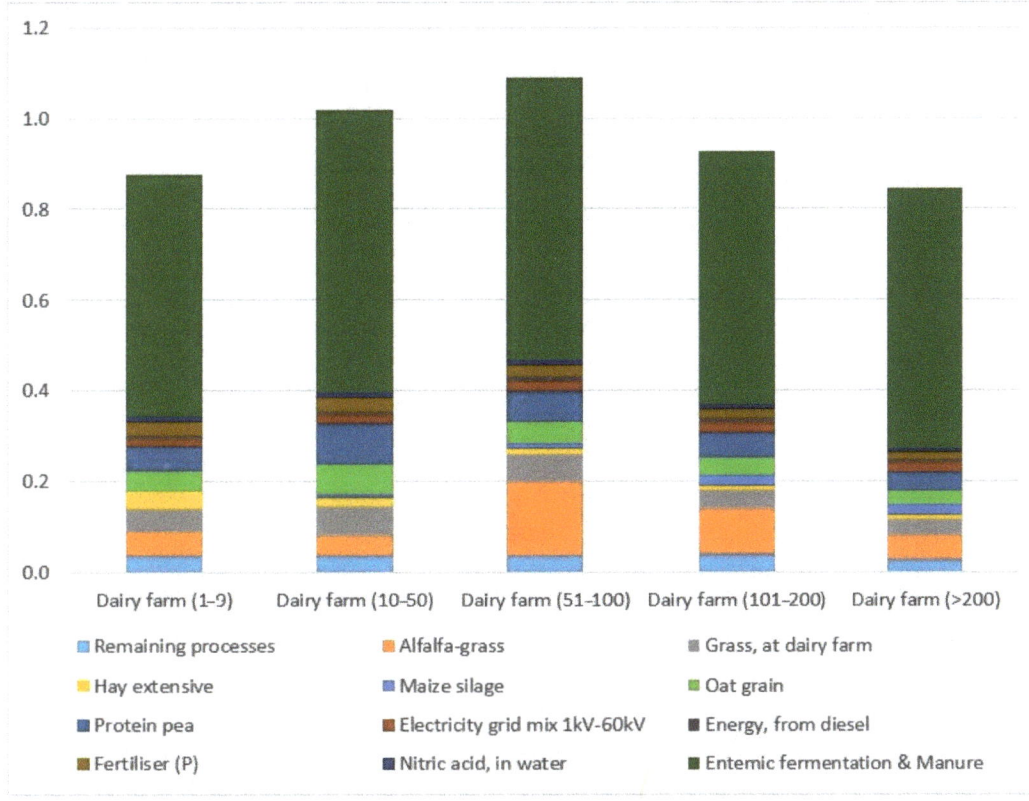

Figure 1. Global warming potential (t CO_{2e}/t milk) for different dairy systems in Latvia.

Table 6. Annual GWP and land use for different management groups.

Impact Category	Units	Dairy Farm (1–9)	Dairy Farm (10–50)	Dairy Farm (51–100)	Dairy Farm (101–200)	Dairy Farm (>200)	Average/Total
GWP—Total	1000 t CO_{2e}	176.0	284.4	160.6	134.7	275.7	1031
GWP per cow	kg CO_{2e}	5219	6248	7736	7654	7568	6696
Land use—Total	million m^2	342.8	396.1	200.1	150.2	248.2	1 337
Land use per cow	m^2	10,167	8703	9640	8532	6812	8683

One of the important impact categories is land use, which from all the parameters analyzed has the highest coefficient of variation among the different farm sizes (26%). Land use linearly decreases from 1.02 ha per cow in the smallest farms to 0.68 ha in the largest farms (see Table 6). To produce 1 kg of raw milk the smallest farms require 1.71 m^2 (including pastures), but the largest farms require only 0.76 m^2 of land. The main contributors to the land use in the smallest farms are grass (50%) and protein feeds (27%), but as the farms intensify the role of grass decreases. Therefore, for the largest farms, the main contribution to land use is from protein pea (29%), grass (21%), alfalfa-grass silage (17.5%), and oat (14%). The method used to assess land use does not specify different land use categories, such as more or less productive agricultural lands.

Water footprint includes green (rainfall) and blue (freshwater stored in lakes, rivers, and aquifers) water use. The water footprint per kg of milk significantly decreases as farms get larger because the main contributor to water footprint is green water for grass production—which ranges from 88.6% in the smallest farms to 59.6% in the largest. Blue-water footprint (mostly used as drinking water for cows) accounts only for 1.7% of the water footprint in the smallest farms, increasing to 6.3% in the largest.

The differences in terrestrial acidification among the different size farms are minor (the coefficient of variation is 8%; see Table 4), but the impacts are highest for the farms with 10–50 cows and decreases for the largest farms. The largest contributor to terrestrial acidification for all the farms is ammonia emissions accounting for 53–67.5% of all impacts, but the next most important process is the growing of grass in pastures on the dairy farms due to manure from cows which is responsible for 17–24%. As ammonia accounts for most of the terrestrial acidification, the rankings are likely related to the amount and ruminal degradability of the protein as well as a level of milk production. As a result, it would seem that the amount of acidification could be constant across herd sizes if diets and cow genetic potential were equalized.

In the case of marine and freshwater eutrophication, the largest environmental impacts are on farms with 10–100 cows. The differences in these parameters among different management groups are large, with the coefficient of variation being 22% and 28%, respectively. Nitrogen flows into marine waters is estimated to be from 0.0037 kg N_e per kg of raw milk from the farms with 51–100 cows to 0.0018 kg N_e in the largest farms (see Table 4). The main contributor to marine eutrophication from milk production on the larger farms is alfalfa-grass silage feed (46% of all the impact from the farms with 51–100 cows to 30% in the largest farms), followed by grass grown at the farm (18–26%). However, for the smaller farms, the main processes contributing to marine eutrophication are grass and oat, which combined contribute more than 70% of all impacts. With regards to freshwater eutrophication, the two management groups which include the smallest farms (up to 50 cows) have the highest impact—0.00023 kg P_e/kg of FPCM. However, the main contribution to freshwater eutrophication from all the farms is from grass, followed by oat and alfalfa-grass silage.

Additionally, for freshwater ecotoxicity the largest environmental impact is from farms with 10–50 cows. The largest contributors are concentrates (oat ~30% and pea (24–31%)) (see Table 4). Alfalfa-grass silage is also responsible for a significant part of freshwater ecotoxicity—contributing 18% of all the terrestrial ecotoxicity impact from the smallest farms to 13% from the largest farms.

The results of our study demonstrate that the production of one kg of milk needs 37–62 g of oil equivalent fossil resources (see Table 4). In this case, the highest fossil resource intensity is for farms with 51–100 cows. The main contributors are production of alfalfa-grass silage (13–26%), protein feeds (17–25%), and fertilizer (P) (10–15%).

Additionally, we looked at changes in the environmental impacts of supplementary protein substitution. In this case, we analyzed substituting pea in the combined feed with soy, as it is increasingly entering the market as a high-protein feed for dairy cows in Latvia. For this, we used global process data on the environmental impacts of soy meal production from the Ecoinvent database [21]. Results demonstrated that most of the environmental impacts would increase, e.g., GWP for the farms with 10–50 cows would increase by 57% and for the largest farms by 32% (see Figure 2), but on average GWP per kg of milk would increase by 43%. As a result of the substitution, the land use in the smallest farms would increase by 12% and for other farms by 19%, on average increasing the land use of raw milk production by 18%. These increasing impacts are mostly linked to the soy production-related land-use change, domestic transport, and industrial processing [28]. However, it can also be assumed that introducing soybean meal to the diets of the dairy cows would decrease the protein supplement fed to keep the diets isonitrogenous.

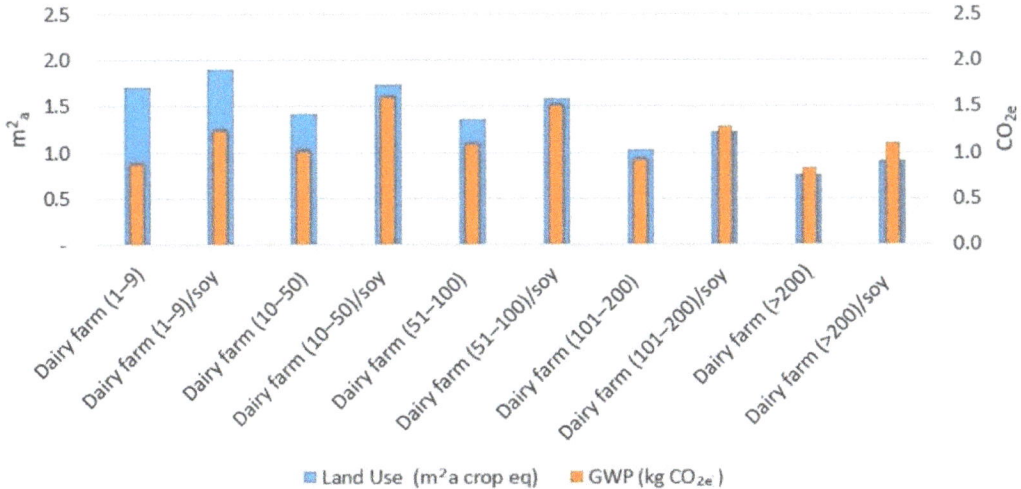

Figure 2. Impact on land use and GWP per kg of FPCM if pea in cow diet was substituted by soy.

4. Discussion and Conclusions

Our results show that the environmental impacts of milking cows are different on farms with different management strategies. For some of the impact categories, e.g., global warming, marine eutrophication, terrestrial acidification, and ecotoxicity, the highest environmental impacts are for mid-sized farms with 10 to 50 cows. In general, the trend of most environmental impacts is decreasing as the farms increase in size, thus partly confirming that many of the environmental impacts assessed here decrease with increasing milk yields [29–31]. However, the smallest farms perform worse than the mid-sized farms with 10–50 cows only in two out of eight environmental impact categories.

The largest farms (>200 cows), through increasing protein content in their feed and choosing more productive breeds, have managed to significantly increase their milk yields and thus, also improve the environmental performance on certain environmental impact categories of production. Our results demonstrate that dairy system intensification generally provides a significant decrease in all of the impact categories assessed, when measured per kg of raw milk. The most significant reduction was observed for water and land use. These results are somewhat different from others that argue that intensification decreases eutrophication and land use per kg of raw milk but has no clear effects on other impact categories.

In our study, for all the farms enteric fermentation had the largest impact on the GWP, followed by feeds (alfalfa, protein feeds, and grass) and electricity consumption. Increasing the consumption of concentrated feeds in the largest herds lead to higher milk yields and thus, helped to decrease relevant environmental impacts. The global warming potential of the raw milk at the farm gate in different LCA studies varies significantly—from 0.9 to 2.4 kg CO_{2e} per kg of milk [27,32]. Results from our study are on the bottom line of this spectrum. This could be explained with the scope of the study excluding heifer rearing, which could increase the global warming potential of the milk by roughly 1/5 [33].

Differences in the environmental impacts of different farms can be explained by differences in feeds and diets of lactating cows as well as by differences in energy use of these farms, as the smallest farms are less energy-intensive in their performance but are more reliant on grazing (grass consumption). Higher milk yields are obtained in the larger farms, which have been achieved by incorporating proteins that are more resistant to ruminal degradation—soy or pea protein may be more efficiently used than proteins in forages like alfalfa.

It should be also noted that dairy cows in traditional systems provide other essential services, e.g., biodiversity protection and the conservation of cultural landscapes [34]. Unfortunately, as in the current study, these grazing benefits (e.g., carbon sequestration, biodiversity) are excluded due to the lack of approved assessment methods, and thus intensive systems are likely favored [35]. As a result, our study highlights the potential conflict between GHG emission and land use efficiencies and other environmental objectives, similar to Edwards-Jones et al. [36], and further emphasizes the need for the development of scientifically robust methods to assess carbon sequestration, biodiversity, and social impacts. Carbon sequestration is not covered by this study, but it is clear that optimized grazing, pasture improvement, and the restoration of degraded pastures are critical in increasing grassland carbon sequestration [37].

In addition, the specification and characterization of land and water use by different areas and crops could provide more insights. Grass production often takes place in areas unsuitable, or at least economically not profitable, for other crop production. Thus, the higher land use of small farms (of which 50% is of grass) may not reflect the higher land use of highly productive agricultural land. Similarly, the method used to assess water use in the paper does not reflect water availability or scarcity. Thus, also regarding water use, the higher impact of small farms may not reflect an actual higher water consumption of scarce water resources.

We can also assume that smaller farms tend to use cultivated local protein feeds which have a relatively lower environmental impact than the soybean byproducts commonly used in larger conventional farms. From this study, the total national pea consumption is estimated to be 157 thousand tons of peas a year. According to the FAO data in 2017 only 29.5 thousand tons of dry pea were produced in Latvia [38]. However, not all of the on-farm production of peas for domestic consumption is included in the national statistics.

It should be also considered that soy (one of the most popular high-protein feeds) production results in other social-ecological impacts [39], e.g., land acquisitions [40] and biodiversity loss [41], which are not properly covered by this study. In assessing soy production-related climate impacts, it would be also important to include LUC-related emissions. Additionally, local characterizations of water use to reflect a scarcity of water would improve the understanding of the water use of different crops from different countries.

One of the main policy recommendations from this study is to differentiate the policy approach to different size dairy farms in Latvia. The largest problems are with the management of farms with a herd size of 10–50 cows as they have some of the highest environmental impacts because their feeding strategy does not deliver higher milk yields. These farms should receive better training in feeding strategies as most likely they are overfeeding their cattle. Farms with a diverse range of cow breeds should also develop diversified feeding strategies for each of the breeds. The use of an automatic feeding system normally linked to milking robots could help to do this, as they increase the possibility of controlling the nutritional value of feed concentrates for each particular cow.

The smallest farms (≤ 9 cows) on another hand should be supported as self-sufficient providers and as important mechanisms for landscape and biodiversity conservation. Small farms with a higher fat content in the milk they produce should be encouraged to focus on differentiated production, e.g., cheese making. The smallest farms also should be supported in the use of cultivated local protein feeds and utilize cow manure effectively to create regenerative benefits in their field area [42]. Small farms are also important in terms of the efficiency of utilizing local sources of dry matter. Their feeding strategies appeared to be the most flexible of all the management groups. Thus, small farms also need support for this purpose. Diverse feeding strategies in small farms also would be a proxy for enhanced biodiversity.

However, we should also acknowledge that this study has several limitations linked to national data availability (e.g., national data on feed production, heifer breeding, differences among farms regards soil type, manure management, and the proximity to marine or aquatic habitats) and methodological shortcomings (e.g., excluding emissions of carbon sequestration, the assessment method of water use, the use of proxy allocation, and the integration of social and biodiversity impacts in LCA). There is a need for the development of a national LCA inventory to further improve the quality and possibilities for environmental impact assessment.

Author Contributions: J.B. led the conceptualization, data collection, model development, calculations, interpretation of results, and writing of the manuscript; S.K. and H.H. contributed to research design, model development, analysis, interpretation of results, and writing of the manuscript. All authors have read and agreed to the published version of the manuscript.

Funding: This research was partly financed by the specific support objective activity 1.1.1.2. "Post-doctoral Research Aid" of the Republic of Latvia (Project No. 1.1.1.2/VIAA/1/16/065 "Developing New Tools for the Sustainability Assessment of the Bioeconomy"), funded by the European Regional Development Fund (Project id. N. 1.1.1.2/16/I/001). The APC was funded by authors.

Institutional Review Board Statement: Not applicable.

Informed Consent Statement: Not applicable.

Data Availability Statement: The data presented in this study are available on request from the corresponding author.

Conflicts of Interest: The authors declare no conflict of interest.

References

1. Mu, W.; van Middelaar, C.; Bloemhof, J.; Engel, B.; de Boer, I. Benchmarking the environmental performance of specialized milk production systems: Selection of a set of indicators. *Ecol. Indic.* **2017**, *72*, 91–98. [CrossRef]
2. Gerber, P.; Vellinga, T.; Opio, C.; Henderson, B.; Steinfeld, H. *Greenhouse Gas Emissions from the Dairy Sector, A Life Cycle Assessment*; FAO Food and Agriculture Organisation of the United Nations. Animal Production and Health Division: Rome, Italy, 2010.
3. Garnett, T.; Appleby, M.C.; Balmford, A.; Bateman, I.J.; Benton, T.G.; Bloomer, P.; Burlingame, B.; Dawkins, M.; Dolan, L.; Fraser, D. Sustainable intensification in agriculture: Premises and policies. *Science* **2013**, *341*, 33–34. [CrossRef] [PubMed]
4. Alvarez, A.; del Corral, J.; Solís, D.; Pérez, J.A. Does Intensification Improve the Economic Efficiency of Dairy Farms? *J. Dairy Sci.* **2008**, *91*, 3693–3698. [CrossRef] [PubMed]
5. Pilvere, I.; Nipers, A.; Krievina, A. Analysis of the factors affecting cost efficiency in the dairy industry in Latvia. *Econ. Sci. Rural. Dev. Conf. Proc.* **2016**, *41*, 251–258.
6. CSB. *Grouping of Farms of All Kinds by the Number of Cattle and Dairy Cows at End of Year (LLG240)*; Central Statistical Bureau of Latvia: Rīga, Latvia, 2019.
7. Pilvere, I.; Nipers, A.; Krievina, A.; Upite, I. Development prospects of milk production in various size farm groups in Latvia. In Proceedings of the 19th International Scientific Conference Engineering for Rural Development, Jelgava, Latvia, 20–22 May 2020.
8. Aubron, C.; Cochet, H.; Brunschwig, G.; Moulin, C.-H. Labor and its productivity in Andean dairy farming systems: A comparative approach. *Hum. Ecol.* **2009**, *37*, 407–419. [CrossRef]
9. Bernard, J.; Le Gal, P.-Y.; Triomphe, B.; Hostiou, N.; Moulin, C.-H. Involvement of small-scale dairy farms in an industrial supply chain: When production standards meet farm diversity. *Animal* **2011**, *5*, 961–971. [CrossRef]
10. Verhees, F.; Malak-Rawlikowska, A.; Stalgiene, A.; Kuipers, A.; Klopčič, M. Dairy farmers' business strategies in Central and Eastern Europe based on evidence from Lithuania, Poland and Slovenia. *Ital. J. Anim. Sci.* **2018**, *17*, 755–766. [CrossRef]
11. Nordborg, M.; Davis, J.; Cederberg, C.; Woodhouse, A. Freshwater ecotoxicity impacts from pesticide use in animal and vegetable foods produced in Sweden. *Sci. Total Environ.* **2017**, *581*, 448–459. [CrossRef]
12. Osītis, U. *Govju ēdināšana (Cow Feeding)*; Latvian Agricultural Advisory and Training Centre: Ozolnieki, Latvia, 2002; p. 45.
13. Nipers, A.; Pilvere, I.; Valdovska, A.; Proskina, L. Assessment of key aspects of technologies and cow farming for milk production in Latvia. In Proceedings of the 15th International Scientific Conference "Engineering for Rural Development", Jelgava, Latvia, 25–27 May 2016; pp. 175–181.
14. ADC. *Pārraudzības Rezultāti Ganāmpulkos 2017/2018 Pārraudzības Gadā (Monitoring Results of Herds in 2017/2018)*; Agricultural Data Center: Rīga, Latvia, 2019.
15. Crosson, P.; Shalloo, L.; O'brien, D.; Lanigan, G.; Foley, P.; Boland, T.; Kenny, D. A review of whole farm systems models of greenhouse gas emissions from beef and dairy cattle production systems. *Anim. Feed Sci. Technol.* **2011**, *166*, 29–45. [CrossRef]
16. Bell, M.; Wall, E.; Russell, G.; Simm, G.; Stott, A. The effect of improving cow productivity, fertility, and longevity on the global warming potential of dairy systems. *J. Dairy Sci.* **2011**, *94*, 3662–3678. [CrossRef]
17. Casey, J.; Holden, N. The relationship between greenhouse gas emissions and the intensity of milk production in Ireland. *J. Environ. Qual.* **2005**, *34*, 429–436. [CrossRef] [PubMed]
18. Audsley, E.; Wilkinson, M. What is the potential for reducing national greenhouse gas emissions from crop and livestock production systems? *J. Clean. Prod.* **2014**, *73*, 263–268. [CrossRef]

19. Yan, M.-J.; Humphreys, J.; Holden, N. Life cycle assessment of milk production from commercial dairy farms: The influence of management tactics. *J. Dairy Sci.* **2013**, *96*, 4112–4124. [CrossRef]
20. 20. IDF. Common carbon footprint approach for dairy: The IDF guide to standard lifecycle assessment methodology for the dairy sector. *Bull. Int. Dairy Fed.* **2010**, *445*, 1–46.
21. Wernet, G.; Bauer, C.; Steubing, B.; Reinhard, J.; Moreno-Ruiz, E.; Weidema, B. The ecoinvent database version 3 (part I): Overview and methodology. *Int. J. Life Cycle Assess.* **2016**, *21*, 1218–1230. [CrossRef]
22. MEPRD. *Latvia's National Inventory Report Submission under UNFCCC and the Kyoto Protocol Common Reporting Formats (CRF) 1990–2017*; Ministry of Environmental Protection and Regional Development: Riga, Latvia, 2019.
23. Degola, L.; Cielava, L.; Trūpa, A.; Aplociņa, E. Feed rations in different size dairy farms. In Proceedings of the Zinātniski praktiskā konference "Līdzvarota Lauksaimniecība", Jelgava, Latvia, 25–26 February 2016; pp. 161–168.
24. Lauku, T. Racionālu Piena Lopkopības Ražošanas Modeļu Rokasgrāmata (Handbook of Rational Dairy Production Models). Available online: http://www.laukutikls.lv/racionalu-piena-lopkopibas-razosanas-modelu-rokasgramata (accessed on 16 July 2020).
25. EDA. *Product Environmental Footprint Category Rules for Dairy Products*; European Dairy Association: Bruxelles, Belgium, 2018.
26. Huijbregts, M.; Steinmann, Z.; Elshout, P.; Stam, G.; Verones, F.; Vieira, M.; Hollander, A.; Zijp, M.; Van Zelm, R. *ReCiPe 2016: A Harmonized Life Cycle Impact Assessment Method at Midpoint and Endpoint Level Report I: Characterization*; National Institute for Public Health and the Environment: Bilthoven, The Netherlands, 2016.
27. Priekulis, J.; Āboltiņš, A. Calculation methodology for cattle manure management systems based on the 2006 IPCC guideline. In *Nordic View to Sustainable Rural Development, Proceedings of the 25th NJF Congress, Riga, Latvia, 16–18 June 2015*; NJF Latvia: Riga, Latvia, 2015; pp. 274–280.
28. Gil, J. Carbon footprint of Brazilian soy. *Nat. Food* **2020**, *1*, 323. [CrossRef]
29. Nemecek, T.; Schmid, A.; Alig, M.; Schnebli, K.; Vaihinger, M. Variability of the global warming potential and energy demand of Swiss cheese. In Proceedings of the SETAC Europe 17th LCA Case Studies Symposium, Budapest, Hungary, 28 February–1 March 2011.
30. Gerber, P.; Vellinga, T.; Opio, C.; Steinfeld, H. Productivity gains and greenhouse gas emissions intensity in dairy systems. *Livest. Sci.* **2011**, *139*, 100–108. [CrossRef]
31. Knapp, J.; Laur, G.; Vadas, P.; Weiss, W.; Tricarico, J. Invited review: Enteric methane in dairy cattle production: Quantifying the opportunities and impact of reducing emissions. *J. Dairy Sci.* **2014**, *97*, 3231–3261. [CrossRef]
32. Thomassen, M.A.; van Calker, K.J.; Smits, M.C.; Iepema, G.L.; de Boer, I.J. Life cycle assessment of conventional and organic milk production in the Netherlands. *Agric. Syst.* **2008**, *96*, 95–107. [CrossRef]
33. Mc Geough, E.; Little, S.; Janzen, H.; McAllister, T.; McGinn, S.; Beauchemin, K. Life-cycle assessment of greenhouse gas emissions from dairy production in Eastern Canada: A case study. *J. Dairy Sci.* **2012**, *95*, 5164–5175. [CrossRef]
34. Plieninger, T.; Höchtl, F.; Spek, T. Traditional land-use and nature conservation in European rural landscapes. *Environ. Sci. Policy* **2006**, *9*, 317–321. [CrossRef]
35. Ripoll-Bosch, R.; De Boer, I.; Bernués, A.; Vellinga, T.V. Accounting for multi-functionality of sheep farming in the carbon footprint of lamb: A comparison of three contrasting Mediterranean systems. *Agric. Syst.* **2013**, *116*, 60–68. [CrossRef]
36. Edwards-Jones, G.; Plassmann, K.; Harris, I. Carbon footprinting of lamb and beef production systems: Insights from an empirical analysis of farms in Wales, UK. *J. Agric. Sci.* **2009**, *147*, 707–719. [CrossRef]
37. Chang, J.; Ciais, P.; Gasser, T.; Smith, P.; Herrero, M.; Havlík, P.; Obersteiner, M.; Guenet, B.; Goll, D.; Li, W. Climate warming from managed grasslands cancels the cooling effect of carbon sinks in sparsely grazed and natural grasslands. *Nat. Commun.* **2020**, *12*, 118. [CrossRef] [PubMed]
38. FAO Stat. *Food and Agriculture Organisation of the UN (FAO) Statistics Database: Production, Trade, Supply*; FAO: Rome, Italy, 2019.
39. Eriksson, M.; Ghosh, R.; Hansson, E.; Basnet, S.; Lagerkvist, C.-J. Environmental consequences of introducing genetically modified soy feed in Sweden. *J. Clean. Prod.* **2018**, *176*, 46–53. [CrossRef]
40. Borras Jr, S.M.; Kay, C.; Gómez, S.; Wilkinson, J. Land grabbing and global capitalist accumulation: Key features in Latin America. *Can. J. Dev. Stud./Rev. Can. D'études Du Développement* **2012**, *33*, 402–416. [CrossRef]
41. Crenna, E.; Sinkko, T.; Sala, S. Biodiversity impacts due to food consumption in Europe. *J. Clean. Prod.* **2019**, *227*, 378–391. [CrossRef]
42. Schreefel, L.; Schulte, R.; de Boer, I.; Schrijver, A.P.; van Zanten, H. Regenerative agriculture–the soil is the base. *Glob. Food Secur.* **2020**, *26*, 100404. [CrossRef]

Article

LCA to Estimate the Environmental Impact of Dairy Farms: A Case Study

Sara Zanni [1], Mariana Roccaro [2,*], Federica Bocedi [3], Angelo Peli [2] and Alessandra Bonoli [4]

1. Department of Management, University of Bologna, 40126 Bologna, Italy; sara.zanni7@unibo.it
2. Department for Life Quality Studies, University of Bologna, 47921 Rimini, Italy; angelo.peli@unibo.it
3. Department of Veterinary Medical Sciences, University of Bologna, 40064 Bologna, Italy; federica.bocedi@studio.unibo.it
4. Department of Civil, Chemical, Environmental and Materials Engineering (DICAM), University of Bologna, 40136 Bologna, Italy; alessandra.bonoli@unibo.it
* Correspondence: mariana.roccaro2@unibo.it; Tel.: +39-051-20-9-7306

Abstract: Intensive farming is responsible for extreme environmental impacts under different aspects, among which global warming represents a major reason of concern. This is a quantitative problem linked to the farm size and a qualitative one, depending on farming methods and land management. The dairy sector is particularly relevant in terms of environmental impact, and new approaches to meeting sustainability goals at a global scale while meeting society's needs are necessary. The present study was carried out to assess the environmental impact of dairy cattle farms based on a life cycle assessment (LCA) model applied to a case study. These preliminary results show the possibility of identifying the most relevant impacts in terms of supplied products, such as animal feed and plastic packaging, accounting for 19% and 15% of impacts, respectively, and processes, in terms of energy and fuel consumption, accounting for 53% of impacts overall. In particular, the local consumption of fossil fuels for operations within the farm represents the most relevant item of impact, with a small margin for improvement. On the other hand, remarkable opportunities to reduce the impact can be outlined from the perspective of stronger partnerships with suppliers to promote the circularity of packaging and the sourcing of animal feed. Future studies may include the impact of drug administration and the analysis of social aspects of LCA.

Keywords: LCA; climate change; agro-livestock sector; GHG emissions; dairy farming

1. Introduction

Climate change, described as the long-term heating of our planet caused by human activities since the pre-industrial period, is among the main challenges on a global scale [1]. As it is well known, one of the main causes of increasing temperatures is greenhouse gas (GHG) emissions in the atmosphere.

GHG concentration affects the global temperature by absorbing strongly radiant electromagnetic energy at wavelengths capable of emitting quantities of elevated heat [2].

According to their global warming potential, the main gases responsible for the greenhouse effect are methane, carbon dioxide, used as a reference for the phenomenon, water vapor, and nitrous oxide, each with different global warming potential and specific sources. Methane, for example, is about 21 times more efficient in trapping heat in the atmosphere compared to CO_2, while nitrous oxide is 296 times more efficient, with persistence in the atmosphere for up to 114 years [3].

The agro-livestock sector is responsible for a relevant share of these gases due to direct and indirect emissions, accounting for about 17% of the global GHG emissions in 2018 [4].

In addition to the release of GHG into the atmosphere, farming contributes in part to the worsening of air quality through the production of mainly nitrogen compounds from manure, particulates obtained from combustion, and volatile organic compounds other than methane (NMVOC), carbon black (BC), heavy metals (i.e., chromium, copper, nickel, selenium, zinc, lead, cadmium, mercury), dioxins, polychlorinated biphenyls (PCBs), polycyclic aromatic hydrocarbons (PAHs), hexachlorobenzene (HCB), and of particulates, both less than 10 microns (PM_{10}) and 2.5 microns ($PM_{2.5}$), deriving from several farm-level operations [5].

Livestock production has a significant environmental impact as it affects several natural resources, including land and soil, water, air, and biodiversity. Nevertheless, growing populations and economies have led to increased demand for animal products and the consequent expansion of the livestock sector over the past decades. At the same time, in developed countries, consumers demand animal products that are both animal welfare- and environment-friendly [3].

In 2019, world milk production grew to about 852 Mt and was forecast to grow at 1.6% per year over the next decade, faster than most other main agricultural products [6]. Globally, cattle are the largest contributors to total livestock greenhouse gas (GHG) emissions, producing about 65% (4.6 Gt CO_2eq annually) of sector emissions, with milk production contributing 20% of total sector emissions [7]. Besides direct emissions, livestock systems are responsible for indirect emissions arising from land-use change, fertilizer use, energy, and transport emissions related to livestock operations and supply chains [8]. The three largest sources of GHG from milk production are emissions from manure management (CH_4 and N_2O), emissions from feed production, processing, and transport (CO_2 and N_2O), and emissions from enteric fermentation (CH_4), the latter accounting for more than half the total of emissions [9].

Although absolute emissions from the dairy sector have increased in the last decades due to production growth in response to the increasing demand, dairy farming is becoming more efficient considering that emissions per unit of product are decreasing [8].

Strategies to reduce GHG emissions include changes in feeding, breeding, and management practices, which essentially lead to the intensification of livestock farming. Strategies that aim to increase productivity are very promising ways to reduce environmental impact; however, in most cases, they are likely to negatively impact animal welfare. For example, intensive housing conditions increase the risk of social stress or hinder the expression of natural behavior [10].

The rising GHG emissions require shifting production systems toward carbon neutrality in order to take action to combat climate change and its impacts, which is one of the United Nations 2030 agenda for sustainable development goals (SDG 13). Moreover, sustainable food production systems and resilient agricultural practices are among the targets to achieve the "zero hunger" goal (SDG 2) and to promote new consumption and production models (SDG 12) by changing the way we produce and consume, mainly in the agri-food sector where these elements are so close, in the farm-to-fork perspective. At the same time, progress toward the sustainable use of natural resources is key to protecting biodiversity and ecosystems (SDG 15). Improving animal health and welfare and reducing GHG emissions through new farming techniques to meet societal demands on safe and sustainable food and contribute to climate change mitigation are among the nine key objectives of the 2023–2027 European common agricultural policy.

The relationship between livestock production and climate change is two-sided. Climate change, deriving from global warming, adversely impacts livestock production, with direct effects on animal health, reproductive efficiency, production performances, and behavior, but also indirect effects deriving from changes in the quality and quantity of feed, water availability, ecosystem alterations leading to changes in the biology and distribution of pathogens and vector-borne diseases [11]. An increase in temperatures negatively affects milk production, especially in cows with higher milk yield and milk quality, with a reduction in casein content [12].

Dairy cattle production systems need to adapt to climate change, but, on the other hand, they must commit to contributing to GHG reduction targets and minimize other negative environmental impacts while continuing to meet society's needs. Agriculture is estimated to generate 11% of all global emissions [13], and dairy farm contribution to the overall impact of milk products is almost as high as 72% [14]. A special focus on this production phase is required to meet the global goals.

The dairy sector is an extremely complex system with numerous interacting components; consequently, determining the best strategies to reduce GHG emissions will depend on each farm's local conditions and objectives. Therefore, a deeper understanding of the underlying driving factors of dairy cattle farming environmental impact is highly relevant to support strategies for limiting adverse effects on the environment and protecting the livestock sector [15].

Studies on the environmental impact assessment of dairy products have increased dramatically in the last decade and have been reviewed by several authors [16–21]. The main goals of these studies were either to assess the potential environmental impact of the product [22–25] or to compare different management systems [26–28]. However, very few of these have focused on Italian production systems [29–32] and even less on Italian Protected Designation of Origin (PDO) dairy products (e.g., Parmigiano Reggiano, Grana Padano) [33]. In Italy, milk production is concentrated in the northern plain regions, characterized by little rain during summer and a highly urbanized territory with low availability of agricultural land; cattle are kept mostly indoors, and maize silage is the main forage crop. Moreover, PDO product specifications impose restrictions on management strategies, sourcing, and type of animal feed.

Considering the peculiarity of the environmental conditions and farming practices, this study aimed to set up an LCA model specific for intensive dairy cattle farms involved in PDO production in Northern Italy. We aimed to evaluate the most relevant items of environmental impact triggered by this activity, considering its ability to impair animal health and welfare. To this end, a preliminary assessment was performed on a dairy cattle farm in the Emilia-Romagna region in Italy. The LCA model was realized using Simapro 8.5.1, including farm processes and main supplies, in terms of animal feed and end-of-life processes of the most relevant waste flows.

With the ultimate aim of integrating farm environmental impact and animal health and welfare, an interdisciplinary research approach combining engineering and veterinary expertise was carried out to evaluate farms operating conditions in terms of animal management and welfare, and specifically, to assess environmental impacts based on an LCA approach [34]. This approach offers grounds for future research and development of targeted sustainability strategies in the sector, specific to the studied geographical area, with a concrete possibility of application in the field.

2. Materials and Methods

The following sections report a detailed description of the site of the study (Section 2.1) and the conceptual model developed for the implementation of LCA (Section 2.2).

2.1. Site Description

The present study was carried out on an Italian dairy cattle farm producing Parmigiano Reggiano and soft cheeses in the Emilia Romagna Region.

Prior to the farm visit, a form for primary data collection was designed. During the farm visit (July 2020), data were collected on paper and subsequently transferred to Microsoft® Excel files. Data sources included the farm management software records, information collected by interviewing the farmer, and invoices.

For data that could not be collected on-farm, default values were defined based on processes available on Ecoinvent v.3.5 [35], with geographical reference to the European context, where possible, or Italian specific data, as in the case of the electricity production mix.

Information on farm size, number of animals, productivity, animal welfare, drug use, feed management, water consumption, fuel consumption, electric energy, manure management, and plastic waste were collected (as summarized in Table 1). It was not possible to quantify the use of chemical fertilizers and bedding material.

Table 1. Farm-specific primary data used to design the conceptual model for the assessment.

Category	Farm-Specific Information	Data
Farm size	Site A	2 ha
	Site B	0.5 ha
	Site C	1 ha
	Site D	1.9 ha
	Cropland	560 ha
Average number of animals	Milking cows	564
	Dry cows	84
	Replacements	622
	Calves	97
	Bulls	1
Animal turnover	Replacement rate	35%
	Longevity	2.5 deliveries
	Adult mortality	3%
	Calf mortality	1.87%
Productivity data	Annual milk production	66,000 quintals
Bought-in feed (source in brackets)	Sunflower seeds (Black Sea)	191,094 kg
	Soybean (Modena province)	9148 kg
	Maize (Modena province)	181,598 kg
	Protein feed (Cremona province)	22,042 kg
Water consumption (annual)	Drinking water	27,963,555.20 L
	Cooling water	72,000 L
Fuel consumption (annual)	Agricultural e livestock farming machines	181,995 L
Electric energy (annual)	Ventilation system	141,849 kWh
	Other operations	516,844 kWh
Manure management	Palatable slurry storage	6 tanks (12,667 m^3) 3 cesspits (1001 m^3)
	Non-palatable slurry storage	9 open pits (12,229 m^3)
Plastic waste	Plastic wrapping and cleaning product containers	163,900 kg

The farm is structured in four sites, located between Bologna and Modena provinces. Site A hosted calves and replacements up to 5 months of age, pregnant heifers (from two months pregnancy to delivery), and dry cows. Replacements from 5 months of age to the first month of pregnancy were hosted in site B. Site C hosted mostly primiparous milking cows, and site D hosted mostly multiparous milking cows. Animals are moved from site A to site C or D twice a week and from site A to site B twice a month.

The most represented breed was Holstein Friesian, but Italian Red Pied, Jersey, and Montbéliarde x Swedish Red and White crossbreds were also present.

Average daily milk production amounts to 35.15 kg for multiparous cows and 30.15 kg for primiparous cows. The milk is destined to produce Parmigiano Reggiano.

Cows are loose housed on deep litter with straw bedding (dry cows and heifers) or in pens with cubicles covered with sand (milking cows). Dry cows and heifers have access to external exercise areas. The facilities are equipped with ventilation systems consisting of ceiling fans for the resting areas and horizontal flow fans for the feeding zones. The waiting parlor is equipped with a ceiling fan and water sprayers.

Calves up to one month of age are reared in individual pens and then moved to group pens. They are fed through an automatic calf feeder and weaned at 70 days of age.

Site C and site D scored 70.33/100 and 71.31/100, respectively, for animal welfare in 2018, as certified by the Italian National Reference Center for Animal Welfare (CReNBA).

The diet consists of 60% fodder and 40% concentrates. Fodder (alfa-alfa, grass hay) and part of sorghum are self-produced; the other feeds are bought in. Given the production requirements for Parmigiano Reggiano, corn silage is only fed to heifers (site B). Daily average dry matter intake is 25 kg for milking cows, 15 kg for dry cows and pregnant heifers, and 8 kg for replacements.

The water comes from the aqueduct. Site C was also supplied from a well, but it was not possible to quantify the water used. There is no water recycling system in place.

The slurry is spread on the farm's land intended for fodder production (470 ha). Figure 1 reports the data collection at the basis of the conceptual model definition.

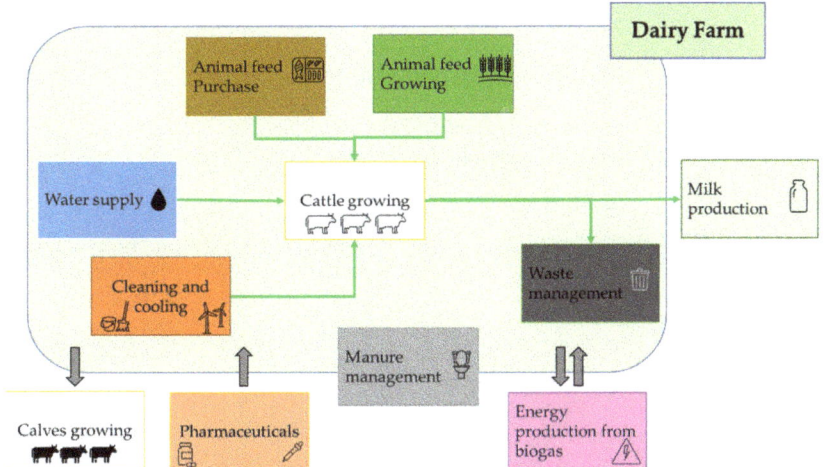

Figure 1. Conceptual model flowchart.

2.2. Conceptual Model for the Assessment

The conceptual model at the basis of the LCA has been defined based on a cradle to farm gate perspective over a calendar year. Considering the specificity of the farm analyses, the model has been developed based on primary data, whenever possible, and previous studies in the field (e.g., Grassauer et al., 2022 [28]).

Several assumptions have been introduced aimed at reducing the uncertainties of the results and at providing a preliminary assessment of the generated impacts, as presented below:

- Water for feeding and cleaning purposes was considered as deriving from the drinking water supply system;
- Electricity consumption was considered entirely covered by energy from the grid, as the biogas generation plant was inoperative at the time of the study. The Italian country mix for medium voltage supply included in the Ecoinvent library was selected as representative for the case;
- The processes related to calf growing were excluded from the model, as meat production, i.e., the co-production line, was out of the scope of the present study;
- As manure management consisted in storage and spreading in the farm fields, it was included exclusively in terms of machinery diesel consumption and equipment involved;
- The use of pharmaceuticals could not be included in the present model due to the lack of primary data about the impact of the specific drugs;
- Carcass disposal was excluded from the boundaries of the present study.

A specific focus was dedicated to animal feed, which has been represented in the following terms, considering each upstream process and relative impacts:

- The growing of sorghum and fodder in local sites was included in terms of machinery diesel consumption and equipment involved;
- Purchased animal feed was detailed by typology, source location, and transportation (mean and distance).

Figure 1 shows the conceptual model in the form of a flowchart. The system boundaries are marked in light blue. Each process included in the analysis is connected through green arrows to the main line of "cattle farming" and "milk-production". The excluded processes, either input (i.e., pharmaceuticals), outputs (i.e., calves growing), or both (i.e., energy production from biogas), are represented out of the system boundaries. Manure management, which lays on the system's boundaries, was included in the analysis exclusively in direct terms, as fuel consumed by the farm's machinery for spreading.

The LCA methodology was applied based on ISO 14040 [34] using the Simapro 8.5.1 software and the Ecoinvent 3.4. database [35]. The Functional Unit (FU) was set as 1 kg of Fat and Protein Corrected Milk (FPCM) produced in 2019, in accordance with similar studies in the field [24,25,33], and it is defined by [36] as

$$\text{FPCM (kg)} = \text{raw milk (kg)} \times (0.337 + 0.116 \times \text{fat content (\%)} + 0.06 \times \text{protein content (\%)})$$

The average Italian national fat and protein contents reported in the year prior to the visit (2019) were used as reference values; hence, all milk was converted to FPCM with 3.8% fat and 3.4% protein [37].

Details about the processes included are reported in Appendix A. Table A1 shows the overall model, Table A2 shows the details regarding animal feed production, and Table A3 shows the details regarding the transport of animal feed.

Considering the relevance of GHG emissions from farming and their impact on the global warming issue, the selected calculation method was IPCC100a, i.e., the method developed by the International Panel on Climate Change, which provides the carbon footprint developed over a time horizon of one hundred years [38,39], but still integrated into an LCA framework. Previous studies in the field have indicated this as the most conservative approach [27].

Uncertainty analysis was performed, applying Monte Carlo simulation with a 95% confidence interval.

3. Results and Discussion

In developed countries, since more and more citizens are concerned about health, environmental, ethical, and animal welfare issues, consumption of animal products has been showing a reduction trend [3].

Despite the negative effects on the environment, our planet faces the challenge of feeding a rapidly growing global population, which is projected to reach 9.73 billion by 2050 and 11.2 billion by 2100 [40], while fulfilling the obligation to reduce greenhouse gas emissions [17].

For these reasons, the involvement of milk and dairy producers at the local level and their commitment to sustainability is strategic to meet the global goals.

The farm included in this case study is part of the nearly 3000 farms active in the Parmigiano Reggiano production area. It uses the produced milk for making both Parmigiano Reggiano and soft cheeses [41]. The production of this PDO cheese follows a specification linked to the characteristics of the production area.

With its 1368 heads, the investigated farm was part of the 4.5% of dairy farms in Emilia-Romagna with a consistency greater than 500 heads, while most dairy cattle farms in this region (39.2%) own between 100 and 499 heads [42]. The farm covers an area of 5.4 ha used for animal rearing and 560 ha for agriculture, compared to the national average of 18 ha, ranking it among the larger farms [43]. In terms of productivity, the farm has an average milk production of 30–35 kg/cow/day, reaching 38 kg/cow/day in winter. These values are significantly higher than the national average, which is around 25 kg of milk per cow per day [42].

Considering the peculiarities of the farm, also within the framework of the Parmigiano Reggiano production area, and the high environmental pressure in the region, we developed a dedicated LCA model, able to valorize the primary data collected.

The LCA applied in this case study allowed us to identify the most relevant items of impact in dairy cattle farming in terms of products, such as animal feed and plastic packaging, and processes, such as machinery fuel consumption for local sourcing of supplements and standard operations of the farm. Table 2 shows the results obtained by LCA in terms of impact generated over a year, considering the farm total production and consumptions reported to the FU (1 kg of FPCM).

Table 2. Impact calculated for the yearly production of the farm and the functional unit, i.e., 1 kg of FPCM, expressed in terms of kg CO_2eq.

Process	Total Yearly Impact (kg CO_2eq)	Impact (kg CO_2eq/1 kg of FPCM)
Farm operating machines	822,797.95	19.97
Electricity consumption	61,470.30	1.49
Water consumption	10,678.27	0.26
Animal feed	244,051.51	5.92
Transport of animal feed	8903.68	0.22
Packaging film, low density polyethylene (LDPE)	443,183.57	10.76
Packaging, for fertilizers or pesticides, cleaning products (PE)	4234.03	0.10
PE recycling	−108,227.27	−2.63
Incineration of waste plastic	175,486.47	4.26
Landfilling of waste plastic	2337.88	0.06
Total	1,664,916.38	40.41

Uncertainty analysis was performed based on the Monte Carlo simulation. With a 95% confidence interval, the results range between 1,506,532.347 and 1,887,999.032 kg CO_2eq/year, with a standard deviation corresponding to about 6% of the mean value. Overall results of the simulation are available in Appendix B, Table A4.

The available literature shows high variability in the environmental impact of milk production due to the different assumptions and models used in the studies. Moreover, differences in the functional units, system boundaries, data sources, characterization factors, and allocation approaches add uncertainty to the comparisons [23,31]. For example, the study by Berton et al. [24] was based on data collected in different dairy farms of the Eastern Alps, therefore characterized by specific cattle breeds and farming and management systems. The tool proposed by Famiglietti et al. [33] outlines different system boundaries, including the cheese production phase and processes from different databases, to provide a comprehensive approach to the analysis.

However, the flows that mainly affect the results are the same, although their contribution to the total results differs due to the different assumptions and models used in each study. These include enteric fermentation, animal feed production, manure management, and spreading [44]. Nevertheless, it was evident that the case study presented discrepancies in terms of energy consumption, packaging, and animal feed compared to similar studies in the field. Consequently, the related impact results are higher than expected and suggest pathways for improvement.

As evident from the results, the diesel consumption for the farm operating machines is the most relevant item of impact, accounting for about 49% of the total impact, followed by the production of plastic packaging (27%) (Figure 2).

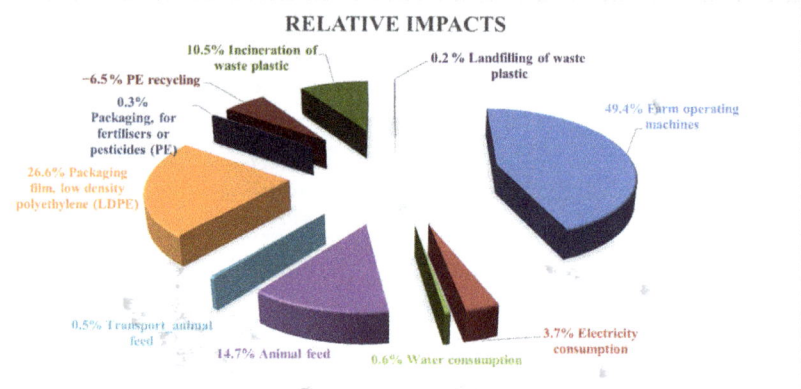

Figure 2. LCA results in terms of the relative impacts of each process.

Considering the possible solutions to tackle these specific issues, we recognized the necessity to involve the supply chain in both cases. The fuel used by the farm operating machines is a factor that is currently difficult to mitigate, as the technologies for the production of electric machines are not sufficiently advanced to ensure the necessary power for the vehicles used in livestock farming and agriculture.

Waste production and management evaluation, particularly in terms of plastic packaging coming from livestock management activities, such as the polyethylene wrapping of hay and straw bales, and cleaning product containers, represent another important item for assessing the overall environmental impact of the farm. A potential solution for reducing the production of plastic waste is almost ready at hand. Figure 2 shows that the disposal of plastic waste through recycling reduces the impact of this item by 7%. The thin polyethylene film, which represents 90% of the plastic present in the farm, can be replaced, for example, with recyclable plastic, thus significantly reducing the environmental impact and working in accordance with the Single-use Plastics Directive (SUPs) [45]. This would require a strong commitment to sustainable farming and partnership with suppliers toward the eco-design of animal feed packaging [46]. In addition to this, clear pathways to make plastic recycling the everyday standard are required to avoid the diffuse practice of incineration or, worse, landfilling [47–49]. Moreover, local initiatives may provide an opportunity for the integration of new business lines [50], as the implementation of sustainability-related practices may boost the overall performance of small and medium enterprises [51].

Animal feed production and the related logistics resulted responsible for about 16% of the total impact, considering only the sourcing from third-party suppliers, as the internal cultivation was accounted for in the consumption items for the site. Figure 3 shows in detail the impact triggered by each component of animal feed. The production phase is evidently the most impactful, and maize represents the main item of impact. The amount supplied each year is comparable with sunflower, which displays an overall impact that is almost 60% lower.

Considering only the transport of purchased animal feed (Figure 4), it is evident how the local sourcing supports limited impacts, while overseas supply triggers about 75% of the overall impact of animal feed transport.

In this regard, improvement actions that could reduce the potential environmental impact include increasing the consumption of grass silage instead of maize silage and lowering the use of concentrates or using locally produced concentrates instead of imported ones [26]. Therefore, a revision of the supply chain is strongly recommended to maintain cost control in case of fluctuating prices of international logistics [52] and to secure the consistency of supplies in uncertain international conditions, as we are experiencing in relationship with the COVID-19 crisis [53]. This would allow keeping the commitment to the SDGs by securing food supply chains (SDG 2) and improving working conditions (SDG 8) as well as the sustainability of production and consumption modes (SDG 9), with the overarching aim of reaching the climate goals (SDG 13), regardless of contingent factors [54].

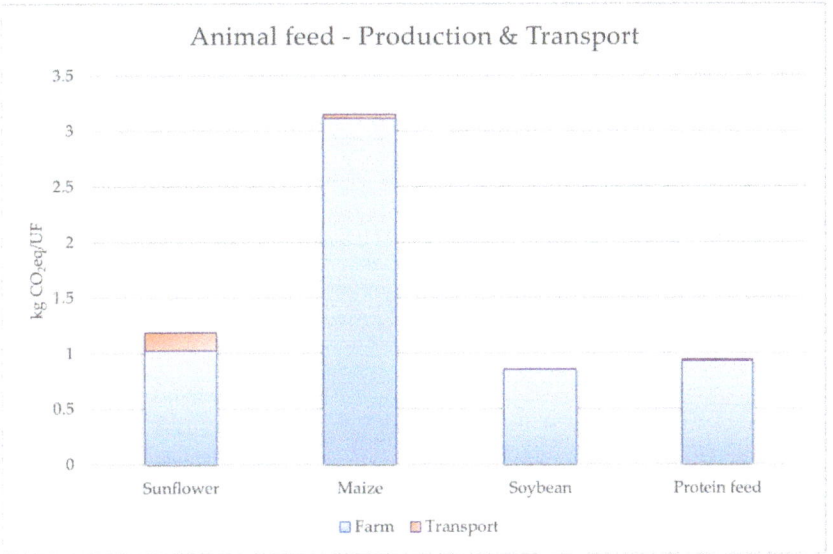

Figure 3. Impact results for external production and transport of animal feed, in terms of kgCO$_2$eq, referred to the FU.

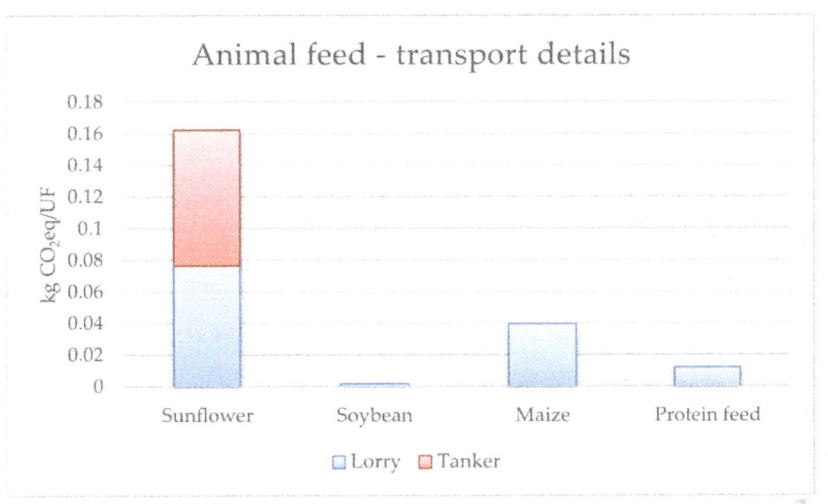

Figure 4. Impact results, in terms of kgCO$_2$eq, referred to the FU.

Table 3 summarizes the impact results for both animal feed production and transport.

Table 3. Impact calculated for the yearly supply of animal feed and the functional unit, i.e., 1 kg of FPCM, expressed in terms of kg CO_2eq.

Process	Total Yearly Impact (kg CO_2eq)	Impact (kg CO_2eq/1 kg of FPCM)
Production		
Sunflower	42,244.29	1.025
Maize	128,295.48	3.114
Soybean	35,211.73	0.855
Protein feed	38,300.00	0.930
Transport		
Sunflower	6686.42	0.162
Maize	1635.06	0.040
Soybean	82.37	0.002
Protein feed	499.83	0.012

As regards energy consumption, accounting for 4% of the total impact, according to the Institut de l'Elevage results [55], in the investigated farm, the following high-consumption areas could be identified: milk collection, refrigeration, and pumping; water heating for washing operations; overall lighting; cleaning and washing equipment.

Global warming is a major problem for livestock farming. It is particularly relevant in the Po Valley, where ensuring an adequate microclimate, which often requires putting in place air conditioning systems, is essential to reduce the risk of heat stress in cows [56]. In this case, about a quarter of the total energy consumed by the farm was related to the ventilation systems. Together with the consumption of water for summer cooling (72,000 L), this value reflects the robust farm commitment to ensuring a good welfare level for the cows.

Manure and sewage management is another fundamental issue for reducing GHG emissions, ensuring the environmental quality of rural areas, and protecting the aquifers and surrounding water basins. In this case, the lack of primary data on manure yearly production forced us to exclude the Tier 1 and Tier 2 emissions [38]. Therefore, the spreading, which the company carries out according to the guidelines of the Regional Agency for Prevention, Environment and Energy of Emilia-Romagna (Arpae), was the only treatment considered. In the modeled farm, sewage storage is carried out in uncovered basins and beds, trying to minimize the surface-volume ratio to avoid the transfer of dissolved organic carbon on the surface and reduce the related CO_2 emissions. Finally, the production of biogas from the anaerobic digestion of sludge represents an important sustainability strategy of the company, but bottom-line production data are still lacking.

4. Implications

The relationship between farm animal welfare and environmental sustainability is complex, and these two research topics have historically been addressed separately [57]. Their integrated study would help to better understand the synergies and antitheses between these two pillars of livestock farming sustainability and may facilitate the identification of coordinated actions for improvement.

The theoretical implications that would derive from such a multidisciplinary approach can be found at a systemic level since environmental impact assessment could be included in an integrated risk-based farm classification system, like the one developed in Italy, which already includes data on animal health and welfare, biosecurity, antimicrobial usage and related antimicrobial resistance [58].

Practical implications would include integrating environmental impacts and animal welfare items into control dashboards for farmers for the smart management of dairy farms. This would benefit livestock farmers as it would enable them to monitor their management strategies, both short-term, related, for instance, to animal feed purchase, and long-term, considering, for example, machinery purchasing or the implementation of infrastructures, in view of reducing the environmental impact of their farms. In perspective, the collection of summary and aggregated data from dairy farms may feed the policy-making process, supporting it with updated information about the state-of-the-art and the impact of sector-specific policies to meet the global goals.

5. Conclusions

Climate change is an established problem, and cattle farming actively contributes by emitting 14.5% of greenhouse gases from human activities [40]. Stricter environmental policies triggered by the diffuse awareness of the pressing urgency of global challenges could lead to innovative solutions that will improve the competitiveness of the livestock sector in the long term [6].

The investigation of the most relevant items of impact in the dairy farm management represents, in this sense, the first and fundamental step to building awareness and measuring the results of sustainability-oriented actions. Considering the variety of farms, management strategies and supply chains involved into the livestock industry, it is now crucial to create a solid benchmark of cases tailored to the specific milk production scenarios of different areas, thus avoiding the "one-size-fits-all" approach and allowing to identify the main items of impact based on the different production approaches.

The study presents some inherent limitations: firstly, the process of growing calves and the disposal of effluents and animal waste were excluded from the analysis; secondly, pharmaceuticals were not accounted for. In the first case, future research development may broaden the system boundaries, considering the multi-output process, i.e., meat production. In the second case, the main constraint is represented by the lack of information about these products regarding environmental impacts. However, drug use has been carefully mapped. It appears to be highly promising as a possible hotspot of impact, both in input (for the production) and output (for the contaminant load carried into animal urine and, consequently, wastewater). Considering the social dimension of sustainability, future studies may also evaluate the impacts of dairy production on workers, local communities, and society at large.

This study outlines pathways for future research at different levels despite its limitations. Firstly, it provides a preliminary LCA that can be extended to include elements so far neglected, evaluating, for example, the impact of different scenarios of drug administration. Secondly, considering the extensive data collection performed through an interdisciplinary approach. It represents the first step towards the integration of LCA with other frameworks for the performance assessment of dairy cattle farming, namely oriented to include animal health and welfare, with the final aim of evaluating livestock farms from both the environmental sustainability and social sustainability point of views.

Author Contributions: Conceptualization, A.B. and A.P.; methodology, A.B. and A.P.; software, S.Z.; validation, M.R. and S.Z.; formal analysis, M.R. and S.Z.; investigation, F.B.; resources, A.B. and A.P.; data curation, F.B. and S.Z.; writing—original draft preparation, M.R. and S.Z.; writing—review and editing, A.B.; visualization, S.Z.; supervision, A.P. All authors have read and agreed to the published version of the manuscript.

Funding: This research received no external funding.

Institutional Review Board Statement: Not applicable.

Informed Consent Statement: The farm manager signed an informed consent before the assessment started.

Data Availability Statement: All study data are presented in the article.

Acknowledgments: The authors would like to acknowledge the farmer for providing the case study and their precious collaboration throughout the project.

Conflicts of Interest: The authors declare no conflict of interest.

Appendix A

Details on the processes included are reported in the following tables.

Table A1. Overall model, process details.

Process	Amount	Unit	Notes
Resources			
Water, well, in ground, IT	72,000	L	
Materials/fuels			
Diesel burned in agricultural machinery	1,097,763.05	kWh	consumption for farm processes (both feed production and animal husbandry)
Tap water {Europe without Switzerland}	market for	APOS, U	28,035,555.2	kg	
Packaging film, low density polyethylene {GLO}	market for	APOS, U	147,510	kg	
Packaging, for fertilizers or pesticides {GLO}	market for packaging, for fertilizers or pesticides	APOS, U	16,390	kg	
Animal feed production	1	p	See Table A2
Electricity/heat			
Electricity, medium voltage {IT}	market for	APOS, U	141,840	kWh	
Animal feed transport	1	p	See Table A3
Final waste flows			
Polyethylene waste	163,900	kg	
Waste to treatment			
PE (waste treatment) {GLO}	recycling of PE	APOS, U	67,199	kg	41% of total waste, considering average regional waste disposal
Waste plastic, mixture {Europe without Switzerland}	treatment of waste plastic, mixture, municipal incineration	APOS, U	73,755	kg	45% of total waste, considering average regional waste disposal
Waste plastic, mixture {Europe without Switzerland}	treatment of waste plastic, mixture, sanitary landfill	APOS, U	22,946	kg	14% of total waste, considering average regional waste disposal

Table A2. Animal feed—production, process details.

Process Detail	Amount	Unit
Soybean, feed {GLO}	market for	APOS, U	9148	kg
Maize grain, feed {GLO}	market for	APOS, U	181,598	kg
Sunflower silage {GLO}	market for	APOS, U	191,094	kg
Protein feed, 100% crude {GLO}	market for	APOS, U	22,042	kg

Table A3. Animal feed—transport, process details.

Process Detail	Distance	Unit	Notes
Transport, freight, sea, transoceanic tanker {GLO} | market for | APOS, U	519,180.2	tkm	Sunflowers' transport—tanker, from Black Sea port (Ukraine) to Ravenna
Transport, freight, lorry 16–32 metric ton, EURO4 {GLO} | market for | APOS, U	21,211.43	tkm	Sunflower—road transport from Ravenna port
Transport, freight, lorry 16–32 metric ton, EURO4 {GLO} | market for | APOS, U	493.99	tkm	Soy—road transport from production site (Modena province)
Transport, freight, lorry 16–32 metric ton, EURO4 {GLO} | market for | APOS, U	9806.29	tkm	Maize—road transport from production site (Modena province)
Transport, freight, lorry 16–32 metric ton, EURO4 {GLO} | market for | APOS, U	2997.71	tkm	Protein feed—road transport from production site (Cremona province)

Appendix B

Results of the uncertainty analysis.

Table A4. Complete results of the Monte Carlo simulation for the overall process and for animal feed production and transport; other processes are integrated directly from Ecoinvent "SD".

	Unit	Mean	Median	Standard Deviation	Coefficient of Variation	2.5%	97.5%	Standard Error of the Mean
Total Yearly Impact	kg CO_2eq/year	1,669,074	1,661,037	102,232.5	6.125107	1,506,532	1,887,999	3232.877
Animal feed production	kg CO_2eq/year	243,847.2	242,761.9	16,982.24	6.964294	215,360.6	279,755.3	537.0255
Animal feed transport	kg CO_2eq/year	8883.801	8831.835	586.796	6.605236	7890.041	10,200.85	18.55612

References

1. Intergovernmental Panel on Climate Change—IPCC. IPCC SPECIAL REPORT: GLOBAL WARMING OF 1.5 °C (2018). Available online: https://www.ipcc.ch/sr15/ (accessed on 12 May 2022).
2. Raven, P.H.; Lasos, J.B.; Singer, S.R.; Mason Kenneth, A.; Jonson, G.B. *Biologia*, 9th ed.; Piccinin Nuova Libreria: Padova, Italy, 2011; Volume 59, pp. 1316–1343.
3. Steinfeld, H.; Gerber, P.; Wassenaar, T.; Castel, V.; Rosales, M.; De Haan, C. *Livestock's Long Shadow: Environmental Issues and Options*; FAO: Rome, Italy, 2006; p. 390.
4. FAO. *Emissions Due to Agriculture. Global, Regional and Country Trends 2000–2018*; FAOSTAT Analytical Brief Series No 18; FAO: Rome, Italy. Available online: https://www.fao.org/3/cb3808en/cb3808en.pdf (accessed on 20 February 2022).
5. Taurino, E.; Bernetti, A.; Caputo, A.; Cordella, M.; De Lauretis, R.; D'Elia, I.; Di Cristofaro, E.; Gagna, A.; Gonella, B.; Moricci, F.; et al. Italian Emission Inventory Report 1990–2020 Informative Inventory Report 2022. Available online: https://www.isprambiente.gov.it/files2022/pubblicazioni/rapporti/iir_2022_italy-stampa-rev.pdf (accessed on 12 May 2022).
6. OECD/FAO. *OECD-FAO Agricultural Outlook 2020–2029*; FAO, Rome/OECD Publishing: Paris, France, 2020; Available online: https://doi.org/10.1787/1112c23b-en (accessed on 8 February 2022).
7. Gerber, P.J.; Steinfeld, H.; Henderson, B.; Mottet, A.; Opio, C.; Dijkman, J.; Falcucci, A.; Tempio, G. *Tackling Climate Change Through Livestock: A Global Assessment of Emissions and Mitigation Opportunities*; FAO: Rome, Italy, 2013.
8. Reisinger, A.; Clark, H. How much do direct livestock emissions actually contribute to global warming? *Glob. Chang. Biol.* **2017**, 24, 1–13. [CrossRef] [PubMed]
9. FAO & Global Dairy Platform. *Climate Change and the Global Dairy Cattle Sector: The role of the Dairy Sector in a Low-Carbon Future*; FAO: Rome, Italy, 2019; Available online: http://www.fao.org/publications/card/en/c/CA2929EN/ (accessed on 10 February 2022).
10. Llonch, P.; Haskell, M.J.; Dewhurst, R.J.; Turner, S.P. Current available strategies to mitigate greenhouse gas emissions in livestock systems: An animal welfare perspective. *Anim. Int. J. Anim. Biosci.* **2017**, 11, 274–284. [CrossRef] [PubMed]
11. Henry, B.K.; Eckard, R.J.; Beauchemin, K.A. Review: Adaptation of ruminant livestock production systems to climate changes. *Animal* **2018**, 12, 445–456. [CrossRef] [PubMed]
12. Nardone, A.; Ronchi, B.; Lacetera, N.; Ranieri, M.S.; Bernabucci, U. Effects of climate changes on animal production and sustainability of livestock systems. *Livest. Sci.* **2010**, 130, 57–69. [CrossRef]
13. Smith, P.; Clark, H.; Dong, H.; Elsiddig, E.A.; Haberl, H.; Harper, R.; House, J.; Jafari, M.; Masera, O.; Mbow, C.; et al. Agriculture, Forestry and Other Land Use (AFOLU). 2014. Available online: https://www.ipcc.ch/site/assets/uploads/2018/02/ipcc_wg3_ar5_chapter11.pdf (accessed on 12 May 2022).

14. Thoma, G.; Popp, J.; Shonnard, D.; Nutter, D.; Matlock, M.; Ulrich, R.; Hellogg, W.; Neiderman, Z.; Kemper, N.P.; Adom, F.X.; et al. Regional analysis of greenhouse gas emissions from USA dairy farms: A cradle to farm-gate assessment of the American dairy industry circa 2008. *Int. Dairy J.* **2013**, *31*, S29–S40. [CrossRef]
15. Rotz, C.A. Modeling greenhouse gas emissions from dairy farms. *J. Dairy Sci.* **2018**, *101*, 6675–6690. [CrossRef]
16. de Vries, M.; de Boer, I.J.M. Comparing Environmental Impacts for Livestock Products: A Review of Life Cycle Assessments. *Live Sci.* **2010**, *128*, 1–11. [CrossRef]
17. Crosson, P.; Shalloo, L.; O'Brien, D.; Lanigan, G.J.; Foley, P.A.; Boland, T.M.; Kenny, D.A. A Review of Whole Farm Systems Models of Greenhouse Gas Emissions from Beef and Dairy Cattle Production Systems. *Anim. Feed. Sci. Technol.* **2011**, *166–167*, 29–45. [CrossRef]
18. Hagemann, M.; Hemme, T.; Ndambi, A.; Alqaisi, O.; Sultana, M.N. Benchmarking of Greenhouse Gas Emissions of Bovine Milk Production Systems for 38 Countries. *Anim. Feed Sci. Technol.* **2011**, *166–167*, 46–58. [CrossRef]
19. Milani, F.X.; Nutter, D.; Thoma, G. Invited Review: Environmental Impacts of Dairy Processing and Products: A Review. *J. Dairy Sci.* **2011**, *94*, 4243–4254. [CrossRef]
20. Yan, M.-J.; Humphreys, J.; Holden, N.M. An Evaluation of Life Cycle Assessment of European Milk Production. *J. Environ. Man.* **2011**, *92*, 372–379. [CrossRef] [PubMed]
21. Laca, A.; Gómez, N.; Laca, A.; Díaz, M. Overview on GHG Emissions of Raw Milk Production and a Comparison of Milk and Cheese Carbon Footprints of Two Different Systems from Northern Spain. *Environ. Sci. Pollut. Res.* **2020**, *27*, 1650–1666. [CrossRef] [PubMed]
22. Castanheira, É.G.; Dias, A.C.; Arroja, L.; Amaro, R. The Environmental Performance of Milk Production on a Typical Portuguese Dairy Farm. *Agri. Sys.* **2010**, *103*, 498–507. [CrossRef]
23. González-García, S.; Castanheira, E.G.; Dias, A.C.; Arroja, L. Using Life Cycle Assessment Methodology to Assess UHT Milk Production in Portugal. *Sci. Total Environ.* **2013**, *442*, 225–234. [CrossRef] [PubMed]
24. Berton, M.; Bovolenta, S.; Corazzin, M.; Gallo, L.; Pinterits, S.; Ramanzin, M.; Ressi, W.; Spigarelli, C.; Zuliani, A.; Sturaro, E. Environmental impacts of milk production and processing in the Eastern Alps: A "cradle-to-dairy gate" LCA approach. *J. Clean Prod.* **2021**, *303*, 127056. [CrossRef]
25. Kumar, M.; Kumar Choubey, V.; Deepak, A.; Gedam, V.V.; Raut, R.D. Life cycle assessment (LCA) of dairy processing industry: A case study of North India. *J. Clean Prod.* **2021**, *326*, 129331. [CrossRef]
26. O'Brien, D.; Shalloo, L.; Patton, J.; Buckley, F.; Grainger, C.; Wallace, M. A Life Cycle Assessment of Seasonal Grass-Based and Confinement Dairy Farms. *Agric. Syst.* **2012**, *107*, 33–46. [CrossRef]
27. Guerci, M.; Knudsen, M.T.; Bava, L.; Zucali, M.; Schönbach, P.; Kristensen, T. Parameters Affecting the Environmental Impact of a Range of Dairy Farming Systems in Denmark, Germany and Italy. *J. Clean Prod.* **2013**, *54*, 133–141. [CrossRef]
28. Grassauer, F.; Herndl, M.; Nemecek, T.; Fritz, C.; Guggenberger, T.; Steinwidder, A.; Zollitsch, W. Assessing and improving eco-efficiency of multifunctional dairy farming: The need to address farms' diversity. *J. Clean Prod.* **2022**, *338*, 130627. [CrossRef]
29. Fantin, V.; Buttol, P.; Pergreffi, R.; Masoni, P. Life Cycle Assessment of Italian High Quality Milk Pro-duction. A Comparison with an EPD Study. *J. Clean Prod.* **2012**, *28*, 150–159. [CrossRef]
30. Pirlo, G.; Carè, S. A Simplified Tool for Estimating Carbon Footprint of Dairy Cattle Milk. *Ital. J. Anim. Sci.* **2013**, *4*, e81. [CrossRef]
31. Guerci, M.; Bava, L.; Zucali, M.; Tamburini, A.; Sandrucci, A. Effect of summer grazing on Carbon Footprint of Milk in Italian Alps: A Sensitivity Approach. *J. Clean Prod.* **2014**, *73*, 236–244. [CrossRef]
32. Pirlo, G.; Lolli, S. Environmental Impact of Milk Production from Samples of Organic and Conventional Farms in Lombardy (Italy). *J. Clean Prod.* **2019**, *211*, 962–971. [CrossRef]
33. Famiglietti, J.; Guerci, M.; Proserpio, C.; Ravaglia, P.; Motta, M. Development and testing of the Product Environmental Footprint Milk Tool: A comprehensive LCA tool for dairy products. *Sci. Total Environ.* **2019**, *648*, 1614–1626. [CrossRef] [PubMed]
34. ISO 14044: Environmental Management—Life Cycle Assessment—Requirements and Guidelines. 2006. Available online: https://www.iso.org/standard/38498.html (accessed on 10 October 2021).
35. Wernet, G.; Bauer, C.; Steubing, B.; Reinhard, J.; Moreno-Ruiz, E.; Weidema, B. The ecoinvent database version 3 (part I): Overview and methodology. *Int. J. Life Cycle Assess* **2016**, *21*, 1218–1230. Available online: http://link.springer.com/10.1007/s11367-016-108 7-8 (accessed on 15 November 2021). [CrossRef]
36. Gerber, P.; Vellinga, T.; Opio, C.; Steifeld, H. Productivity gains and greenhouse gas emissions intensity in dairy systems. *Livest. Syst.* **2011**, *139*, 100–108. [CrossRef]
37. CLAL.it. Available online: https://www.clal.it/index.php?section=quadro_europa&country=IT (accessed on 12 January 2022).
38. Intergovernmental Panel on Climate Change—IPCC. IPCC Fifth Assessment Report. The Physical Science Basis. 2013. Available online: http://www.ipcc.ch/report/ar5/wg1/ (accessed on 20 February 2022).
39. PRé. SimaPro Database Manual—Methods Library, 4.15. 2020. Available online: https://simapro.com/wp-content/uploads/20 20/06/DatabaseManualMethods.pdf (accessed on 20 September 2021).
40. FAO. *The Future of Food and Agriculture: Trends and Challenges*; Food and Agriculture Organization of the United Nations: Rome, Italy, 2017; Available online: http://www.fao.org/publications/fofa/en/ (accessed on 12 May 2022).
41. ISTAT. 2017. Available online: http://dati.istat.it/Index.aspx (accessed on 21 September 2020).
42. CLAL. 2020. Available online: https://teseo.clal.it/?section=produttivita_capi (accessed on 21 September 2020).
43. ISTAT. 2016. Available online: http://dati.istat.it/Index.aspx (accessed on 21 September 2020).

44. Roma, R.; Corrado, S.; De Boni, A.; Forleo, M.B.; Fantin, V.; Moretti, M.; Palmieri, N.; Vitali, A.; Camillo, D.C. Life cycle assessment in the livestock and derived edible products sector. In *Life Cycle Assessment in the Agri-food Sector*; Springer: Cham, Switzerland, 2015; pp. 251–332.
45. Directive (EU) 2019/of the European Parliament and of the Council of 5 June 2019 on the Reduction of the Impact of Certain Plastic Products on the Environment. Available online: https://eur-lex.europa.eu/legal-content/EN/TXT/PDF/?uri=CELEX:32019L0904 (accessed on 12 May 2022).
46. Foschi, E.; Zanni, S.; Bonoli, A. Combining Eco-Design and LCA as Decision-Making Process to Prevent Plastics in Packaging Application. *Sustainability* **2020**, *12*, 9738. [CrossRef]
47. Pubblicazioni: PlasticsEurope. Available online: https://plasticseurope.org/wp-content/uploads/2021/10/2019-Plastics-the-facts.pdf (accessed on 12 May 2022).
48. Plastic Fantastic—Solving the Problem We Created for Ourselves | Heinrich Böll Foundation | Southeast 359 Asia Regional Office. Available online: https://th.boell.org/en/2019/10/31/plastic-fantastic-solving-problem-we-created-ourselves (accessed on 12 May 2022).
49. Jambeck, J.R.; Geyer, R.; Wilcox, C.; Siegler, T.R.; Perryman, M.; Andrady, A.; Narayan, R.; Law, K.L. Plastic 357 waste inputs from land into the ocean. *Science* **2015**, *347*, 768–771. [CrossRef]
50. Bonoli, A.; Zanni, A.; Awere, E. Organic waste composting and sustainability in low-income communities in Palestine: Lessons from a pilot project in the village. *Intl. J. Rec. Org. Waste Agric.* **2018**, *8*, 253–262. [CrossRef]
51. Mura, M.; Longo, M.; Zanni, S. Circular economy in Italian SMEs: A multi-method study. *J. Clear Prod.* **2020**, *245*, 118821. [CrossRef]
52. Kumar Tarei, P.; Kumar, G.; Ramkumar, M. A Mean-Variance robust model to minimize operational risk and supply chain cost under aleatory uncertainty: A real-life case application in petroleum supply chain. *Comp. Ind. Eng.* **2022**, *166*, 107949. [CrossRef]
53. Anderson, J.D.; Mitchell, J.L.; Maples, J.G. Invited Review: Lessons from the COVID-19 pandemic for food supply chains**Presented as part of the ARPAS Symposium: Building a Resilient Food Production System in the US: What COVID-19 and Other Black Swan Events Have Exposed About Modern Food Production, July 2021. *Appl. Animal Sci.* **2021**, *37*, 738–747. [CrossRef]
54. UN: SHARED RESPONSIBILITY, GLOBAL SOLIDARITY: Responding to the socio-economic impacts of COVID-19 March 2020. Available online: https://unsdg.un.org/sites/default/files/2020-03/SG-Report-Socio-Economic-Impact-of-Covid19.pdf (accessed on 12 May 2022).
55. Institut de l'Elevage. Le Consommations d'Energie en Batiment d'Elevage Laitier. Repères de Consommations et Pistes D'économies. Collection: Synthèse, Janvier. 2009. Available online: http://www.bretagne.synagri.com/ca1/PJ.nsf/TECHPJPARCLEF/19296/$File/CONSOMMATION%20BAT%20ELEVAGE%20LAITIER%20ADEME.pdf?OpenElement (accessed on 28 March 2022).
56. Bernabucci, U.; Biffani, S.; Buggiotti, L.; Vitali, A.; Lacetera, N.; Nardone, A. The effects of heat stress in Italian Holstein dairy cattle. *J. Dairy Sci.* **2014**, *97*, 471–486. [CrossRef] [PubMed]
57. Arvidsson Segerkvist, K.; Hansson, H.; Sonesson, U.; Gunnarsson, S. Research on Environmental, Economic, and Social Sustainability in Dairy Farming: A Systematic Mapping of Current Literature. *Sustainability* **2020**, *12*, 5502. [CrossRef]
58. Istituto Zooprofilattico Sperimentale della Lombardia e dell'Emilia-Romagna "BRUNO UBERTINI", Classyfarm. Available online: http://www.classyfarm.it (accessed on 22 April 2022).

Article

Organizational Forms and Agri-Environmental Practices: The Case of Brazilian Dairy Farms

Tiago Teixeira da Silva Siqueira [1,2,*], Danielle Galliano [2], Geneviève Nguyen [2,3] and Ferenc Istvan Bánkuti [4]

[1] INRAE UMR 1114 EMMAH, Department Agroecosytem, Dom St Paul, 84914 Avignon, France
[2] INRAE UMR 1248 AGIR, 24, Department Actions and Transitions, Chemin de Borde-Rouge, 31326 Castanet Tolosan, France; danielle.galliano@inrae.fr (D.G.); genevieve.nguyen@toulouse-inp.fr (G.N.)
[3] INP-ENSAT, Department of Social Sciences, Université Fédérale Toulouse Midi-Pyrénées, 31326 Castanet-Tolosan, France
[4] Department of Animal Science, State University of Maringá, Maringá, PR 87020-900, Brazil; fibankuti@uem.br
* Correspondence: tiago.teixeira.dasilva.siqueira@gmail.com

Citation: Siqueira, T.T.d.S.; Galliano, D.; Nguyen, G.; Bánkuti, F.I. Organizational Forms and Agri-Environmental Practices: The Case of Brazilian Dairy Farms. *Sustainability* **2021**, *13*, 3762. https://doi.org/10.3390/su13073762

Academic Editor: Rajeev Bhat

Received: 21 February 2021
Accepted: 24 March 2021
Published: 28 March 2021

Publisher's Note: MDPI stays neutral with regard to jurisdictional claims in published maps and institutional affiliations.

Copyright: © 2021 by the authors. Licensee MDPI, Basel, Switzerland. This article is an open access article distributed under the terms and conditions of the Creative Commons Attribution (CC BY) license (https://creativecommons.org/licenses/by/4.0/).

Abstract: Understanding the relationship between the organizational characteristics of a farm and its environmental performance is essential to support the agro-ecological transition of farms. This is even more important as very few studies on the subject have been undertaken and as there is a growing diversity of organizational forms of farms that differ from the traditional family model. This paper proposes a comprehensively integrated approach of dairy farms in Brazil. A case study of six archetypes of farms with contrasted organizational characteristics is developed to explore the relations between, on the one hand, farms' organizational structure and governance, and on the other hand, the adoption of agri-environmental practices. Results show that the adoption of agri-environmental practices varies across the wide range of farm's organizational forms—from the family to the industrial models. Farms with limited internal resources depend more specifically on external sectoral or territorial resources to implement environmental practices. If the environment is conducive to the creation of incentives and coordination mechanisms underlying learning processes, farms will adopt agri-environmental practices, regardless of they are organized. The creation of local cooperatives, farmer's networks and universities extension programs can strengthen farmers' absorption, adaptation and transformation capacities and boost the adoption of environmental practices. Finally, considering farms as heterogeneous organizational forms in terms of human capital, resources, market, and informational access is essential to accelerate the agroecological transition.

Keywords: farm; organization; governance; adoption; agroecology; practices; regulation; Brazil

1. Introduction

Understanding the interconnections between agricultural activities and ecosystems is essential to build sustainable agriculture. Various authors represent these complex interactions by describing the services, both positive and negative, that agriculture provides, as well as those from which it benefits [1,2]. Indeed, agriculture benefits from services generated by ecosystems (climate regulation, pollination, soil conservation, etc.) can itself contribute to the provisioning of ecosystems services (maintenance of biodiversity, carbon storage, water purification, etc.) through agri-environmental practices employed in the farms.

This dual interaction exists at the farm level and for a given production. It raises the question of understanding the relations between the farms' organization and the choice of agri-environmental practices that can at the same time reduce negative externalities and increase positive externalities. However, there are very few studies on this issue, while the diversity of organizational models of farm (peasant farms, family farms, entrepreneurial farms, family business farms, etc.) has been well documented and discussed in the scientific literature [3,4]. The majority of existing studies tend to look at only two archetypes of farms, the large agro-industrial farms, and the traditional small family farms, recognizing that the latter is more efficient in achieving sustainability objectives [3,4]. A few others are mainly interested in the relation between farm's organizations and environmental choices [5,6], but they are mainly applied to European farming systems. The issue has not yet been addressed for the newly emerged farm forms, such as the entrepreneurial farms or family business farms, and in contexts other than Europe. Do the latter perform better or worse than the traditional family farms in terms of the adoption of environmental practices? What are the organizational factors, if any, that drive the farmer's adoption of environmental practices?

This article develops a qualitative analysis of the relations between a farm's organizational forms and its agri-environmental practices. A farm's organizational form is defined by the farm's structural characteristics (size, family and employed labor force, capital structure, etc.) and its governance system. Our main hypothesis is that these internal factors, as well as the farm's external environment, do influence the adoption of environmental practices. We provide empirical evidences based on the case study of different archetypes of dairy farms in Brazil. First, dairy farms are known to generate both negative and positive externalities. Second, Brazil appears to be particularly suitable for studying such an issue. In addition to being the fourth biggest milk producer in the world, Brazil has a regulatory environment that recognizes the existence of different farm's forms of organization (Law number 11,326, 24 July 2006 established the criteria for the definition of family farming related do the size of the farm, the labor force used and the percentage of family income coming from the farming activity), ranging from the peasant forms to the large agro-industrial ones. Brazilian environmental law (Law number 12,651, 25 May 2012, known as the "Forest Code" established the rules about rivers' margins preservation with native vegetation on the width of the riverbed. The law also established the protection of 20% to 80% of farmland with native vegetation according to biome and allows exceptions to family farms) also gives special attention to family farms. Another interesting fact is that the number of family farms reduced by 9.5% between 2006 and 2017 while milk production increased by 62 percent in the same period [7]. These facts make relevant the question of the environmental impact of farm's organizational forms, on the one hand, and that of the efficiency of the measures implemented in Brazil to support the adoption of agri-environmental schemes by family farms, on the other hand. Several studies focusing on the study of Brazilian farm's organizational forms [8,9] but they usually do not consider the relation between farm's organizational forms and the adoption of agri-environmental practices [10]. Our study is based on six semi-directive face-to-face interviews with farm owners and a visit to their farms conducted in the states of Paraná and São Paulo, the 2nd and 6th largest milk producing states respectively [7].

The objective of this paper is to add to the existing literature by focusing on the process of adoption of agri-environmental practices, based on an in-depth and integrated approach to the farm's organizational forms. First, we examine the influence of farm's internal factors specific to each organizational form. Second, we study the influence of farm's regulatory, sectoral, spatial, and market environments in the adoption. This paper uses concepts from the economics of innovation [11] and concepts of evolutionary economics to build up the analytical framework. The first allows us to better understand the factors influencing the adoption. The last helps to go deeper into the learning processes associated with the adoption of agri-environmental practices on farms [12,13]. We believe that these results can provide a better understanding about how different farming organizational forms can influence the agri-environmental practices adopted by farmers. Besides that, the results can support private and public strategies to generate environmental policies that fit better the diversity of farming models in the Brazilian context.

2. Theoretical Framework

The emergence of new farm models different from the traditional family farm leads us, firstly, to consider the farm as a company similar to any other, which supports the joint production of marketed goods and non-marketed environmental goods/bad, and secondly, to study the farmer's adoption of environmental practices through the lens of an organizational innovation process undergone by its company. Our main hypothesis is that thanks to their internal characteristics, some farming organizational forms are more favorable to environmental innovation than other is. However, any organization operates in a given socio-economic environment. As such, external factors do also influence the farm's capacity to innovate. The theoretical framework developed to address these hypotheses is detailed below.

2.1. Environmental Externalities and Farm's Agri-Environmental Practices

The concept of externality has been widely used in the study of environmental problems [1,2,14,15]. An environmental externality is defined as the effects that some agents cause on the wellbeing of others and can be generated as "joint products" of the production of a good or service [1]. Indeed, given those complex interdependencies between agricultural socio-technical and environmental systems exist, we can consider that the production of marketable agricultural goods cannot be considered independently of the production of (non-marketable) environmental externalities [14]. Studies therefore converge on the fact that agricultural practices constitute satisfactory proxies to analyze the environmental externalities produced by farms [15]. These externalities are either unintended positive (carbon storage) or negative (air pollution) [2].

Dairy farms in particular are known to yield environmental externalities, both positive and negative, depending on the breeding practices. The preservation of native vegetation on farmland for grazing (especially in areas adjacent to rivers or water bodies) produces positive externalities in terms of biodiversity preservation, maintenance of water stocks, support to water quality, animal well-being, etc. [2,3,16]. Another example is the use of appropriate animal waste management to reducing microbiological and chemical pollution [17,18]. Animal effluents and used water storage and treatment facilities, especially when placed far from watercourses, can be built to minimize farms' negative externalities into the water, air as well biodiversity [6,17,18]. Feeding animals with a balanced diet is also related to the reduction of negative environmental externalities. More generally, less input intensive farming systems (i.e., that use no synthetic fertilizers and chemicals), in which permanent grasslands are grazed throughout the year, can produce less negative environmental externalities and can contribute to the production of positive environmental externalities [1,2,6,16,19]. Assessing a farm's agri-environmental practices helps indeed to better understand the environmental externalities produced through a specific farm socio-technical system.

2.2. What Organizational Factors Drive Farm's Adoption of Environmentally Friendly Practices?

The innovation economy approach proposes to study the environmental innovation process by analyzing not only the incentive and regulatory mechanisms but also the organizational characteristics of the adopter and its specific technical and sectoral systems [11]. Evolutionary approaches consider meanwhile the processes of interaction and co-evolution between the factors composing a company, and between the company and its environment [20]. Based on these two sets of theoretical literature, we propose the following framework for analyzing the determinants of farm's agri-environmental practices for different farm's organizational forms (Figure 1). Inspired by the framework developed by [6], we distinguish in particular two sets of variables: the internal factors related to farm's organizational forms (Section 2.2.1) and those related to the coordination between the organization and external actors, and more specifically those related to market, regulatory, sectoral, or spatial dimensions (Section 2.2.2).

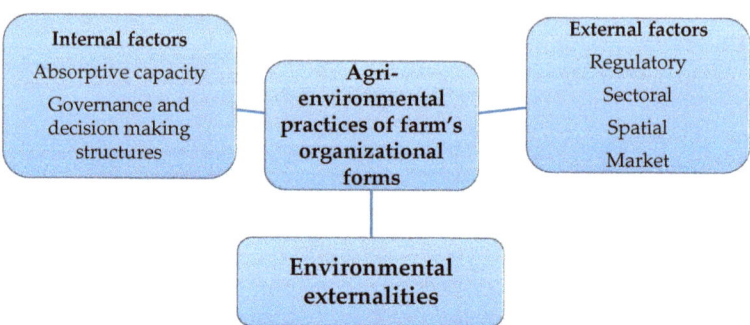

Figure 1. Framework to study the relations between farm's organizational forms and agri-environmental practices.

Reference [11] developed a framework proposed to study the environmental innovation in the industrial sector. Inspired by the latter, we intend to analyze the role of farm's absorptive capacity, governance, and decision-making structure, on the learning processes associated with the adoption of agri-environmental practices. We also study the sectoral, spatial, and market factors [21,22] related to the farms' adoption of agri-environmental practices learning processes. This theoretical framework allows us to analyze the interactions, over time, between the farm's internal and external factors in the consideration of environmental externalities by the different farm's form of organization.

2.2.1. The Internal Factors: The Role of Farm's Organizational Forms

As shown by several studies [5,6,9,12,13,23,24], some forms of organization are better able to manage certain types of environmental externalities involving different assets. They pointed out, in particular, the effect of the organization's absorptive capacity, of the manager's characteristics, and the farm's governance structure, on the adoption of agricultural practices.

The farm's absorptive capacity is related to its ability to assimilate and apply new practices [13,24]. It refers to the set of organizational strategies, routines, and processes through which a firm or system acquires, assimilates, transforms, and exploits knowledge [20,24]. This capacity is identified by factors such as investment capacity, training, quality and diversity of human and managerial skills, and the experience gained within organizations along their learning trajectories [20,24]. Thus, learning is not only a technological issue but also an organizational one. This refers to the notion of techno-organizational learning, which refers to the inseparable and progressive construction of technological and organizational capabilities of a firm in the innovation process [20]. The learning trajectories associated with an adoption process enable the creation of an organizational memory [12,20] that is often associated with the firm's dynamic capacity to adapt to a changing environment [13].

In terms of the manager's characteristics on the adoption behavior, studies stress the effect of the following ones: level of education, believes and representations [5,25], sense of belonging to a community [25,26], reliability on his advisors [26], and perceptions related with the transmission issue of the farm [27]. Literature also highlighted purely mimetic or epidemic behaviors in practices adoption behavior [6]. The role of the governance structures seems to be less explored and the few studies that have been conducted tend to oppose the traditional family structures to the so-called industrial structures on the adoption of environmental practices [3,6,12].

In fact, the literature recognizes the coexistence of different farm's governance structures: family farms with emerging forms that deviate from the traditional family farming structure [3,28]. The traditional family structure refers to a productive entity in which all assets (land, capital, labor) are under family ownership, in which agricultural production and family life are intertwined, and which is characterized by a high rate of intergenerational succession and transfer of agricultural know-how [3,4,12]. Farm structures that deviate from the family structure are characterized by a partial to total separation of ownership and asset management rights, by the development of wage employment, and by non-farmers capital investors. This phenomenon results in a high variety of farms' organizational forms described in the literature, such as: corporate-style farms [29], "factory-style corporate farms" [28] (p. 175), family business farms [4], partnership farms [28] (p. 172), entrepreneurial farms [30] or family and peasant farms [3,4,12].

2.2.2. The External Factors: The Role of Farm's Environment

The external factors are associated with mechanisms of interaction and the co-evolution of the organization with its environment [20,31]. The effect of regulation on agents' behaviors is the most discussed dimension in the standard literature in environmental economics. Trust relations and reputation as well market pull effect are also identified as key external factors in the establishment of incentive and coordination mechanisms in the process of adoption of environmental practices [11].

Concerning regulation, the empirical literature highlights the role of the mechanisms of regulatory compliance (voluntary or mandatory) and the anticipation of future regulations in the adoption of environmental practices [20,24]. The degree of stringency of the regulation, the level of implementation, and the effectiveness of the associated control mechanisms [32,33] as well as the legitimacy ascribed to the regulation [25] are also important factors of adoption. The effects of the regulatory environment on farmers' adoption of agricultural production practices were studied in European countries [32], in the United States [33], and particularly Brazil [34]. In the latter case, studies show that the combination of public control policies and incentive mechanisms for encouraging farmers to adopt environmental practices has significantly contributed to reducing impacts.

The demand for products with eco-friendly attributes is constantly increasing. Yet, it is difficult to identify eco-friendly attributes ex-ante or even ex-post [23]. Private actors seeking to respond to this demand generally do so as part of a strategic initiative to develop a brand image [22] and/or to maintain their reputation [35]. They must therefore implement incentive and coordination mechanisms for reducing uncertainty about the characteristics of their products and information asymmetry between producers and consumers [22,36]. This requires specific investments in implementing standardized processes and practices (specifications, etc.), in developing a technical mastery of all production stages, and in setting up a traceability system [21]. It is to ensure a return on these specific investments that economic actors resort to vertical integration and/or arrangements with partners based on various coordination (contracts, hierarchy, and reputation) and incentive (financial and non-financial) mechanisms [21]. Production contracts between farmers and upstream actors are the instruments traditionally used to facilitate the creation of incentive mechanisms for the adoption of low-input production practices.

The adoption of environmental practices can also be the result of a deliberative process combined with knowledge sharing between the actors of a collective [26]. Indeed, collective learning processes can encourage farmers to re-examine their shared knowledge, values, and beliefs and, in turn, convince them of the legitimacy of adopting alternative agricultural practices [26]. Finally, the participation of farmers in arrangements facilitating investments for the reduction of negative environmental externalities depends on its specific environment.

3. Materials and Methods

3.1. Criteria for Identifying the Multiple Case Studies of Dairy Farms Organizational Forms

We selected the multiple case studies representing a diversity of farm's organizational forms, which we shall present below without any prior hypothesis as to their environmental profile in terms of agri-environmental practices. Four main criteria were used to identify a diversity of farm's organizational forms studied [29]: (i) the methods of governance and operational management of the farm; (ii) the characteristics and management of the workforce; (iii) the farm's capacity for innovation and, (iv) the degree of integration of the farm in the local area, supply chains, and markets. Regarding governance, some farms pursue goals that have nothing to do with family heritage (land is seen as a production tool rather than a family asset, priority is given to short-term profitability, etc.) [3,4,29]. Some types of farms also differ from the traditional family model by the involvement of several decision-makers, by capital that may belong to investors from outside the family, and by the division of the farm into operational and decision-making units [28–30]. Regarding labor management, non-family farms may also differ from family ones. They are predominantly managed by wage-earning, generally skilled workers to whom responsibilities can be delegated [28–30].

Thus, governance and labor management are key dimensions for characterizing the different forms of organization and were central in our choice of the farms to study. These dimensions enabled us to define the main organizational profiles to be studied, ranging from the family farm forms [3,4] to more business-like forms [28–30]. Among the hybrid forms highlighted in the literature are the "family business model" [4], the "family entrepreneurial model" [7,30], or the "factory-style" industrial model [28] (p. 175). "Family farms", according to the Law number 11,326, are the most important form and represent about 80% of Brazilian dairy farms in 2006 [7]. The diversity of farm's organizational forms cannot be clearly identified in the Brazilian agricultural census [7]. However, the figures show that only 3% of farms produce more than 200 L of milk per day and it represents 35% of the milk produced in Brazil [7].

Contact with local organizations (cooperatives, universities, and farmer unions) was useful to get access to the interviewed farmers. They provided us a list of farms potentially considering a diversity of farm's organizational forms following the four main criteria established before. Then, we randomly contacted farmers and ask if they were interested in participating in the study and able to welcome us on their farms. The selected farm's organizational forms are consistent with the literature presented in Section 2.2.1. Again, we do not pretend to do a representative study of all the forms of organization presenting in the Brazilian dairy sector. We aim to illustrate contrasted farms organizational forms existing in the literature to explore the relations between, on the one hand, farms' organizational structure and governance, and on the other hand, the adoption of agri-environmental practices.

3.2. The Survey of the Selected Sample of Farms

Our methodological approach was based on a series of semi-structured interviews with dairy farm owners and a visit to their farms. We conducted the interviews in 2016 in the states of São Paulo and Paraná (Figure 2), the 6th and 2nd largest milk producing states respectively [7]. Farm visits are useful when studying sensitive topics such as environmental issues. It allow us to cross farmers' declarations and researcher's on-site observations. This also helps to draw-up the links between practices and the environmental externalities observed generated on-site.

The interview guide was structured around three key sections corresponding to the three main points presented in our theoretical framework:

1. The first section enabled us to identify the forms of organization characterizing traditional family farms, on the one hand, and other types of farms, on the other. We also identify the internal factors relative to the farms' governance structures, the socio-economic characteristics, and how it can influence the adoption of farm's agri-environmental practices.
2. The second section collects information about farm's agri-environmental practices to assess the degree of environmental externalities potentially produced by the system. It is organized around three criteria:
    (i) Farmer's compliance to the Law of May (2012) in terms of the protection of native vegetation in farmland. We used the statements made by our interviewees to assess four levels of positive externalities potentially produced: very high, if they go beyond the law; high, if they fully implemented the protection;

medium, if they partially implemented the protection; low, if they have not implemented the protection.

(ii) Production practices used (tillage method, fertilization methods, crop rotation, pesticide use, permanent grasslands are and grazing system, etc.). Based on these practices we analyze farm's input intensity and grazing systems to define the negative externalities potentially produced in three levels: high, input intensive system and no grazing; medium, semi-intensive use of inputs and permanent grassland areas where animals can graze; low, low use of inputs (use of agroecological practices: no pesticides, no synthetic fertilizers, no tillage, etc) and grazing all the year.

(iii) Animal waste management system (treatment, storage, and spreading of effluents) and distance from watercourses we can define the potential negative externalities of farms in three levels: high, no waste management system and close to watercourses; medium, partially implemented waste management system and are closed to watercourses; low, fully implemented waste management system and far from watercourses

3. The third section identifies the farm's external factors. We collect information to analyze the influence of the commercial factors on farms' agri-environmental profile, to characterize the regulatory and market environment (production standards) of the farm as well as the innovation networks in which they are involved.

Figure 2. Map of Brazil and states where data was collected.

For each section, we asked the owner to describe not only the current state of their farm organization and the agri-environmental practices used but also the main stages and the processes of changes, when relevant. In other words, this historical analysis informed us about the manager's perception regarding the various topics discussed and helped us to better characterize the mechanisms that influence the adoption of practices by farmers. Farmer's interviews and farms visits took from 3 to 5 h. During the farm visits, observations are made of the farmer's practices, and crosschecking are done with the farmer's declaration. When it is possible, discussions are conducted with people present on the farm other than the farmer interviewed. This helps to evaluate the relative validity and reliability of the interviews' content [37]. All the interviews were transcribed in full being translated from Portuguese into English. Selected verbatim excerpts were then used for the analysis and illustration of the farmer's discourse.

4. Results

4.1. Farm's Organizational Forms and Its Internal Characteristics

The dairy farms studied represent different forms of organization. The first farm (EA1) is a family farm (it meets the criteria established by the Brazilian law of 2006), while the other five farms differ in various degrees from this model. Farm EA1 is in the State of São Paulo. It has family governance. Indeed, the capital belongs to the family and the farm management has been centralized in the hands of the family (father and son) for three generations. The land property, as well as farming skills, are passed on from father to son. In terms of management, farming activities and family life are closely intertwined. Production for self-consumption is a family tradition. The interviewee's discourse also shows that he is guided by an objective of long-term profitability, and by a desire to pass on land and farming legacy through the next generations. The family derives almost all income from farming. The farm is characterized by low internal absorptive capacity, due mainly to its small size (20 milking cows) and low investment capacity; furthermore, the education level of its owners is low. Its capacity for innovation is highly dependent on external resources (advice, financing, etc.).

The second farm (EA2) is in the State of Paraná. This farm is characterized by what can be called an entrepreneurial governance style. It means that the family owns only a small portion of the farmland. The farm entrepreneur leases the land, the herd, and the farm infrastructure from his neighbor. The latter delegates the operational and financial management of his farm, as well as the decision-making power to the farm entrepreneur. Two employees work on the farm, one full-time and one part-time. Part of the production activities (maize and silage production) is outsourced to an agricultural contracting company. The farm entrepreneur, who recently graduated from a veterinary school, derives 100% of his income from managing the farm. The entrepreneur considers the activity profitable but wishes to stop it once his contract with his neighbor expires showing a short-term profitability strategy. This farm has a low absorptive capacity due to its owner's lack of experience and managerial skills (25 years-old), its low investment capacity, and small size (30 milking cows). Its capacity for innovation is highly dependent on external resources.

The third farm (EA3) is also located in Paraná. The farm governance is qualified as a "family business form", with an owner who does not work on the farm and delegates the work to an employee but has all the decision-making power. The farm accounts for only 2% of his income. He is a lawyer, and the farm is one of the properties in his investment portfolio. The owner wants to keep the farm in the family because he feels a strong attachment to the land and to dairy farming. He has two intertwined objectives: to keep the farm for leisure and personal enjoyment on the one hand, and on the other hand, to possibly pass it on to one of his daughters, who recently graduated from a veterinary school. Three employees work on the farm, one of whom supervises the other two. Only the maize harvesting and silage production operations are outsourced to another company. The farm has a limited absorptive capacity. However, despite its relatively small size (38 milking cows) and its owner's lack of farming skills and experience, the farm has a high investment capacity. Its capacity for innovation is also highly dependent on external resources.

The fourth farm (EA4) is also in Paraná State. In terms of governance, it is what we call a family partnership farm. Two brothers co-own and co-manage the family farm. One of them is a veterinarian and was the director of a multinational company in the meat industry. The other is an occupational safety consultant. They have off-farm activities, but the farm now accounts for most of their income. Decision-making is more decentralized than on the first three farms. The two brothers make strategic decisions jointly. The latter seldom perform farming tasks and delegate all the operational management to an employee. Thus, there is one manager, who supervises six other employees. The sowing and harvesting activities are outsourced. The owners' focus is on ensuring the short-term profitability of the farm, but they are also concerned about the succession. This issue is a matter of concern for the brothers, as none of their children seems to be interested in farming. The farm has a high absorptive capacity due to its large size (190 milking cows), to the owners' 30 years of experience, and considerable investment capacity. Its capacity for innovation is partly dependent on external resources.

The fifth farm (EA5) is also located in the State of Paraná. Its governance is close to an industrial or corporate farm style. Six shareholders (two from the family and four from outside the family) manage the farm. Several decision-making bodies exist and are structured hierarchically. The Board of Directors makes strategic decisions. One of the shareholders is the administrative and financial director and another is the director of production operations. In addition to being shareholders, they are also employees. Finally, there are three waged managers supervising 16 workers. All the activities related to food production and the construction of facilities are outsourced. The primary objective is financial profitability and milk production on the farm started recently (less than 5 years prior to the survey). This farm has a high absorptive capacity, due for the most part to its large size (730 milking cows), the high investment capacity, and because the shareholders have complementary skills (one animal scientist, one veterinarian specialized in reproduction, two agronomist-farmers, and dairy cow breeders, one corporate administrator, and dairy cow breeder). Their capacity for innovation depends very little on external resources.

The last farm (EA6) is in the State of São Paulo. This farm is an agribusiness corporation whose capital stock is owned exclusively by one family and its governance is that of a family-owned industrial corporation, with a board of directors composed of four family shareholders and chaired by the father. The farm has been in the family for three generations. Two of the family members work on the farm. The first is an agricultural engineer and oversees the operational management of the production and processing activities. The second is a business school graduate and acts as a financial and commercial director. The operation employs 230 wageworkers, including 12 team managers, in charge of the different stages of production, processing, and marketing of the farm's products. They outsource very few of their activities. The governance is guided by a financial as well patrimonial rationale, with a focus on profit maximization through intensification and the creation of benefit on the products. This farm has a high absorptive capacity thanks to its large size (1600 milking cows), a significant investment capacity, the diversified and competent human capital (continuous employee training), and the skills acquired through experience in managing large-scale farming operations for several generations. Their adoption of the practices depends very little on external resources.

4.2. The External Environment of the Dairy Farms

Regarding the regulatory environment, all farms' managers reported that they comply with the 2012 Brazilian law for the protection of areas covered with indigenous vegetation on farm holdings (Native Vegetation Protection Law). However, all the interviewees testify that although environmental protection is considered a major issue, the State has disengaged itself entirely from the provision of support to farming communities in their transition to compliance with regulations and has withdrawn from its role of enforcing regulations. All interviewees express a feeling of unfairness, a sense that the law is unclear and that the State does not support farmers.

Thus, for EA1, the technical support provided by a local university has been key in bringing the farm into compliance with regulatory requirements. "A professor came with students to help us with this new regulation and they even planted trees". The farm manager also expressed a sense of injustice and feel that the State is illegitimate in its application of environmental regulations. "Instead of punishing the big ones, they are going to punish the small farms of 10 to 20 hectares ... There's a big farm in the area that doesn't comply with environmental regulations and they have never been punished". The farm managers of EA2 and EA3 reported that they are aware of the importance of compliance but that the State has not adopted any compliance monitoring measure. The managers of EA2 consider that the lack of information and clarity in the laws as well as the lack of state support are major obstacles to regulatory compliance.

In EA6, our respondent concurs with this assessment "the law is not very clear and the obligation for farmers to implement conservation measures or not depends very much on the demands of the controlling officer". The managers in EA 4, 5, and 6 reports that they have had no problems in bringing their operations into compliance and that there is increasingly stringent monitoring of farms' compliance with environmental regulations in their region. The owner of EA4 underlines the key role played by the cooperative through knowledge sharing into the compliance process: "The cooperative has organized collective training sessions, provided technical support to its members to facilitate their transition to compliance with environmental standards".

Regarding milk production, due to the lack of environmental regulations governing the treatment, storage, and spreading of animal waste, some farmers are not necessarily aware that poor waste management can result in severe environmental externalities (e.g., water contamination, loss of biodiversity, etc.). Consequently, perceptions, motivations, and behaviours associated with the impacts of animal effluents vary from farmer to farmer. Due to the small volume of effluent produced on his farm, the owner of EA1 does not perceive animal waste as a major source of pollution. For the owners of EA2 and EA3, this source of pollution is not perceived as a problem either. The owners of EA4, EA5, and EA6 are more aware of the environmental impact of animal waste. According to the owner of EA4 "In a region like ours, which has a large concentration of dairy farms, with many animals confined in stalls, animal effluents is becoming a concern . . . ". As for the owner of EA6, he believes that "with the big amount of effluents produced in the farm, I can't flush it all down the river like people used to do in the old days".

We observed that the managers of EA4, EA5, and EA6 have been more proactive in anticipating future regulations but that there is also some dissatisfaction them regarding the lack of support they receive in the process towards compliance, as well as the lack of compensation for complying with the requirements. According to the owner of EA4, "regulations similar to those imposed in the swine production industry will soon apply to the dairy sector". For the owner of EA6, "environmental laws are constantly changing, and the tendency is to pay more attention to the issue of dairy effluents". Despite the cost incurred by the producer, he also states that he is vigilant and stays ahead of future regulatory changes: "In our production planning, we had already considered the question of the environmental impact and kept the recycling of effluents in mind. We have only minor adjustments. For us, compliance has not required any major structural changes".

Spatial, sectoral, and market environments of the farms are different and have an important impact on farming practices. EA1 produces raw milk (normative instruction number 62, 29 December 2011, defines production, packaging and processing criteria that allow milk to be classified into 2 categories: cooled raw milk and type A milk. Chilled raw milk concerns all volumes that cannot be qualified as type A milk. Type A milk must meet specifications with requirements for more stringent microbiological and sanitary processes and parameters. This milk costs more and is intended for consumers with greater purchasing power) and sells it through different channels and market segments (with low to high-value-added). Most of the milk is collected and processed by a small local cooperative (100 members) founded with strong involvement of the local University and which the farmer has been a member since its creation. There are no written contracts, but there is a strong sense of belonging among the members and of satisfaction with the cooperative: "The cooperative has played a central role in changing the lives of small milk producers in this region and we have always learned by working together . . . prices at the cooperative are more advantageous and stable . . . we no longer pay for the collection . . . The farmers are paid for quality and farmers are aware of the importance of producing quality milk". Some of the milk is processed into cheese that is sold at the local producers' markets. The farmer perceives this marketing channel as 'ideal'.

EA2 and EA3 produce raw milk, all of which is then sold to a private processing business. The dairy products manufactured by this business are intended for a "low-end, low added value" market that extends beyond the State of Paraná. There are no written contracts. As the EA2 farmer explains: "I can stop delivering milk overnight without getting any penalty". This can lead to disputes. The farmer adds: "They didn't pay me for 3 months in a row, their cheque bounced. I changed buyers". Processing companies set the prices, on par with the prices of the competition, but with the possibility of negotiating them. Traceability and quality standards are low or even non-existent and the farmer. The farmer reports: "they say they pay for quality but in practice, they only pay for the volume of milk we deliver to the factory ... So, there is no point in investing money and effort in improving quality". This form of opportunistic behaviour is an obstacle to risk pooling and makes it difficult to share the costs incurred in implementing quality standards.

Farms EA4 and EA5 have a contract (with an exclusivity clause) to deliver milk, to the local cooperative. If a farmer wishes to stop supplying the cooperative, he is required to give the latter at least 6 months' notice. This cooperative is larger than that mentioned above. It processes part of the milk under its own brand, supplies the national market, and more specifically the country's main consumer centres (São Paulo, Rio de Janeiro, etc.). The cooperative operates in a high value-added segment of the dairy market. The cooperative, in partnership with an international dairy company, put in place mechanisms of price incentives and technical support to encourage farmers to adopt environmentally sustainable farming practices. These mechanisms reduce uncertainties related to quality and the practices employed by farmers. They add value and help the cooperative develop a reputation for its products and brand.

EA6 farm produces type A milk, most of which is processed on-site and marketed under its own brand. "We started producing type A milk thanks to a joint venture with a well-known domestic brand. After ten years of operating as a joint venture, we had a good knowledge of this market and so we decided to start producing under our own brand". The regulations to produce this type of milk require, among other things, full traceability of the production process. The milk produced on the farm is also certified Kosher (milk produced, conserved, and processed according to dietary criteria established by the Torah). Among the products of this farm: "Type A dairy products are high-quality-differential products. They have a distinct freshness. Our customers are diverse, but most have a strong purchasing power. Direct producer-to-consumer delivery is a fast-developing marketing channel". Other processing companies also buy farm's raw milk. The strategy of producing this type of milk and selling it under their own brand requires creating a brand image and provide quality guarantees to the consumer. Provide these guarantees to build up a brand image requires the adoption of production practices complying with standards related to the welfare of workers and animals, and environmentally sustainable practices.

4.3. The Agri-Environmental Practices and the Environmental Externalities of the Dairy Farms

The degree of environmental externalities potentially produced by a farm is the result of dairy farm's agri-environmental practices. It can vary from low to very high and is assessed by three main criteria (see Section 3.2): the compliance with environmental laws related to the preservation of areas with native vegetation cover, production practices used (input intensity and grazing systems), and animal waste management (treatment, storage and spreading effluents). These practices combined allow us to define the farm's agri-environmental profile (Table 1).

Table 1. Agri-environmental profile of practices and farm's organizational forms.

	EA1 Family	EA2 Entrepreneurial	EA3 Family Business	EA4 Family Partnership	EA5 Corporate Farm	EA6 Agro-Industrial
Degree of positive externalities potentially produced associated with the farmers' protection of areas with native vegetation *	Very high, thanks to the preservation of areas with native vegetation, in full compliance with the law, and beyond.	Medium, thanks to the partial preservation of areas with native vegetation, within the limits defined by the law; drainage of the ponds for irrigation	Medium, thanks to the partial preservation of areas with native vegetation, within the limits defined by the law	Very high, thanks to the preservation of areas with native vegetation, in full compliance with the law	High, thanks to the preservation of areas with native vegetation, within the limits defined by the law	High, thanks to the reservation of areas with native vegetation, within the limits defined by the law
Degree of negative externalities potentially produced by farming practices (level of inputs and permanent grassland)	Low, due to the set of agroecological practices **	Medium, due to large acreage in permanent grassland and semi-intensive use of inputs; conventional practices	Medium due to large acreage in permanent grassland. Semi-intensive use of inputs, and use of conventional practices	Medium due to percentage of the land is in permanent grassland; But high input systems, and use of conventional practices	High, due to intensive input systems, Conventional practices	High, due to intensive input systems, Conventional practices
Degree of negative externalities potentially produced by livestock waste management	Low, due to sufficient distance between milking facilities and watercourses. Animal waste used in the family gardens	Medium, due to partially implemented waste storage facilities and proximity to watercourses	Medium, due to partially implemented waste storage facilities are proximity to watercourses	Low, due to waste storage and treatment facilities (compost barn)	Low, due to waste storage and treatment facilities (methanation)	Low due to very well managed: waste storage and treatment facilities, regular monitoring of the quality of surface and underground water
Agri-environmental profile	Agroecological with low potential externalities	Semi-intensive with medium potential externalities	Semi-intensive with medium potential externalities	(Semi)intensive with medium to low potential externalities	Intensive with low to high potential externalities	Intensive with low to high potential externalities

* Brazilian Environmental Preservation Law. Farmers are required to maintain large areas of native vegetation under protection, to maintain biodiversity. Farmers are also required to take measures to protect water sources, riverbanks, ponds, etc. In the regions studied, farmers must keep approximately 30% of their surface area with native vegetation under protection. ** Direct seeding, without using any synthetic fertilizers nor pesticides, rotational crops, year-round pasture, no silage corn production, use of grass-legume mixtures, rational rotational grazing on small plots.

Regarding the preservation of areas with native vegetation cover, the representatives of all the farms declare that they have complied with the regulations and have partially or fully protected vegetation close to riverbanks, following the criteria established by law. The owners of EA1 and EA4 report: "We have taken all the necessary measures—and more—to comply with environmental protection laws". The owners of EA2 and EA3 admit that they have not taken all the protection measures stipulated by the law. According to the owner of EA2 "In the past, the animals use to have free access to the river. Now, the areas along the riverbanks, with native vegetation cover are all protected from the livestock". The owner of EA3 reported that he had already initiated the compliance process: "a large part of the areas to be protected have already been brought up to environmental standards. The rest will be done soon". The owners of EA5 and EA6 claim that they protected the areas with native vegetation cover according to the criteria stipulated by the law. Finally, the degree of positive externalities potentially produced due to the farmers' protection of areas with native vegetation is considered as very high in the EA1 and EA4, high in the EA5 and EA6, and medium in the EA2 and EA3.

In terms of production practices, EA1 stands out from the other farms. It has adopted agroecological practices: direct seeding, no pesticides or synthetic fertilisers are used, use of grass-leguminous in the grassland, use of different species of grass, use of hardier livestock, absence of grain concentrates and corn silage in the ration, rotational grazing on 70 parcels throughout the year and no irrigation. All these practices, combined with the fact that the farm produces a low volume of milk (around 10 L/milking cow/day), lower the risk of producing negative externalities on this farm. It is also interesting to note that the extensive farming method used is suitable on a farm in which little family labour is available. "We tried, but it didn't work . . . My cousin does it, but his wife does the milking and operates a tractor too, my son's wife does not do that".

EA2, EA3, and EA4 produce corn silage and buy feed (mainly concentrates) but have a larger acreage in permanent pasture grazed by the animals. They use synthetic fertilisers, and pesticides. They use dairy cattle specialized breeds, but the cows' milk productivity per day varies between the three farms 28 L for EA4, 22 for EA2, and 18 for EA3. Therefore, farm EA4 has a slightly more intensive system than the others do. The three farms have in common genetic improvement, artificial insemination, and direct seeding practices as well as the absence of irrigation. EA2, EA3, EA4 use conventional practices and use much more inputs than EA1. Because of these practices and productivity figures, the level of environmental externalities potentially produced by farms' practices is considered as medium. This leads us to classify their production methods as semi-intensive.

EA5 outsources all food production activities. One of the farm's shareholders produces part of the feed (corn silage and grass). The latter uses conventional practices (he does, however, use direct seeding and crop rotation), including input-intensive techniques (synthetic fertilisation, pesticides, etc.). EA6 uses irrigation and conventional, input-intensive production practices (but he also uses direct seeding and crop rotation). Most of the farm's acreage is used for corn silage production, the other part being used for grass production. Part of the feed is outsourced. On both farms, animals have no access to pastures. The farmers use specialized dairy breeds with very high production potential and apply genetical improvement techniques. The daily milk output per lactating cow is approximately 40 L. Because of these intensive practices and the use of inputs, the degree of environmental externalities potentially produced by both these farms is high.

The negative externalities potentially caused by waste management vary from farm to farm. EA1 has no animal waste management system, but its potential generation of negative externalities is very low due to the production practices it uses (year-round grazing and exclusively grass-based feeding), low level of productivity, and the distance of the milking facilities from watercourses. In EA2, EA3, and EA4, the lactating animals are fed in feeding facilities but also have access to grazing paddocks throughout the year, although there is no rotational grazing. EA2 and EA3 are equipped with a rather inadequate milking and waste management facilities located close to watercourses. Because of this partial management of animal wastes, the risk of externalities potentially produced by the farms is classified as medium. EA4 manages livestock waste by collecting it into settling ponds and composting it in a compost barn (a method of treating excreta by composting excreta under confined or semi-confined animals in a building; agricultural by-products are added to the soil such as: rice husks, coffee straw, sawdust, etc.), which helps to reduce externalities. EA5 and EA6 generate a large quantity of animal waste. However, the farms have efficient waste management systems. EA5 uses methanation as a waste treatment solution and in EA6 the solid effluents are composted while the liquid waste is used for spreading. EA6 regularly monitors the quality of surface and underground water within the farm's boundaries. Because of these reasons, we consider the degree of negative externalities potentially produced by livestock waste management in EA4, EA5, and EA6 as low.

5. Discussion

5.1. Organizational Forms and Agri-Environmental Practices

On the one hand, we have observed the influence of internal factors structural characteristics, governance, and absorptive capacity. On the other hand, we have observed the influence of factors external to farms such as the regulatory, sectoral, spatial, and market environment. We also identified incentive or learning mechanisms playing in the adoption processes.

The results relative to the traditional family farm (EA1) show first that the form of organization and governance influence the farmer's consideration of environmental externalities. In accordance with [13,24], we show that farmer's choices of production methods are intrinsically linked to a strategy of adaptation to the capacities, skills, and preferences of the family as the on the farm available workforce. In line with these studies, our results also show that the farm's trajectory of adoption of agri-environmental practices is intrinsically linked to family-oriented objectives, including that of the transfer to future generations of farming traditions, knowledge, and lifestyle. Regarding external factors, the results show that with the low absorptive capacity (low levels of education and skills and low investment capacity) the sectoral and spatial factors play a major role in the adoption. The interaction with the university (for more than 20 years) has contributed to the farmer's learning and adopting agri-environmental practices. The university provided the technical and operational support necessary to bring the farm up to environmental regulations. Thus, the role of educational and research institutions is important for the definition of environmental actions in the analyzed production systems. In this sense, public policy should give more attention to regions where there is a lack of research and extension agencies. In these regions, the role of industry or collective forms of production can be an important alternative. The main role of the Universities and NGO's in the adoption of agroecological practices in Brazilian family farms is also showed in other study [9]. The reinforcement of informative networks is a key point on the development of family farm's resilience [13]. The market environment of the farm seems to have very little influence on the adoption of practices.

About the entrepreneurial farm (EA2) and the family farm business (EA3), the results first show that internal factors associated with the form of organization such as low/medium absorptive capacity (the managers' low level of experience, a lack of diversity among low skilled employees) have a limiting effect on the adoption. We also showed that limited awareness of the impacts of the practices employed on the farm is an obstacle to the adoption of agri-environmental practices [38]. The fact that the farmers (EA2) see land as "a production tool" and prioritizes short-term profitability can also explain the negative effect on the adoption of greener practices. Other authors [5] also identified this kind of environmental attitude associated with a "yield optimizer" farm governance profile. Because of the complexity of EA3's short and long-term strategies, the links between the governance of the farm and environmental strategies are less clear for this farm. The absence of incentives associated with the lack of standards compliance monitoring has a negative impact on adoption and is a major barrier to the application of Brazilian Forest

Code [16,34]. Sectoral and market characteristics also make it difficult to set up contractual and incentive arrangements for sharing the value-added, which could contribute to the adoption of agri-environmental practices. The interviewees explain that overly opportunistic behavior, combined with the absence of quality standards and compensation payments, makes difficult for relations of trust between the farmers and processing companies of the region to develop. These problems of coordination to put in place quality standards are mentioned in many studies [22,35]. The difficulty to precisely measure the agricultural environmental externalities make it harder to set up incentive arrangements for environmental practices adoption along agro-food value chains [23]. Especially in the case of livestock effluents, the lack of specific legislation, perception of its impacts, and high cost of waste treatment facilities are the main obstacles to the adoption of best waste management practices. The policy should pay more attention to dairy farms effluent pollution, mainly in the case of the intensification of agriculture practices happening in Brazil now. More than regulation, it seems important to designing incentives to the adoption of low-cost dairy waste management solutions.

In the case of the family partnership farm (EA4), we first find that the farm's high absorptive capacity (the managers' high level of training and experience and their investment capacity) is an important factor promoting adoption. Manager's awareness of the environmental impacts also drives the adoption of the farm's practices [38]. Our results corroborate other studies [26] showing that collective arrangements have a significant and positive effect on the adoption of environmental practices. In fact, by contributing to the construction of a common reputation and the development of common values between the members, the cooperative has played a key role. It facilitates the dissemination of knowledge, the distribution of value added in their production, the pooling of specific resources, learning, as well as the implementation of bonuses for the adoption of agri-environmental practices. Selling milk in a high value-added market, with standards governing quality and production practices, is a factor contributing to the reduction of environmental externalities.

Internal factors appear to be the main drivers of adoption agri-environmental practices by the corporate farm (EA5). Its large absorptive capacity (large size, high investment capacity, highly qualified human resources, and diversified skills) contributes to the establishment of coordination mechanisms in the organization that facilitate the adoption of the practices. In addition to these factors, the farmers' ability to anticipate possible stricter regulations also influences adoption. Moreover, the farm's participation in a cooperative network and the fact that it operates in a high value-added market with standards governing quality and production practices seem to be factors contributing to the adoption.

The adoption of agri-environmental practice by the industrial farm (EA6) seems to be linked to a brand image strategy. Indeed, to produce dairy products for the high value-added market in which it operates, and to be able to sell under its own brand, the farm must use the incentive and coordination mechanisms associated with the construction of an environmentally friendly" image. For this purpose, it relies on highly structured internal coordination mechanisms based on knowledge acquisition (continuous employee training). This explains the farm's high absorptive and innovation capacity (investment capacity, high level of organizational experience, highly qualified human resources, diversified skills), which positively influences the adoption of agri-environmental practices. The organizational memory developed over a three-generation long process of learning the ins and outs of industrial farming also seems to explain the implementation of agri-environmental practices, as part of a strategy of anticipation of stricter environmental regulations

5.2. What Explains Farm's Adoption of Agri-Environmental Practices? A Synthesis of the Main Sights

First, the results illustrate the internal factors associated with the farms' organizational choices and help to better understanding the adoption of agri-environmental practices. In line with the literature, the study shows that structural [6,9] and governance factors [3,5,9,12,13], the managers' perception [9,25,38], and the organization's absorptive capacity [9,13,24] influence farmer's choices in terms of environmental practices.

About governance, we observe that in the case of the family farm, the organization of production and family life are closely intertwined, which has an impact on the farmer's adoption of agri-environmental practices. We show that the involvement of the family members in farm activities is related to the practices used [27]. In fact, the availability, skills and wishes of the family labor force are important drivers of agri-environmental practices choices [27]. These corroborate the studies stressing that the adoption of practices on farms is closely related to family dynamics and changes in family preferences [4,13,21]. Our results show that technical learning processes are inseparable from organizational learning processes. The influence of the land ownership status and the farmer's objective to pass on the farm to future generations on the reduction of environmental externalities seems less clear. As for farmers' perception of the impacts of practices on the environment, behaviors vary. The limited awareness by some farm managers seems to be a major obstacle to the reduction of externalities. Some studies also highlight farmers' perceptions of environmental risk is the most important determinant of the adoption of good practices [39].

In line with the literature, we observe that the internal incentive and coordination mechanisms, as well as the organizational memory of organizations, are important determinants of adoption [9,12,13]. Indeed, we show that it is thanks to their significant investment and managerial capacities, the quality, and diversity of their human resources, and their organizational memory that these organizations can adopt agri-environmental practices.

Secondly, the results illustrate the role of external factors in the adoption process. They show that organizations with low absorptive capacity and limited internal resources rely strongly on their regulatory, spatial, and market environment to be able to implement environmental practices. Indeed, if the environment is not conducive to the creation of incentive and coordination mechanisms nor the implementation of learning processes, farms will not adopt agri-environmental practices. We find that environmental regulations must be accompanied by incentive and support policies to push the implementation of those practices in dairy farms.

In situations where the State is deficient in this respect, regulations can be fully respected only if they are accompanied by incentive mechanisms and mimetic or learning processes arising from the farm's interaction with its spatial and market environments [6,22,26]. In fact, our results show that local cooperative networks and interactions with the university play key roles in the process of adoption of agri-environmental practices by farmers. These networks allow for the emergence and development of collective learning and knowledge-sharing processes [26]. Indeed, factors such as trust, reputation, and the sharing of common values achieved through these networks all have positive effects on the adoption of agri-environmental practices [22].

The study shows that in parallel, operating in a high value-added market also gives rise to market price-based incentive mechanisms and mechanisms associated with brand image (reputation) building strategies [21,22,35]. The study shows that anticipating stricter environmental regulations is a factor that influences the choice of practices. Nevertheless, similar to other studies show, the implementation of stricter regulations and standards generates controversy concerning the cost-benefit impact of such regulations, whether in economic or social terms [16,34,40]. Indeed, our results also reveal that the adoption of quality standards, especially environmental standards, can benefit some actors while excluding others. Furthermore, as showed in the literature [39], our case studies illustrate that the spatial heterogeneity of human capital and resources makes compliance with quality and environmental standards particularly difficult in Brazil.

6. Conclusions

This article has aimed to contribute to existing knowledge on the processes of adoption of agri-environmental practices by conducting a more in-depth study of the internal organization factors and those related to farms' external environment. We contribute to make an empirical progress in the analysis of the links between models of agricultural organizations of choice of practices. For this purpose, we conducted case studies that has helped us to highlight the decision-making processes, incentive, coordination mechanisms, and learning processes on which the environmental profile of farms is based. We have used innovation and organizational economics approaches and evolutionary economics concepts to better understand the decision-making and learning processes associated with farms' adoption of agri-environmental practices.

About the internal factors, the study has highlighted the role in the adoption of agri-environmental practices of farms' structural and governance characteristics, absorptive capacity, and managers' perceptions. In the case of operations with high absorptive capacity, the farms' investment capacity, the quality, and diversity of their human resource, their organizational experience, and learning seem to play a key role in the implementation of agri-environmental practices. At the same time, we find that organizations with limited internal resources depend more specifically on external sectoral or territorial resources to be able to implement environmental practices. We have highlighted the important role of local cooperative networks and the partnership with the University in the adoption of agri-environmental practices, particularly for farms with lower absorptive capacity. Indeed, the implementation of arrangements for encouraging farmers to adopt agri-environmental practices involves the pooling of technical, informational, and financial resources as well as values related to trust and reputation. More generally, a manager's perception of the environmental externalities generated by productive practices is also a factor in the adoption of practices. The links between environmental profiles, land ownership status and farmer's succession issues seem more complex and call for further exploration.

Regarding the role of regulatory factors, the study highlights that more than setup environmental regulations, it seems important to designing incentives to push farms to preserve native vegetation and adopt waste management measures. Policy, market, and sectoral environment should provide these incentives and support mechanisms. Operating in a high value-added market also gives rise to price-based and reputation (brand image building) incentives promoting the adoption of environmental practices. The study also shows that anticipating stricter regulations is also a factor that influences the choice of practices.

This study has also provided theoretical and methodological insights. First, it seems important to use a multifactorial approach (internal structure, governance, and external environment) to understand a farm's environmental profile. It seems useful and relevant to apply the analytical framework generally used when studying eco-innovations in the industrial sector to the study of farms. Moreover, in the context of the case studies, the semi-directive interviews combined with visits and observations in the fields and on the farms enabled us to collect original, detailed, and reliable information. This approach has enabled us, not only to highlight the complex relationship between the form of organization and farm's environmental externalities by exploring the decision-making and learning processes associated with the adoption of practices. However, it is important to stress that the results of this study should be generalized with caution. Conducting interviews with a larger sample of farms would allow for a wider generalization of the results.

Finally, to promoting the adoption of better agri-environmental practices, policies should better consider farms as heterogeneous organizational forms. This heterogeneity can be related to factors composing farm's structure and governance but also to the sectoral, spatial, and market environments. Considering the constraints and needs of these different organizational forms to strengthening the farm's absorption, adaptation, and transformation capacities seems to be good insights to accelerate the agroecological transition.

Author Contributions: Conceptualization, T.T.d.S.S., D.G. and G.N.; methodology, T.T.d.S.S., D.G. and G.N.; software, T.T.d.S.S.; validation, T.T.d.S.S., D.G., G.N. and F.I.B.; formal analysis, T.T.d.S.S.; investigation, T.T.d.S.S.; resources, T.T.d.S.S., D.G. and G.N.; data curation, T.T.d.S.S.; writing—original draft preparation, T.T.d.S.S., D.G., G.N. and F.I.B.; writing—review and editing, T.T.d.S.S., D.G., G.N. and F.I.B.; visualization, T.T.d.S.S., D.G., G.N. and F.I.B.; supervision, D.G. and G.N.; project administration, T.T.d.S.S., D.G. and G.N.; funding acquisition, T.T.d.S.S. and D.G. All authors have read and agreed to the published version of the manuscript.

Funding: This research was funded by French LABEX "SMS: Structuring social worlds" (ANR-11-LABX-0066) and the PSDR4 – Repro-Innov program (INRAE—Conseil Regional Occitanie). This study was financed in part by the Coordenação de Aperfeiçoamento de Pessoal de Nível Superior—Brasil (CAPES)—(Ph.D. fellowship 0900-13-3).

Institutional Review Board Statement: Not applicable.

Informed Consent Statement: Not applicable.

Acknowledgments: We would like to thank you all the farmers that kindly welcomed me and shared their knowledge and time.

Conflicts of Interest: The authors declare no conflict of interest.

References

1. Wossink, A.; Swinton, S.M. Jointness in production and farmers' willingness to supply non-marketed ecosystem services. *Ecol. Econ.* **2007**, *64*, 297–304. [CrossRef]
2. Zhang, W.; Ricketts, T.H.; Kremen, C.; Carney, K.; Swinton, S.M. Ecosystem services and dis-services to agriculture. *Ecol. Econ.* **2007**, *64*, 253–260. [CrossRef]
3. Van der Ploeg, J.D.; Laurent, C.; Blondeau, F.; Bonnafous, P. Farm diversity, classification schemes and multifunctionality. *J. Environ. Manag.* **2009**, *90*, 124–131. [CrossRef] [PubMed]
4. Bélières, J.-F.; Bonnal, P.; Bosc, P.-M.; Losch, B.; Marzin, J.; Sourisseau, J.-M.; Baron, V.; Loyat, J. *Les Agricultures Familiales du Monde. Définitions, Contributions et Politiques Publiques*; Thirion, M.-C., Bosc, P.-M., Eds.; AFD: Paris, France, 2014; 276p.
5. Schmitzberger, I.; Wrbka, T.; Steurer, B.; Aschenbrenner, G.; Peterseil, J.; Zechmeister, H.G. How farming styles influence biodiversity maintenance in Austrian agricultural landscapes. *Agric. Ecosyst. Environ.* **2005**, *108*, 274–290. [CrossRef]
6. Galliano, D.; Siqueira, T.T.S. Organizational design and environmental performance: The case of French dairy farms. *J. Environ. Manag.* **2021**, *278*, 111408. [CrossRef]
7. IBGE—Brazilian Institute of Geography and Statistics. Censo Agropecuário 2017. 2019. Available online: https://censos.ibge.gov.br/agro/2017 (accessed on 10 November 2020).
8. Vennet, B.V.; Schneider, S.; Dessein, J. Different farming styles behind the homogenous soy production in southern Brazil. *J. Peasant. Stud.* **2015**, *43*, 396–418. [CrossRef]
9. Teixeira, H.M.; Van den Berg, L.; Cardoso, I.M.; Vermue, A.J.; Bianchi, F.J.J.A.; Peña-Claros, M.; Tittonell, P. Understanding Farm Diversity to Promote Agroecological Transitions. *Sustainability* **2018**, *10*, 4337. [CrossRef]
10. Bánkuti, F.I.; Prizon, R.C.; Damasceno, J.C.; De Brito, M.M.; Pozza, M.S.S.; Lima, P.G.L. Farmers' actions toward sustainability: A typology of dairy farms according to sustainability indicators. *Animal* **2020**, *14*, 417–423. [CrossRef]
11. Horbach, J.; Rammer, C.; Rennings, K. Determinants of eco-innovations by type of environmental impact—The role of regulatory push/pull, technology push and market pull. *Ecol. Econ.* **2012**, *78*, 112–122. [CrossRef]
12. Wilson, G.A. From 'weak' to 'strong' multifunctionality: Conceptualising farm-level multifunctional transitional pathways. *J. Rural Stud.* **2008**, *24*, 367–383. [CrossRef]
13. Darnhofer, I. Strategies of family farms to strengthen their resilience. *Environ. Policy Gov.* **2010**, *20*, 212–222. [CrossRef]
14. Kremen, C.; Miles, A. Ecosystem services in biologically diversified versus conventional farming systems: Benefits, externalities, and trade-offs. *Ecol. Soc.* **2012**, *17*, 1–25. [CrossRef]
15. Vatn, A. Markets in environmental governance—From theory to practice. *Ecol. Econ.* **2014**, *105*, 97–105. [CrossRef]
16. Siqueira, T.T.S.; Duru, M. Economics and environmental performance issues of a typical Amazonian beef farm: A case study. *J. Clean. Prod.* **2016**, *112*, 2485–2494. [CrossRef]
17. Hamoda, M.F.; Al-Awadi, S.M. Wastewater management in a dairy farm. *Water Sci. Technol.* **1995**, *32*, 1–11. [CrossRef]
18. Fyfe, J.; Hagare, D.; Sivakumar, M. Dairy shed effluent treatment and recycling: Effluent characteristics and performance. *J. Environ. Manag.* **2016**, *180*, 133–146. [CrossRef] [PubMed]
19. Ryschawy, J.; Disenhaus, C.; Bertrand, S.; Allaire, G.; Aznar, O.; Plantureux, S.; Josien, E.; Guinot, C.; Lasseur, J.; Perrot, C.; et al. Assessing multiple goods and services derived from livestock farming on a nation-wide gradient. *Animal* **2017**, *11*, 1–12. [CrossRef] [PubMed]
20. Van den Bergh, J.C.J.M. Evolutionary thinking in environmental economics. *J. Evol. Econ.* **2007**, *17*, 521–549. [CrossRef]
21. Ménard, C.; Valceschini, E. New institutions for governing the agri-food industry. *Eur. Rev. Agric. Econ.* **2005**, *32*, 421–440. [CrossRef]
22. Raynaud, E.; Sauvée, L.; Valceschini, E. Aligning branding strategies and governance of vertical transactions in agri-food chains. *Ind. Corp. Chang.* **2009**, *18*, 835–868. [CrossRef]
23. Hagedorn, K. Particular requirements for institutional analysis in nature-related sectors. *Eur. Rev. Agric. Econ.* **2008**, *35*, 357–384. [CrossRef]
24. Darnhofer, I. Resilience and why it matters for farm management. *Eur. Rev. Agric. Econ.* **2014**, *41*, 461–484. [CrossRef]
25. Davies, B.B.; Hodge, I.D. Farmers' Preferences for New Environmental Policy Instruments: Determining the Acceptability of Cross Compliance for Biodiversity Benefits. *J. Agric. Econ.* **2006**, *57*, 393–414. [CrossRef]

26. Del Corso, J.-P.; Kephaliacos, C.; Plumecocq, G. Legitimizing farmers' new knowledge, learning and practices through communicative action: Application of an agro-environmental policy. *Ecol. Econ.* **2015**, *117*, 86–96. [CrossRef]
27. Burton, R.J.F.; Walford, N.S. Multiple succession on family farms in the South East of England: A counterbalance to agricultural concentration? *J. Rural Stud.* **2005**, *21*, 335–347. [CrossRef]
28. Allen, D.W.; Lueck, D. *The Nature of the Farm: Contracts, Risk, and Organization in Agriculture*; MIT Press: Cambridge, MA, USA, 2003; 258p.
29. Nguyen, G.; Purseigle, F. The Corporate Challenge to the Family Farm. *Etudes Rurales* **2012**, *190*, 99–118. [CrossRef]
30. Pritchard, B.; Burch, D.; Lawrence, G. Neither "family" nor "corporate" farming: Australian tomato growers as farm family entrepreneurs. *J. Rural Stud.* **2007**, *23*, 75–87. [CrossRef]
31. Del Rio Gonzalez, P. The empirical analysis of the determinants for environmental technological change: A research agenda. *Ecol. Econ.* **2009**, *68*, 861–878. [CrossRef]
32. Falconer, K. Farm-level constraints on agri-environmental scheme participation: A transactional perspective. *J. Rural Stud.* **2000**, *16*, 379–394. [CrossRef]
33. Kara, E.; Ribaudo, M.; Johansson, R.C. On how environmental stringency influences adoption of best management practices in agriculture. *J. Environ. Manag.* **2008**, *88*, 1530–1537. [CrossRef] [PubMed]
34. Nepstad, D.; McGrath, D.; Stickler, C.; Alencar, A.; Azevedo, A.; Swette, B.; Bezerra, T.; DiGiano, M.; Shimada, J.; Seroa da Motta, R.; et al. Slowing Amazon deforestation through public policy and interventions in beef and soy supply chains. *Science* **2014**, *344*, 1118–1123. [CrossRef] [PubMed]
35. Carriquiry, M.; Babcock, R.A. Reputation, Market structure, and the choice of quality assurance systems in the food industry. *Am. J. Agric. Econ.* **2007**, *89*, 12–23. [CrossRef]
36. Barbieri, C.; Mahoney, E. Why is diversification an attractive farm adjustment strategy? Insights from Texas farmers and ranchers. *J. Rural Stud.* **2009**, *25*, 58–66. [CrossRef]
37. Meyer, C.B. A case in a case study methodology. *Field Methods* **2001**, *13*, 329–352. [CrossRef]
38. Hadrich, J.C.; Winkle, A.V. Awareness and proactive adoption of surface water BMPs. *J. Environ. Manag.* **2013**, *127*, 221–227. [CrossRef]
39. Toma, L.; Mathijs, E. Environmental risk perception, environmental concern and propensity to participate in organic farming programmes. *J. Environ. Manag.* **2007**, *83*, 145–157. [CrossRef]
40. Miranda, S.H.G. Evaluating WTO institutions for solving trade disputes involving non-tariff measures: Four cases involving Brazil. In *The Economics of Regulation in Agriculture: Compliance with Public and Private Standards*; Browuer, F.M., Fox, G., Jongeneel, R.A., Eds.; CABI Publishing: Oxfordshire, UK, 2012; 296p.

Article

Effect of SOP "STAR COW" on Enteric Gaseous Emissions and Dairy Cattle Performance

Elizabeth G. Ross [1], Carlyn B. Peterson [1], Angelica V. Carrazco [1], Samantha J. Werth [1], Yongjing Zhao [2], Yuee Pan [1], Edward J. DePeters [1], James G. Fadel [1], Marcello E. Chiodini [3], Lorenzo Poggianella [4] and Frank M. Mitloehner [1,*]

1. Department of Animal Science, University of California, Davis, Davis, CA 95616-8521, USA; eghumphreys@ucdavis.edu (E.G.R.); cbpeterson@ucdavis.edu (C.B.P.); acarrazco@ucdavis.edu (A.V.C.); sjwerth@ucdavis.edu (S.J.W.); yepan@ucdavis.edu (Y.P.); ejdepeters@ucdavis.edu (E.J.D.); jgfadel@ucdavis.edu (J.G.F.)
2. Air Quality Research Center, University of California, Davis, Davis, CA 95616-8521, USA; yjzhao@ucdavis.edu
3. Department of Agricultural and Environmental Sciences (DiSAA), University of Milan, Via Festa del Perdono, 7, 20122 Milano MI, Italy; marcello.chiodini@guest.unimi.it
4. Department of Land, Air and Water Resources, University of California, Davis, Davis, CA 95616-8521, USA; lpoggianella@ucdavis.edu
* Correspondence: fmmitloehner@ucdavis.edu

Received: 13 October 2020; Accepted: 4 December 2020; Published: 8 December 2020

Abstract: Feed additives have received increasing attention as a viable means to reduce enteric emissions from ruminants, which contribute to total anthropogenic methane (CH_4) emissions. The aim of this study was to investigate the efficacy of the commercial feed additive SOP STAR COW (SOP) to reduce enteric emissions from dairy cows and to assess potential impacts on milk production. Twenty cows were blocked by parity and days in milk and randomly assigned to one of two treatment groups ($n = 10$): supplemented with 8 g/day SOP STAR COW, and an unsupplemented control group. Enteric emissions were measured in individual head chambers over a 12-h period, every 14 days for six weeks. SOP-treated cows over time showed a reduction in CH_4 of 20.4% from day 14 to day 42 ($p = 0.014$), while protein % of the milk was increased (+4.9% from day 0 to day 14 ($p = 0.036$) and +6.5% from day 0 to day 42 ($p = 0.002$)). However, kg of milk protein remained similar within the SOP-treated cows over the trial period. The control and SOP-treated cows showed similar results for kg of milk fat and kg of milk protein produced per day. No differences in enteric emissions or milk parameters were detected between the control and SOP-treated cows on respective test days.

Keywords: feed additive; methane mitigation; enteric emissions; greenhouse gas; climate change

1. Introduction

Animal-sourced foods (ASF) have been under increased scrutiny due to public awareness and concern over environmental impacts. Animals are vital in many regions of the world and represent the foundation of the human food system. Animal-sourced foods can also improve national agricultural alignment to several UN Sustainable Development Goals by providing nutritious food to the population and stable livelihoods for rural communities [1], where the lack of arable land makes it possible only for ruminants to convert non-edible plants into food.

Nevertheless, the agricultural livestock sector (i.e., ASF) has been identified for its contributions to greenhouse gas (GHG) production. According to the Intergovernmental Panel on Climate Change (IPCC) [2], agriculture contributes 10 to 12% of anthropogenic CO_2, 40% of methane (CH_4), and 60% of nitrous oxide (N_2O) emissions. Methane and N_2O are the most significant greenhouse gases produced by livestock production. While N_2O originates mainly from nitrogen (N) fertilizers and manure application to agricultural soils [3], CH_4 comes from enteric fermentation in ruminants [2] and manure decomposition during storage.

In the United States, the livestock sector is estimated to contribute 35% of the anthropogenic CH_4, 72% of which originates from enteric fermentation and 28% from manure management [4]. In California, where 19% of US milk is produced [5], the California Air Resource Board inventory estimated that the dairy sector is responsible for 55% of anthropogenic CH_4 emissions, 45% of which come from enteric fermentation [6].

In rumen, feedstuffs are digested and converted through the process of microbial fermentation primarily into volatile fatty acids (VFA), including propionate, butyrate, and acetate. Methane is also produced via archaea present in the rumen. Methane constitutes a loss of approximately 5.8% of dietary gross energy intake for U.S. dairy cattle [7]. Energy loss in the form of CH_4 as well as the environmental impacts associated with enteric CH_4 production give rise to a need for CH_4 mitigation strategies in dairy production. Finding economically feasible options for dairy producers to reduce emissions is paramount because California Senate Bill 1383 requires a reduction in CH_4 emissions from California dairies by 40% from 2013 levels by 2030. California is the first state with a methane mitigation law and is setting the standard for how this reduction can be achieved in the U.S. and throughout other regions in the world.

Several enteric CH_4 mitigation strategies for dairy cattle have been investigated, including: CH_4 inhibitors such as bromochloromethane [8,9] and 3-Nitrooxypropanol [10]; electron receptors (e.g., nitrate [11]); ionophores (e.g., monensin [9]); and plant bioactive compounds such as tannins [12], essential oils [13], and bromoform found in certain seaweeds [14]. Although some of these strategies have shown promising mitigation potential, they have also manifested issues, including toxicity to the animal or the environment, short-term effects due to rumen adaptation, inconsistent results, or a negative effect on production.

SOP SQC233-005A-SQE034 (commercial name: SOP® STAR COW; SOP Srl, VA, Italy) is a feed additive containing minerals, deactivated yeast, condensed tannins from carob flour, and bentonite clay. SOP STAR COW (SOP) is processed using proprietary technology with the aim of improving feed efficiency and reducing production of CH_4, and its subsequent eructation, resulting in reduced energy loss. SOP STAR COW has been commercially available for several years and its individual components are widely used and commercialized. Over a year testing period, SOP was found to increase milk yield on seven commercial dairy farms in Italy [15]. SOP has not been previously studied for the efficacy of reducing enteric CH_4 production from lactating dairy cows. This study aims to evaluate the efficacy of the feed additive SOP on enteric gaseous emissions and the impact on milk production from lactating dairy cows. It was hypothesized, given the combination of ingredients included in SOP and the previous in vivo work conducted for milk production, that when fed to lactating dairy cows SOP will reduce CH_4 emissions and improve milk production.

2. Materials and Methods

2.1. Study Design

This study was conducted at the UC Davis Dairy Teaching and Research facility (Davis, CA, USA) with the approval of the Institutional Animal Care and Use Committee, protocol number 20601. Twenty lactating Holstein dairy cows in mid to late lactation (DIM = 153 ± 17) were randomly assigned to one of two treatment groups: treatment (SOP) or control, with 10 cows per group ($n = 10$). The study was arranged as a randomized complete block design with cows blocked by parity and days in milk. Within each treatment group, half of the animals were first lactation cows and the other half were multiparous animals either in their second or third lactation, to be representative of a typical commercial dairy operation in California.

Animals were fed an industry-standard total mixed ration (TMR) containing corn silage as the main forage component (Table 1). Diets were formulated to contain approximately 17% crude protein. Corn silage was sampled daily for dry matter (DM) with the SCiO, a handheld micro spectrometer (Consumer Physics, Inc; St. Cloud, Minnesota) to determine the correct inclusion amount for the TMR, in addition to weekly DM samples that were collected. Feed samples were dried in an oven at 100 °C for 14 h in triplicate and averaged to determine DM. All cows were adapted to the basal control diet without SOP supplementation for 14 days prior to the beginning of treatments (acclimation period, day −14 to day 0). At the end of the acclimation period, cows were fed either the control diet or the SOP treatment diet. Treatment was supplemented for a 42-day period, with the first 14 days per each cow considered as an acclimation period to the SOP feed.

Table 1. Ingredients of basal total mixed ration on an as fed and dry matter basis (kg/d/cow).

Feed Ingredients	As Fed (kg/cow Daily)	Dry Matter Basis (kg/cow Daily) [1]
Corn Silage	21.97	6.15
Corn, Steam Flaked	4.08	3.58
Soybean Meal	3.36	2.99
Alfalfa Hay	2.72	2.42
Almond Hulls	2.51	2.26
Cottonseed, Linted	1.99	1.84
Soybean Hulls	1.36	1.24
Mineral [2]	0.39	0.38
EnerGII [3]	0.27	0.26
Strata [4]	0.09	0.09
Limestone, Ground	0.09	0.09

[1] The diet was formulated using a linear program for an average milk yield of 36.5 kg daily at 3.6% milk fat that assumed an intake of 21.3 kg DM daily of the formulated diet. [2] Custom mineral mix containing: calcium, 12.56%; phosphorus, 5.33%; magnesium, 4.3%; sulfur, 2.17%; iron, 1985.36 ppm; manganese, 2664.5 ppm; zinc, 4519.78 ppm; copper, 668.8 ppm; iodine, 58.54 ppm; cobalt, 25.06 ppm; selenium, 22.79 ppm; vitamin A, 553.00 KIU/kg; vitamin D, 185.19 KIU/kg; vitamin E, 4188.79 IU/kg; biotin 58.80 mg/kg; sodium bicarb, 33.33%; magnesium oxide, 7.14%; Ethylenediamine dihydroiodide, 29.34 mg/kg; yeast, 29.32 BCFU/kg; diflubenzuron, 0.0197%; Zinpro 120, 0.88% (Nutrius, Kingsburg, CA). [3] A calcium salt of fatty acids containing 50% palmitic and 35% oleic fatty acids (Virtus Nutrition, Corcoran, CA, USA). [4] A calcium salt of fatty acids containing a blend of palmitic, stearic, and oleic fatty acids with a 16% eicosapentaenoic acid (EPA)/docosahexaenoic acid (DHA) omega-3 fatty acids (Virtus Nutrition, Corcoran, CA, USA).

The SOP additive was mixed with ground corn and fed as a top dress to deliver a total of 8 g of SOP fed per cow per day, according to the manufacturer's specifications. The treatment cow top dress included 92 g of ground corn mixed with 8 g of SOP, for a total 100 g of top dress per day. Control cows received a total of 100 g of ground corn as a top dress daily. Animals received 67 g of the top dress at the morning feeding and 33 g of the top dress in the evening, as the morning intake contained, on average, 2/3 of the cows' daily feed. Cows were individually fed their respective diets using the Calan Broadbent Feeding System (Calan gate; American Calan, Northwood, NH, USA).

Prior to the acclimation phase, each cow was trained to use their respective Calan gate. Feed was administered twice daily after the morning and evening milkings and diets were offered on an ad libitum basis, with a target of 5% daily feed refusals. Refusals were weighed before each morning feeding and sampled for DM analysis to determine daily dry matter intake (DMI). Weekly feed samples of corn silage and TMR were collected and analyzed for chemical composition and DM to ensure correct diet formulation. Chemical composition was determined by proximate analysis conducted by Denele Analytical, Inc (Woodland, CA). Dry matter was determined by drying samples in triplicate in an oven for 14 h at 100 °C and averaging the three sub samples. The feeding schedule and treatment periods for the cows were staggered to allow for gaseous emission sampling of two cows per day (one control and one treatment) in the head chamber system, with animal pairs randomly assigned to their respective treatment start time.

2.2. Emissions Measurements

Enteric emission measurements were collected using head chambers (HC). Both chamber construction and sampling procedures were based on the work of Place et al. (2011) [16]. Each head chamber was 151 cm × 104 cm × 76.2 cm (H × W × D) with polycarbonate sheeting on all sides to allow a full view of the cows during the enteric emission data collection. The chambers were equipped with head hoods specially made from Cordura waterproof fabric (Cordura Advanced Fabrics, USA) to fit the chamber opening and secure around the animal's neck. A vacuum was attached to the HC to pull air from inside the chamber and pump it outside the chamber (Peerless Blowers, Hot Springs, North Carolina, USA). Cattle were secured in the head chamber using quick-release neck chains. Emissions were collected over a 12-h period (approximately 0600 to 1800 h) and animals were sampled at 14-day intervals. HC sampling occurred on each cow's respective days 0, 14, 28, and 42.

The HC sampling system has the advantage of allowing continuous enteric emission data collection over an extended time period (12-h in the current study) and therefore reduces the cow-to-cow variability, which would be lost with shorter measurement periods. Eructated emissions were analyzed for CH_4, CO_2, N_2O, and NH_3.

Gas samples were measured in rounds of 15 min from each chamber, followed by a 15-min ambient air sampling period to correct chamber emissions from ambient emissions, for 12 h. Gas samples were collected in a mobile trailer that housed an Innova 1412 photo-acoustic multi-gas analyzer (LumaSense Technologies Inc., Ballerup, Denmark), a computer, and other support equipment. A full list of gases analyzed and their respective detection ranges are reported in Table 2.

Table 2. Gases analyzed, detection limits, and detection ranges used to measure emissions from heads chambers.

Gases	Detection Limits (µg/L)	Upper Range (g/L)
CO_2	2.75	1.83
NH_3	0.71	0.71
CH_4	0.06	0.57
N_2O	0.05	1.83

2.3. Milk Sampling

Cows were milked immediately before entering and after exiting the head chambers and had ad libitum access to their respective diet and water for the 12-h sampling period. Milk yields were collected at each milking for all animals. Milk samples were collected every 14-days and analyzed for fat, true protein, milk urea nitrogen (MUN), and solids-not-fat (SNF). Samples were sent to Central Counties DHIA (Atwater, CA, USA) for analysis and used to establish treatment period averages for ECM.

2.4. Calculations

2.4.1. Emissions Calculations

Data regarding the concentrations of the outlet air samples from the heads chamber over each 15-min period were truncated to remove the first five minutes and last two minutes of the sample to prevent carry-over effects. The following equation was used to calculate the emission rate in mg/h/head of gases from the head chambers:

$$\text{Emission Rate (mg/h/head)} = \{[(\text{MIX}) \times (\text{FL}) \times (60)]/\text{MV}\} \times (\text{MW}) \times (\text{Conv})/\text{Head} \qquad (1)$$

where MIX is the net concentration (inlet concentration—outlet concentration) in either ppm (parts per million) or ppb (parts per billion), FL is the continuous ventilation rate of the head chambers (2300–2500 L/min), 60 is the conversion from minute to hour, MV is the volume of one molar gas and equals to 24.04 (liter/mole) at temperature 20 °C and one atmosphere pressure, MW is the molecular weight of the gas in grams per mole, and Conv is a conversion factor of 10^{-3} for concentration in ppm and 10^{-6} for concentration in ppb. Head is the number of animals in the head chamber. In this experiment, Head = 1.

2.4.2. Energy-Corrected Milk

Energy-corrected milk (ECM) values were an average ECM for each two-week interval during the treatment period and calculated as follows [17]:

$$(0.327 \times \text{Milk Yield (kg)}) + (12.95 \times \text{Fat (kg)}) + (7.2 \times \text{Protein (kg)}) \tag{2}$$

Energy-corrected milk values were established for the AM and PM milkings. To establish a 24-h ECM, the AM and PM values were added together and averaged over a two-week period.

2.4.3. Corrected Dry Matter Intake

The corrected dry matter intake equation was developed from data reported in van Lingen et al. (2017), showing that approximately 25% of the CH_4 being produced from dairy cattle at any given time is coming from the previous 24 h DMI [18]. The following equation therefore accounts for the contribution of CH_4 coming from the previous day's intake:

$$\text{cDMI (kg)} = 0.25 \times \text{DMI (previous days feed (kg))} + 0.75 \times \text{DMI (in head chamber(kg))} \tag{3}$$

2.5. Data Analysis

All data were analyzed using the lmerTest package in R [19]. Least square means (LSM) and contrast between treatment by day p-values were determined using the emmeans package in R [20]. Pairwise comparisons of treatment by day interaction LSM were determined by a Tukey test using the multcompView package in R [21]. Differences were declared significant at $p \leq 0.05$ and showed a trend at $0.05 < p \leq 0.10$. p-values reported in the tables are from the ANOVA table while p-values reported in the text are from pairwise comparisons of the interaction. The model used to evaluate emissions data is:

$$Y_{ijkl} = \mu + C_i + T_j + D_k + P_l + T_j{:}D_k + e_{ijkl} \tag{4}$$

where Y_{ijkl} is the dependent variable for the ith cow in the jth treatment on the kth test day (0, 14, 28, 42) and in the lth parity. μ is the overall mean, C_i is the experimental unit (cow), T_j is the treatment, D_k is the test day (0, 14, 28, 42), P_l is the parity of the cow, $T_j{:}D_k$ is the interaction between treatment and test day, and e_{ijkl} is the error term associated with the model $\sim N(0, \sigma_e^2)$. Days in milk was initially included in the model as a continuous variable and was removed as it was not significant. Parity was included in the model as a categorical variable. Cow was a random effect, with all other variables as fixed effects.

3. Results

3.1. Enteric GHG Emissions

Table 3 shows uncorrected gas emissions for animals in the head chambers. No differences for CH_4, CO_2, N_2O, and NH_3 were detected between SOP-treated cows and control cows on respective treatment days. The analysis of CH_4 data showed that the emissions from within the SOP group had a significant decrease from day 14 to day 42 with a reduction of 20.4% (Table 3; $p = 0.014$). While the emissions from within the control group did not show significant differences over time there was still approximately a 10% reduction from day 14 to 42 (Table 3). Additionally, there was no significant differences for CH_4 seen from day 0 (prior to treatment administration) to days 14 or 42 within the SOP treatment or the control groups, meaning CH_4 emissions before SOP treatment administration were similar to CH_4 emissions after 14 and 42 days of treatment. Carbon dioxide emissions, within the

SOP-treated cows showed a decrease from day 14 to day 42 (−18.4%, $p = 0.011$), while the emissions from within the control group fluctuated without significant variations throughout the test days (Table 3). The N_2O emissions within both the control and within the SOP group increased when compared with day 0. After the SOP STAR COW supplementation, the SOP group did not show significant variations, while the control group emitted significantly ($p < 0.016$) larger amounts of N_2O at day 28 compared with day 14 (+40.6%; Table 3). Ammonia emissions decreased greatly for both SOP-treated cows and control cows after the initial measurements (day 0 of trial period; Table 3).

Table 3. Gaseous emissions from head chambers for control and treatment groups ($n = 10$) on days 0, 14, 28, and 42 with least square means, pooled standard errors (SEM), and p-values. Measured gases include methane, carbon dioxide, nitrous oxide, and ammonia.

Trait	LSM								SEM	p-Value		
	C	SOP	C	SOP	C	SOP	C	SOP		Trt	Day	Trt:Day
	d0		d14		d28		d42					
CH_4 (g/h)	24.70[ab]	21.87[ab]	24.04[ab]	25.10[b]	23.71[ab]	24.73[b]	21.59[ab]	19.98[a]	1.14	0.55	<0.001	0.014
CO_2 (g/h)	718[abc]	659[abc]	672[abc]	713[bc]	675[abc]	728[c]	593[ab]	582[a]	29.93	0.81	<0.001	0.041
N_2O (mg/h)	12.69[ab]	11.38[a]	18.00[bc]	24.45[cd]	25.31[d]	28.54[d]	28.43[d]	27.71[d]	1.62	0.065	<0.001	0.033
NH_3 (mg/h)	21.32[b]	21.51[b]	5.93[a]	5.99[a]	4.78[a]	5.15[a]	2.72[a]	1.83[a]	1.47	0.94	<0.001	0.96

Means with the same letter ([abcd]) are not significantly different ($p > 0.05$); SOP = Star Cow Treatment; C = Control; d = day; d0 = day 0; d14 = day 14; d28 = day 28; d42 = day 42; Trt = Treatment; CH_4 = methane; CO_2 = carbon dioxide, N_2O = nitrous oxide; NH_3 = ammonia.

Table 4 reports gaseous emissions standardized for DMI while animals were housed in head chambers. No differences for CH_4, CO_2, N_2O, and NH_3 were detected between SOP-treated cows and control cows on respective treatment days. The reduction seen for CH_4 in uncorrected emissions from day 14 to 42 for SOP was not seen when corrected for DMI. However, the control group does show a reduction in DMI standardized CH_4 emissions from day 0 to day 42 ($p = 0.003$), while no reduction is seen in the treatment group. A similar reduction is seen for CO_2 from day 0 to 42 ($p = 0.001$).

Table 4. Gaseous emissions corrected for dry matter intake (DMI) from head chambers (12 h period) for control and treatment groups ($n = 10$) on days 0, 14, 28, and 42 with least square means, pooled standard errors (SEM), and p-values. Emission measurements reported are on a per cow basis in either mg or g/h/kg DMI.

Trait	LSM								SEM	p-Value		
	C	SOP	C	SOP	C	SOP	C	SOP		Trt	Day	Trt:Day
	d0		d14		d28		d42					
CH_4 (g/h/kg DMI)	1.90[c]	1.63[abc]	1.61[abc]	1.73[bc]	1.43[ab]	1.63[abc]	1.28[a]	1.37[ab]	0.10	0.64	<0.01	0.027
CO_2 (g/h/kg DMI)	55.37[c]	49.15[abc]	45.08[abc]	50.99[bc]	39.36[ab]	48.25[bc]	35.27[a]	40.00[abc]	3.22	0.17	<0.01	0.020
N_2O (mg/h/kg DMI)	1.02[ab]	0.84[a]	1.19[ab]	1.42[abc]	1.55[bc]	1.90[c]	1.74[bc]	2.08[c]	0.16	0.067	<0.001	0.22
NH_3 (mg/h/kg DMI)	1.46[b]	1.44[b]	0.37[a]	0.42[a]	0.39[a]	0.40[a]	0.30[a]	0.21[a]	0.14	0.86	<0.001	0.94

Means with the same letter ([abc]) are not significantly different ($p > 0.05$); SOP = Star Cow Treatment; C = Control; d = day; d0 = day 0; d14 = day 14; d28 = day 28; d42 = day 42; Trt = Treatment; CH_4 = methane; CO_2 = carbon dioxide, N_2O = nitrous oxide; NH_3 = ammonia, DMI = dry matter intake.

Table 5 reports gaseous emissions standardized for corrected dry matter intake (cDMI) from head chamber DMI and the previous 24-h DMI [18]. No differences for CH_4, CO_2, N_2O, and NH_3 standardized for cDMI were detected between SOP-treated cows and control cows on respective treatment days. Both SOP-treated cows and control cows showed an increase from day 0 over the treatment period for N_2O.

Table 6 reports gaseous emissions corrected for energy-corrected milk values established from morning milk samples yield, fat percent, and protein percent. No differences for CH_4, CO_2, N_2O, and NH_3 standardized for ECM were detected between SOP-treated cows and control cows on respective treatment days. Both SOP-treated cows and control cows showed an increase from day 0 over the treatment period for N_2O.

Table 5. Gaseous emissions corrected for corrected dry matter intake (cDMI) from head chambers and the previous 24-h DMI for control and treatment groups ($n = 10$) on days 0, 14, 28, and 42 with least square means, pooled standard errors (SEM), and p-values. Emission measurements reported are on a per cow basis in either mg or g/h/kg cDMI. Corrected DMI was determined by the following equation: cDMI (kg) = 0.25 × DMI (previous days feed (kg)) + 0.75 × DMI (in head chamber (kg)) [18].

Trait	LSM								SEM	p-Value		
	C	SOP	C	SOP	C	SOP	C	SOP		Trt	Day	Trt:Day
	d0		d14		d28		d42					
CH$_4$ (g/h/kg cDMI)	1.13^{ab}	0.97^{ab}	1.11^b	1.08^b	1.04^{ab}	1.04^{ab}	0.94^{ab}	0.85^a	0.06	0.15	<0.01	0.28
CO$_2$ (g/h/kg cDMI)	32.95^b	29.47^{ab}	31.00^{ab}	31.84^b	28.71^{ab}	30.79^{ab}	25.71^{ab}	24.90^a	1.69	0.79	<0.01	0.18
N$_2$O (mg/h/kg cDMI)	0.53^a	0.48^a	0.80^{ab}	0.91^{bc}	1.13^{bc}	1.23^c	1.18^c	1.23^c	0.09	0.36	<0.001	0.71
NH$_3$ (mg/h/kg cDMI)	0.86^b	0.81^b	0.26^a	0.25^a	0.29^a	0.24^a	0.15^a	0.11^a	0.07	0.45	<0.001	0.98

Means with the same letter (^{abc}) are not significantly different ($p > 0.05$); SOP = Star Cow Treatment; C = Control; d = day; d0 = day 0; d14 = day 14; d28 = day 28; d42 = day 42; Trt = Treatment; CH$_4$ = methane; CO$_2$ = carbon dioxide, N$_2$O = nitrous oxide; NH$_3$ = ammonia, DMI = dry matter intake.

Table 6. Gaseous emissions corrected for morning milking's energy-corrected milk values from head chambers for control and treatment groups ($n = 10$) on days 0, 14, 28, and 42 with least square means, pooled standard errors (SEM), and p-values. Emission measurements reported are on a per cow basis in either mg or g/h/kg ECM. Energy-corrected milk was established by the following equation: (0.327 × Milk Yield (kg)) + (12.95 × Fat (kg)) + (7.2 × Protein (kg)) [17].

Trait	LSM								SEM	p-Value		
	C	SOP	C	SOP	C	SOP	C	SOP		Trt	Day	Trt:Day
	d0		d14		d28		d42					
CH$_4$ (g/h/kg ECM)	1.43	1.07	1.32	1.13	1.37	1.19	1.08	1.03	0.12	0.11	0.021	0.096
CO$_2$ (g/h/kg ECM)	41.47	32.26	36.86	33.23	37.86	35.11	29.28	29.90	3.55	0.26	0.020	0.051
N$_2$O (mg/h/kg ECM)	0.69^a	0.54^a	0.99^{abc}	0.94^{ab}	1.46^c	1.40^{bc}	1.45^{bc}	1.39^{bc}	0.14	0.44	<0.001	0.96
NH$_3$ (mg/h/kg ECM)	1.12^b	1.02^b	0.33^a	0.29^a	0.27^a	0.24^a	0.09^a	0.05^a	0.08	0.30	<0.001	0.98

Means with the same letter (^{abc}) are not significantly different ($p > 0.05$); SOP = Star Cow Treatment; C = Control; d = day; d0 = day 0; d14 = day 14; d28 = day 28; d42 = day 42; Trt = Treatment; CH$_4$ = methane; CO$_2$ = carbon dioxide, N$_2$O = nitrous oxide; NH$_3$ = ammonia, ECM = energy-corrected milk.

3.2. Milk Parameters and Intake

The cows enrolled on trial were mid to late lactation (approximately 153 ± 17 days in milk). Over the 42-day treatment period milk yield, ECM, kg of milk fat, milk fat %, kg of milk protein, milk protein %, MUN, dry matter intake from 12 h in head chambers (DMI HC), and average DMI consumed in Calan gate pens outside of head chambers for each 14-day study period (DMI AVG) were not significantly different for the treatment by day interaction (Table 7). Day is representative of the average over the 14-day study period for milk yield, ECM, milk fat %, milk protein %, MUN, and DMI AVG. There was one missing data point for milk component analysis for a milk sample on day 0 during the morning milking. Data for milk yield, and DMI were complete.

No significant variations were observed within or between groups for DMI HC or for DMI AVG (Table 7). There was no difference between the control and SOP-treated cows on respective test days for % milk protein. Within the groups, the SOP treatment resulted in a significant increase in % milk protein, with higher % protein levels throughout the study period (+4.9% from day 0 to day 14 ($p = 0.036$) and +6.5% from day 0 to day 42 ($p = 0.002$; Table 7). No changes were detected in the % milk protein within the control. However, the control and SOP-treated cows showed similar results for kg of milk fat and kg of milk protein produced per day (Table 7).

Table 7. Least square means (LSM), pooled standard errors (SEM), and *p*-values for the control (C) and treated (SOP) groups on study days for milk yield, ECM, kg milk fat, milk fat %, kg milk protein, and milk protein %, milk urea nitrogen (MUN), and dry matter intake in the head chambers (DMI HC) and DMI averaged over the 14 day period (DMI AVG).

Trait	LSM								SEM	*p*-Value		
	C	SOP	C	SOP	C	SOP	C	SOP		Trt	Day	Trt:Day
	d0		d14		d28		d42					
Milk yield (kg/day)	35.8	35.4	34.1	36.2	34.8	34.0	35.6	34.4	0.94	0.92	0.51	0.16
ECM (kg/day)	38.8	39.8	37.4	39.9	38.7	38.5	39.0	39.0	1.19	0.54	0.87	0.38
Fat (kg/day)	1.43	1.59	1.42	1.52	1.46	1.52	1.51	1.42	0.06	0.37	0.71	0.047
Milk Fat (%)	4.13	4.40	4.17	4.22	4.36	4.32	4.19	4.23	0.14	0.46	0.17	0.68
Protein (kg/day)	1.08	1.12	1.09	1.16	1.09	1.16	1.15	1.11	0.04	0.41	0.37	0.004
Protein (%)	3.12[ab]	3.08[a]	3.22[ab]	3.23[b]	3.27[ab]	3.27[b]	3.18[ab]	3.28[b]	0.06	0.66	<0.001	0.21
MUN (mg/100 mL)	12.68[b]	13.41[b]	8.64[a]	8.26[a]	12.11[b]	12.17[b]	9.20[a]	8.63[a]	0.55	0.93	<0.01	0.38
DMI HC (kg/12 h)	12.50[a]	12.82[ab]	15.11[ab]	13.96[ab]	16.82[b]	15.91[ab]	17.29[ab]	15.26[ab]	1.10	0.36	<0.01	0.20
DMI AVG (kg/day)	24.54	25.29	24.24	24.91	24.77	25.50	26.22	26.88	0.79	0.38	<0.001	0.99

Means with the same letter ([abc]) are not significantly different (*p* > 0.05); SOP = STAR COW treatment; C = control; d = day; d0 = day 0; d14 = day 14; d28 = day 28; d42 = day 42, Trt = treatment; ECM = energy-corrected milk; MUN = milk urea nitrogen.

4. Discussion

The use of feed additives to mitigate enteric emissions has received growing attention in recent years since feed additives have the potential to satisfy regulations requiring the dairy sector to reduce its environmental footprint. The present study focused on the possible effects of the commercial feed additive, SOP STAR COW, on enteric emissions and dairy cattle performance.

4.1. Effects on Enteric Emissions

There were no pairwise comparison differences detected for any measured parameter between SOP treatment and controls on respective treatment days. There was a day effect showing a reduction in uncorrected CH_4 emissions and an increase in milk protein within the SOP-treated group over time, which was not measured in the control group. As control and SOP-treated cows did not show significantly different data on respective test days, the efficacy of using SOP STAR COW as an effective means of reducing enteric CH_4 could not be completely validated.

Correcting emissions for DMI in the HC can be problematic as some animals tend to consume less while in the head chambers than they normally would. This can be seen in Table 7, where there is minimal numeric differences in the average DMI of the animals; however, when in the HC, the SOP-treated animals—after day 0—were consistently eating between 1 to 2 kg less feed on a dry matter basis than control cows. While the difference in DMI in the HC was not significant, this can have an effect on standardizing emissions for DMI. Additionally, not all of the CH_4 being measured in the HC is attributable to the feed being consumed in the HC. Van Lingen et al. (2017) showed that up to 25% of measured CH_4 emissions from cattle are from feed consumed in the previous 24 h [18]. A respiratory chamber study using sheep found that approximately 50% of CH_4 emissions could be attributed to the previous 48 hours' DMI [22]. Further research is needed to establish a more precise model for a DMI correction specific for dairy cattle in head chambers. However, Equation 3, used in this study, helps account for some of the variation in intake while in the HCs and likely gives a more accurate representation of standardized CH_4 emissions than just using the HC DMI correction.

Some of STAR COW's components, such as bentonite, tannins, and yeast have previously been shown to individually reduce enteric emissions. Bentonite clay was toxic to some protozoa as it interfered with cilia motion and this has been shown in vitro, when applied at 10% in the feed, to cause an increase in bacterial populations compared with control samples, as well as a reduction in NH_3 production due to its ability to bind NH_3 [23]. Wallace and Newbold (1991), utilizing a Rusitec in vitro design, and Abdullah et al. (1995), using sheep in an in vivo experiment, found an inhibitory effect of bentonite on holotrich protozoa [23,24]. Abdullah et al. (1995) additionally found that a 2% DMI supplementation of sodium bentonite increased the entodinia protozoal population [24].

A large portion of the methanogen population have an endosymbiotic relationship with protozoa, with holotrichs and entodinia supporting up to 526 and 96 methanogens internally. This helps explain why defaunation, in some cases, can result in CH_4 mitigation [25]. However, it is unlikely that the small quantity of bentonite in the SOP dosage would have this effect on the rumen. It is possible that a higher quantity of bentonite may be more effective at mitigating CH_4 emissions.

Research has determined two possible mechanisms to achieve a reduction in enteric CH_4 emissions in cattle after tannin supplementation, including (1) decreasing hydrogen production through a reduction in fiber digestibility, and (2) the inhibition of methanogens [12]. Previous research on tannins as feed additives focused on their ability to improve nutrient utilization efficiency, in particular nitrogen (N), and reduce nutrient loss via NH_3 emissions into the environment [26,27]. A recent in vitro study found a 20 to 27% decrease in CH_4 emissions, as well as a decrease in the total VFA and the acetate to propionate ratio, by injecting both hydrolyzed (HT) and condensed tannins (CT) at a 1:1 ratio into the rumen volume [28]. Reducing total VFA content is not ideal as this indicates a reduction in overall rumen fermentation, which would reduce feed efficiency and production performance. However, these trials were including tannins at a much higher dose than the current study and in vitro trials are not always representative of the effect that will be seen in vivo, largely due to the lack of time microbial populations have to adjust, and are more indicative of short-term results. Further in vivo research is needed to determine if a higher dose of SOP can be more effective at mitigating CH_4 emissions and if VFA concentrations in the rumen or DMI are altered.

Similarly, significant decreases in the molar % of acetate, acetate to propionate ratio, and crude protein digestibility were noted when CT were fed to Angus cattle at 2% DM; however, no differences in CH_4 or BW were seen [29]. Usually, a decrease in acetate to propionate concentrations is consistent with other methods of decreasing enteric CH_4 emissions as it is likely a result of decreased levels of hydrogen being available as a substrate for methanogens in the rumen.

The SOP treatment used low quantities of material (8 g/animal per day, approx. 0.04% DM of the complete feed). Other additives, including tannins, usually need to be included at 20 g/kg diet DM to have a reliable reduction in CH_4 [30]. However, a synergistic effect of the components in SOP has never been researched for CH_4 mitigation. Borgonovo et al. (2019) and Peterson et al. (2020) [31,32] found that gypsum processed with SOP's proprietary technology reduced NH_3 and GHG emissions in liquid manure with a much lower dosage of gypsum than reported previously [33]. However, this same effect with low doses was not seen in the current study.

Given that the uncorrected gas emissions results showed a reduction in CH_4 over time for SOP-treated cows, it is possible that SOP STAR COW has some CH_4 mitigation potential. However, to determine a true reduction potential, this change would also need to be seen when standardized for DMI. Likewise, SOP STAR COW might have better CH_4 mitigation responses if fed at higher amounts, as most effective feed additives with similar compounds are fed at a much higher percentage of DMI [28–30].

4.2. Effects on Milk Production

SOP STAR COW-treated cows showed similar results to control cows for the treatment by day interaction for all milk parameters and intake data. Within the SOP treatment group, the cows showed an increase in milk protein percent over the course of trial period; however, kg of milk protein remained similar within the SOP-treated cows over the trial period.

SOP STAR COW contains tannins, which have the ability to bind proteins in the rumen, thus reducing protein degradation by rumen microorganisms and making proteins available for digestion in the small intestine. This likely increased the availability of amino acids (AA) for the animal to absorb from the feed. Previous research on tannins as a feed additive focused on their ability to improve nutrient utilization efficiency, in particular nitrogen (N) [28,29], though these studies did not investigate their ability to reduce enteric emissions. Aguerre et al. (2016) found that feeding CT at 0.45% of diet DM resulted in an increase in milk protein yield; however, at CT levels higher than 0.45%, milk protein yield and percentage were decreased [29]. While an increase in % milk protein was seen in the current study within the SOP-treated cows over time, there was no difference in kg of milk protein produced, so synthesis of milk protein remained unchanged.

SOP STAR COW contains deactivated yeast cells, which act as a prebiotic for microbiota in the digestive system. Yeast cultures provide soluble growth factors such as organic acids, B vitamins, and amino acids that stimulate the growth of ruminal bacteria populations that utilize lactate and digest cellulose [34]. The supplementation of diets with yeasts was used to increase the final protein content in the milk, by providing probiotic and prebiotic materials to the ruminal flora. Several studies have confirmed the effect of yeast on milk protein percentage, but these studies used live yeast cultures [35–37]. Both deactivated yeast and CT in SOP STAR COW potentially explain the increased protein percent over time in the milk of SOP-treated cows, while the % protein in the control cows' milk remained unchanged.

Previous research has shown that supplementing dairy cows with yeast cultures increased DMI and milk yield and decreased the acetate to propionate ratio in the rumen [38,39]. Additionally, yeast supplementation altered the amino acid profile of bacterial protein and the flow of methionine from the rumen to the small intestine, which could potentially increase milk protein synthesis [39]. Further research is needed to determine if SOP increases the post-ruminal flow of methionine in support of milk protein synthesis.

Since tannins and deactivated yeast comprise only 5% of SOP content, coupled with the small feeding inclusion (8 g/head/day), the suggested mode of action to increase protein content might be related to an increase and/or a shift in the rumen microbial population.

As recent studies have investigated the role of predominant clusters of ruminal microbes in milk production and CH_4 formation [40], further investigations should determine the potential impact of SOP STAR COW on the microbial populations in the rumen as an approach to explain its potential modes of action.

5. Conclusions

No differences were detected for enteric emissions, standardized enteric emissions, milk parameters, or intake between control and SOP-treated cows on respective test days. Analyzing the two groups separately, within SOP-treated cows over time, showed a significant reduction in CH_4 of 20.4% from day 14 to day 42, while the protein % of the milk was increased (+4.9% from day 0 to day 14 and +6.5% from day 0 to day 42). Over time, within the control group, there was no reduction in CH_4 or increase in milk protein. Within the SOP-treated cows, the kg of milk protein remained similar throughout the duration of the study. Tannins and yeast, present in SOP STAR COW, may be effective compounds that enable a reduction in enteric CH_4 emissions, and should be researched further. Future research should investigate the effects of long-term supplementation or higher doses of SOP STAR COW, in order to determine if greater mitigation effects on CH_4 emissions and increases in milk production can be established. Increasing pressure from legislation and consumers is being put on the dairy industry to reduce the environmental impact of dairy production, especially as it relates to climate change. Determining feed additives that both reduce emissions and improve the production of lactating dairy cows is both essential for producers to meet current CH_4 reduction regulations and is an important step towards a sustainable food system.

Author Contributions: Conceptualization, E.G.R., C.B.P., E.J.D. and F.M.M.; data curation, E.G.R., C.B.P., A.V.C., S.J.W., Y.Z., Y.P. and L.P.; formal analysis, E.G.R. and J.G.F.; Funding acquisition, F.M.M.; investigation, E.G.R., C.B.P., A.V.C., S.J.W. and F.M.M.; methodology, E.G.R., C.B.P. and F.M.M.; project administration, E.G.R., Y.Z. and F.M.M.; writing—original draft, E.G.R. and M.E.C.; writing—review and editing, C.B.P., A.V.C., S.J.W., Y.Z., Y.P., E.J.D., J.G.F., L.P. and F.M.M.; All authors have read and agreed to the published version of the manuscript.

Funding: This study was funded by SOP Srl (SOP Srl, VA, Italy).

Acknowledgments: We kindly thank: Doug Gisi, the manager of the UC Davis Dairy, and his staff for their help and support with this trial, the undergraduate student interns that were instrumental in the day to day work, as well as our sponsors at SOP.

Conflicts of Interest: The authors declare no conflict of interest.

References

1. Adesogan, A.T.; Havelaar, A.H.; McKune, S.L.; Eilittä, M.; Dahl, G.E. Animal source foods: Sustainability problem or malnutrition and sustainability solution? Perspective matters. *Glob. Food Secur.* **2020**, *25*, 100325. [CrossRef]
2. IPCC. *IPCC Emissions from Livestock and Manure Management (Chapter 10)*; Eggleston, H.S., Buendia, L., Miwa, K., Ngara, T., Tanabe, K., Eds.; IGES: Kanagawa, Japan, 2006; Volume 4.
3. Syakila, A.; Kroeze, C. The global nitrous oxide budget revisited. *Greenh. Gas Meas. Manag.* **2011**, *1*, 17–26. [CrossRef]
4. USEPA. Inventory of U.S. Greenhouse Gas Emissions and Sinks: 1990–2015. 2017. Available online: https://www.epa.gov/sites/production/files/2017-02/documents/2017_chapter_5_agriculture.pdf (accessed on 17 June 2020).
5. United States Department of Agriculture Economic Research Service. Available online: https://www.ers.usda.gov/webdocs/DataFiles/48685/milkcowsandprod.xlsx?v=1518.6 (accessed on 17 June 2020).
6. CARB. California Air Resources Board. Short-Lived Climate Pollutant Reduction Strategy. 2017. Available online: https://ww2.arb.ca.gov/sites/default/files/2020-07/final_SLCP_strategy.pdf (accessed on 15 June 2020).
7. Johnson, K.A.; Johnson, D.E. Methane emissions from cattle. *J. Anim. Sci.* **1995**, *73*, 2483–2492. [CrossRef] [PubMed]
8. Knight, T.; Ronimus, R.S.; Dey, D.; Tootill, C.; Naylor, G.; Evans, P.; Molano, G.; Smith, A.; Tavendale, M.; Pinares-Patiño, C.; et al. Chloroform decreases rumen methanogenesis and methanogen populations without altering rumen function in cattle. *Anim. Feed. Sci. Technol.* **2011**, *166*, 101–112. [CrossRef]
9. Adesogan, T.; Yang, W.; Lee, C.; Gerber, P.J.; Henderson, B.; Tricarico, J.M. Mitigation of methane and nitrous oxide emissions from animal—SPECIAL TOPICS. *J. Anim. Sci.* **2013**, *91*, 5045–5069.
10. Hristov, A.N.; Oh, J.; Kindermann, M.; Duval, S.; Giallongo, F.; Frederick, T.W.; Harper, M.T.; Weeks, H.L.; Branco, A.F.; Moate, P.J.; et al. An inhibitor persistently decreased enteric methane emission from dairy cows with no negative effect on milk production. *Proc. Natl. Acad. Sci. USA* **2015**, *112*, 10663–10668. [CrossRef] [PubMed]
11. Van Zijderveld, S.; Gerrits, W.; Dijkstra, J.; Newbold, J.; Hulshof, R.; Perdok, H. Persistency of methane mitigation by dietary nitrate supplementation in dairy cows. *J. Dairy Sci.* **2011**, *94*, 4028–4038. [CrossRef]
12. Tavendale, M.H.; Meagher, L.P.; Pacheco, D.; Walker, N.; Attwood, G.T.; Sivakumaran, S. Methane production from in vitro rumen incubations with Lotus pedunculatus and Medicago sativa, and effects of extractable condensed tannin fractions on methanogenesis. *Anim. Feed. Sci. Technol.* **2005**, *123*, 403–419. [CrossRef]
13. Benchaar, C.; Greathead, H. Essential oils and opportunities to mitigate enteric methane emissions from ruminants. *Anim. Feed. Sci. Technol.* **2011**, *166*, 338–355. [CrossRef]
14. Roque, B.M.; Brooke, C.G.; Ladau, J.; Polley, T.; Marsh, L.J.; Najafi, N.; Pandey, P.; Singh, L.; Kinley, R.; Salwen, J.K.; et al. Effect of the macroalgae Asparagopsis taxiformis on methane production and rumen microbiome assemblage. *Animal Microbiome* **2019**, *1*, 3. [CrossRef]
15. Luparia, P.; Rota, N.; Poggianella, M.; Bronzo, V. Preliminary results of a feed additive for rumen functionality through in-field monitoring on 7 Italian commercial dairy farms. In Proceedings of the 19th European Society of Veterinary and Comparative Nutrition (ESVCN) Congress, Tulouse, France, 17–19 September 2015.
16. Place, S.; Pan, Y.; Zhao, Y.; Mitloehner, F.M. Construction and operation of a ventilated hood system for measuring greehouse gas and volatile organic compound emissions from cattle. *Animals* **2011**, *1*, 433–446. [CrossRef]
17. Dairy Records Management Systems DHI Glossary. Available online: http://www.drms.org/PDF/materials/glossary.pdf (accessed on 12 December 2019).
18. Van Lingen, H.J.; Edwards, J.E.; Vaidya, J.D.; Van Gastelen, S.; Saccenti, E.; Bogert, B.V.D.; Bannink, A.; Smidt, H.; Plugge, C.M.; Dijkstra, J. Diurnal Dynamics of Gaseous and Dissolved Metabolites and Microbiota Composition in the Bovine Rumen. *Front. Microbiol.* **2017**, *8*, 425. [CrossRef]
19. Kuznetsova, A.; Brockhoff, P.B.; Christensen, R.H.B. lemerTest: Tests in Linear Mixed Effects Models. R package version 2.0-30. *J. Stat. Softw.* **2017**, *82*, 1–26. [CrossRef]
20. Lenth, R.; Singmann, H.; Love, J.; Buerkner, P.; Herve, M. Emmeans: Estimated Marginal Means, Aka Least-Squares Means; R package version 1.3.2; 2019. Available online: https://CRAN.R-project.org/package=emmeans (accessed on 18 March 2019).
21. Graves, S.; Piepho, H.P.; Selzer, L. MulticompView: Visualizations of Paired Comparisons; R package version 0.1-7. 2017. Available online: https//github.com/rvlenth/emmeans/issues (accessed on 18 March 2019).

22. Robinson, D.L.; Goopy, J.P.; Donaldson, A.J.; Woodgate, R.T.; Oddy, V.H.; Hegarty, R.S. Sire and liveweight affect feed intake and methane emissions of sheep confined in respiration chambers. *Animal* **2014**, *8*, 1935–1944. [CrossRef]
23. Wallace, R.J.; Newbold, C.J. Effects of bentonite on fermentation in the rumen simulation technique (Rusitec) and on rumen ciliate protozoa. *J. Agric. Sci.* **1991**, *116*, 163–168. [CrossRef]
24. Abdullah, N.; Hanita, H.; Ho, Y.W.; Kudo, H.; Jalaludin, S.; Ivan, M. The effects of bentonite on rumen protozoal population and rumen fluid characteristics of sheep fed palm kernel cake. *Asian Australas. J. Anim. Sci.* **1995**, *8*, 249–254. [CrossRef]
25. Finlay, B.J.; Esteban, G.; Clarke, K.J.; Williams, A.G.; Embley, T.M.; Hirt, R.P. Some rumen ciliates have endosymbiotic methanogens. *FEMS Microbiol. Lett.* **1994**, *117*, 157–161. [CrossRef] [PubMed]
26. Dschaak, C.; Williams, C.; Holt, M.; Eun, J.-S.; Young, A.; Min, B. Effects of supplementing condensed tannin extract on intake, digestion, ruminal fermentation, and milk production of lactating dairy cows. *J. Dairy Sci.* **2011**, *94*, 2508–2519. [CrossRef] [PubMed]
27. Aguerre, M.; Capozzolo, M.; Lencioni, P.; Cabral, C.; Wattiaux, M. Effect of quebracho-chestnut tannin extracts at 2 dietary crude protein levels on performance, rumen fermentation, and nitrogen partitioning in dairy cows. *J. Dairy Sci.* **2016**, *99*, 4476–4486. [CrossRef]
28. Jayanegara, A.; Makkar, H.P.S.; Becker, K. Addition of Purified Tannin Sources and Polyethylene Glycol Treatment on Methane Emission and Rumen Fermentation in Vitro. *Media Peternak.* **2015**, *38*, 57–63. [CrossRef]
29. Beauchemin, K.A.; McGinn, S.M.; Martínez, T.F.; McAllister, T.A. Use of condensed tannin extract from quebracho trees to reduce methane emissions from cattle. *J. Anim. Sci.* **2007**, *85*, 1990–1996. [CrossRef]
30. Jayanegara, A.; Leiber, F.; Kreuzer, M. Meta-analysis of the relationship between dietary tannin level and methane formation in ruminants from in vivo and in vitro experiments. *J. Anim. Physiol. Anim. Nutr.* **2011**, *96*, 365–375. [CrossRef]
31. Borgonovo, F.; Conti, C.; Lovarelli, D.; Ferrante, V.; Guarino, M. Improving the Sustainability of Dairy Slurry by A Commercial Additive Treatment. *Sustainability* **2019**, *11*, 4998. [CrossRef]
32. Peterson, C.B.; El-Mashad, H.M.; Zhao, Y.; Pan, Y.; Mitloehner, F.M. Effects of SOP Lagoon Additive on Gaseous Emissions from Stored Liquid Dairy Manure. *Sustainability* **2020**, *12*, 1393. [CrossRef]
33. Yang, F.; Li, G.; Shi, H.; Wang, Y. Effects of phosphogypsum and superphosphate on compost maturity and gaseous emissions during kitchen waste composting. *Waste Manag.* **2015**, *36*, 70–76. [CrossRef] [PubMed]
34. Callaway, E.; Martin, S. Effects of a Saccharomyces cerevisiae Culture on Ruminal Bacteria that Utilize Lactate and Digest Cellulose. *J. Dairy Sci.* **1997**, *80*, 2035–2044. [CrossRef]
35. Moallem, U.; Lehrer, H.; Livshitz, L.; Zachut, M.; Yakoby, S. The effects of live yeast supplementation to dairy cows during the hot season on production, feed efficiency, and digestibility. *J. Dairy Sci.* **2009**, *92*, 343–351. [CrossRef] [PubMed]
36. Rossow, H.A.; Riordan, T. Effects of addition of a live yeast product on dairy cattle performance. *J. Appl. Anim. Res.* **2018**, *46*, 159–163. [CrossRef]
37. Dias, A.; Freitas, J.; Micai, B.; Azevedo, R.; Greco, L.; Santos, J. Effect of supplemental yeast culture and dietary starch content on rumen fermentation and digestion in dairy cows. *J. Dairy Sci.* **2018**, *101*, 201–221. [CrossRef]
38. Erasmus, L.; Botha, P.; Kistner, A. Effect of Yeast Culture Supplement on Production, Rumen Fermentation, and Duodenal Nitrogen Flow in Dairy Cows. *J. Dairy Sci.* **1992**, *75*, 3056–3065. [CrossRef]
39. Williams, P.E.; Tait, C.; Innes, G.M.; Newbold, C.J. Effects of the inclusion of yeast culture (Saccharomyces cerevisiae plus growth medium) in the diet of dairy cows on milk yield and forage degradation and fermentation patterns in the rumen of steers. *J. Anim. Sci.* **1991**, *69*, 3016–3026. [CrossRef]
40. Wallace, R.J.; Sasson, G.; Garnsworthy, P.C.; Tapio, I.; Gregson, E.; Bani, P.; Huhtanen, P.; Bayat, A.; Strozzi, F.; Biscarini, F.; et al. A heritable subset of the core rumen microbiome dictates dairy cow productivity and emissions. *Sci. Adv.* **2019**, *5*, eaav8391. [CrossRef]

Publisher's Note: MDPI stays neutral with regard to jurisdictional claims in published maps and institutional affiliations.

© 2020 by the authors. Licensee MDPI, Basel, Switzerland. This article is an open access article distributed under the terms and conditions of the Creative Commons Attribution (CC BY) license (http://creativecommons.org/licenses/by/4.0/).

Review

A Global Review of Monitoring, Modeling, and Analyses of Water Demand in Dairy Farming

Philip Shine [1,*], Michael D. Murphy [1] and John Upton [2]

1. Department of Process, Energy and Transport Engineering, Cork Institute of Technology, Cork, Ireland; michaelD.murphy@cit.ie
2. Animal and Grassland Research and Innovation Centre, Teagasc Moorepark Fermoy, Co., Cork, Ireland; john.upton@teagasc.ie
* Correspondence: philip.shine@cit.ie

Received: 13 August 2020; Accepted: 31 August 2020; Published: 3 September 2020

Abstract: The production of milk must be balanced with the sustainable consumption of water resources to ensure the future sustainability of the global dairy industry. Thus, this review article aimed to collate and summarize the literature in the dairy water-usage domain. While green water use (e.g., rainfall) was found to be largest category of water use on both stall and pasture-based dairy farms, on-farm blue water (i.e., freshwater) may be much more susceptible to local water shortages due to the nature of its localized supply through rivers, lakes, or groundwater aquifers. Research related to freshwater use on dairy farms has focused on monitoring, modeling, and analyzing the parlor water use and free water intake of dairy cows. Parlor water use depends upon factors related to milk precooling, farm size, milking systems, farming systems, and washing practices. Dry matter intake is a prominent variable in explaining free water intake variability; however, due to the unavailability of accurate data, some studies have reported moving away from dry matter intake at the expense of prediction accuracy. Machine-learning algorithms have been shown to improve dairy water-prediction accuracy by 23%, which may allow for coarse model inputs without reducing accuracy. Accurate models of on-farm water use allow for an increased number of dairy farms to be used in water footprinting studies, as the need for physical metering equipment is mitigated.

Keywords: dairy; water; review; modelling; water footprint; agriculture

1. Introduction

Of the global water supply, 97% is saline, and thus unsuitable for human consumption. Freshwater, encompassing the remaining 3%, therefore represents earths most valuable natural resource, essential for human and animal consumption, agriculture, ensuring biodiversity and ecosystems, as well as offering hygiene services and a vital commodity used by industry [1]. However, only 0.4% of the world's total freshwater is readily available as surface water in rivers, lakes, and wetlands, with the vast majority embedded in glaciers, polar ice caps, or groundwater aquifers [2]. In conjunction with a growing population and the resulting increase in agricultural, industrial, and energy requirements, the per capita availability of freshwater is reducing [2]. Thus, water policy is an increasingly topical subject globally, with the introduction of the EU Water Framework Directive (2000/60/EC) and EU Groundwater Directive (2006/118/EC) aiming to improve and protect the water quality of rivers, lakes, groundwater, and transitional coastal waters [3,4]. The EU Groundwater Directive states, *"water is not a commercial product like any other but, rather, a heritage which must be protected, defended and treated as such"*.

The agricultural sector represents the largest consumer of freshwater globally, accounting for approximately 70% of all global freshwater use [5]. The agricultural sector therefore requires substantial improvements in water-use efficiency and productivity, as the production of agricultural products to meet the needs of a growing population is limited by not only land availability, but also the availability of freshwater [2]. The three largest product categories contributing to the global water footprint are cereals (27%), meat (22%), and dairy products (7%) [6]. Although the production of dairy products is the third largest contributor, its share may increase over the coming decades, as the global consumption of milk and dairy produce is forecasted to increase by 19% by 2050 compared to 2005–2007 levels [5]. The increase of dairy herds and consequent milk production comes with its own unique challenges regarding the consumption of freshwater. The magnitude and efficiency (liters of water per dairy cow, liters of water per liter of milk produced, etc.) of water use on dairy farms varies according to a number of factors including, but not limited to, irrigation requirements, type of production system (e.g., grazing or confinement), level of milk production, type of milking system (e.g., conventional or automatic milking system (AMS)), geographical location, and environmental conditions. Thus, understanding variances in water consumption may offer on-farm benefits regarding the protection of public water systems and local freshwater supplies, and allow optimization of infrastructural equipment for a cost-effective water system.

Water consumption incorporates (1) green water, which is the water required for soil moisture due to evapotranspiration; (2) grey water, which is the volume of freshwater required to mitigate a given pollutant load via dilution; and (3) blue water, which is fresh surface and groundwater usage. Direct use of freshwater on dairy farms may be required for yard wash down, milk pre-cooling, hot washing of milking equipment, and cow drinking water as well as miscellaneous usage throughout the farm, and is therefore the most controllable water use by the farmer. An adequate freshwater system is important to ensure the sufficient hydration of dairy cows for good milk production as well as for the sanitation of infrastructural equipment [7]. The direct water demand required will increase along with expanding milk production and may increase to unsustainable levels, potentially depleting groundwater borehole supplies and placing additional pressure on the public water supply [8]. In particular, the water demand may rise dramatically in line with milk production, which may cause local water shortages during periods of scarce rainfall.

It is clear that the production of milk and dairy produce globally must be carried out with considerations regarding water consumption in order to ensure the future sustainability of the dairy industry. Thus, research in this domain will become increasingly important as researchers aim to identify new technologies and methods to improve the water sustainability of dairy farming. However, there has been a lack of secondary research undertaken in this domain to date, particularly related to direct freshwater use. At this point in time, it is important to understand the current state of the research pertaining to dairy-farm water consumption, and to allow for the identification of areas in which future research efforts should be focused to support the future monitoring and sustainability of global milk production. Therefore, this review focused on critically assessing published literature related to the monitoring, prediction modeling, and analysis of water consumption in dairy farming. Although overall water footprinting and life-cycle assessments in the literature are covered, this review places specific emphasis on the water consumption on dairy farms that can be controlled by the farmer, i.e., direct freshwater consumption for plant washing and cow drinking.

This review incorporates three primary components. Section 2 explores research studies focused on dairy water consumption, prediction modeling, and analysis, while Section 3 provides a discussion overview highlighting common trends throughout the dairy water literature prior to Section 4, which provides a concise conclusion to this review.

2. Dairy Water Consumption

Water consumption may be categorized as green water, grey water, or blue water. Green water is defined as the infiltrated rainfall in the unsaturated soil layer, while not incorporating green-water run-off recharging groundwater supplies [9,10]. Consumption of green water refers to the usage of precipitation (rainwater), required for the production of pasture, forages, and/or crops (e.g., wheat, barley, soybeans), through evapotranspiration, necessary for the manufacturing of feed for animal livestock (i.e., dairy cows) [11,12].

Grey-water consumption refers to the virtual volume of freshwater required to dilute pollutants present in water (due to the manufacturing process) to the required ambient water-quality standards [11,12]. From a dairy production perspective, one source of grey-water consumption might be the dilution of water utilized (and mixed with cleaning products) for cleaning and sanitizing milk bulk tanks and milk harvesting equipment [13]. Thus, grey water is an indicator of water pollution, and is often excluded from water footprint studies as it represents a virtual quantity of water as opposed to an actual water use [14–16]. Lastly, blue water refers to the consumption of freshwater (from groundwater sources, lakes, rivers, or mains supply) that is not immediately returned to the same catchment [11,12]. Demand for blue water therefore results in a loss of water from the available freshwater source in a catchment area through evaporation, and returning to either a different catchment area or the sea, or is incorporated into a product. Sources of blue-water demand on dairy farms include the requirement of freshwater for irrigation, cow drinking water, and water for cleaning plant equipment.

The quantities of green, grey, and blue water required in dairy production vary from region to region depending on the frequency of precipitation and the availability of groundwater sources, which will influence both the ability to grow pasture and thus the type of dairy system employed (e.g., grazing or confinement). Although water may be considered a renewable resource, its availability may be considerably limited in certain months of the year due to meteorological and geographical constraints [17]. Thus, understanding the volume of water consumed in dairy farming and the type of water consumed in conjunction with the water availability is significantly important when quantifying the sustainability of dairy production (and all agricultural activities) in a particular region.

There are three primary methods generally employed across agricultural water literature to quantify water sustainability across the entire supply chain (direct and indirect), as described by Ran et al. [18]. These include (1) water-footprint assessments, which quantify the volume of water required to produce a good or service (e.g., liters of water per kg fat and protein-corrected milk (FPCM)). Generally used for global assessments, calculating the water footprint allows for the identification of the point on the supply chain at which water is required and consumed. A synopsis of dairy water-footprint studies is shown in Table 1; (2) life-cycle assessments, which calculate the volume of water required within a defined system boundary may also be adjusted using a water-stress index (WSI) to account for localized water stress where it is consumed. Although very data-intensive, life-cycle assessments allow for the local impact of water usage to be calculated in relation to potential water scarcity during certain periods using a country-/region-specific WSI [19]. The WSI is a dimensionless value ranging from 0 to 1, measuring the ratio between total annual water withdrawal and total annual water availability, while accounting for seasonal variability in precipitation and flows in a specific country/region; (3) water productivity relates to quantifying water consumption relative to physical or economic outputs. Generally, water-productivity calculations do not separate green- and blue-water consumption, but can help to reduce water footprints substantially by improving overall water productivity, which may be necessary in regions with scarce water supplies.

Global water footprint studies carried out by Mekonnen and Hoekstra [20] estimated country-specific green-, grey-, and blue-water consumption values for a range of farm animals and animal products, distinguishing between different production systems (grazing, industrial, and mixed) while considering the country-specific conditions related to feed-conversion efficiency of the animal, origin of the feed, and feed composition. They estimated a total water-footprint value of 1299 L of water (L_w) kg^{-1} FPCM, whereby 91% was attributed to green-water use, 5% was attributed to grey-water use, and 4% was attributed to blue-water consumption (Table 1). Water-footprint calculations were carried out across countries worldwide. Estimated water-footprint values for China, the Netherlands, the USA, and India are also shown in Table 1, in conjunction with values from six additional research studies found in the literature [13,15,16,21–23]. Two of these additional studies estimated blue-water use values of 66 L_w kg^{-1} FPCM and 13 L_w kg^{-1} FPCM in the Netherlands [16] and South Gippsland, Australia [21], respectively. In the Netherlands case-study farm, 76% of blue-water consumption was utilized for irrigation, 15% was required for the production of concentrates, and 8% for drinking and cleaning services [16]. In Australia, 83% of total blue-water consumption was utilized on-farm for drinking and cleaning of plant equipment, while the remaining 13% was associated with the production of farm inputs such as concentrated feed. The considerably larger blue-water footprint in the Netherlands was due to irrigation. Interestingly, without irrigation, the calculated blue-water footprint in the Netherlands was equal to 16 L_w kg^{-1} FPCM, which was only 23% greater than that in Australia without any irrigation system. De Boer et al. [16] also calculated a stress-weighted water-footprint value of 33 L_w kg^{-1} FPCM, implying that 1 kg of FPCM produced on a case-study farm in the Netherlands had an impact on water deprivation equivalent to the direct consumption of 33 L of water by an average world citizen.

Table 1. Dairy water-footprint assessment (L_w kg^{-1} fat and protein-corrected milk (FPCM)) studies found in the literature.

Study	Year	Region	WF *	GW	Grey	BW
Mekonnen and Hoekstra [20] [1]	2012	Global Avg.	1299	1185	61	53
Mekonnen and Hoekstra [20] [1]	2012	IND	1592	1425	41	126
Mekonnen and Hoekstra [20] [1]	2012	CHN	1651	1438	117	96
Mekonnen and Hoekstra [20] [1]	2012	NTL	781	683	38	60
Mekonnen and Hoekstra [20] [1]	2012	USA	1414	1237	100	77
De Boer et al. [16]	2013	NTL	–	–	–	66
Murphy et al. [15]	2016	IRE	690	684	–	6
Ridoutt et al. [21]	2010	AUS	–	–	–	13
Sultana et al. [22] [1]	2014	Global Avg.	1643	1423	103	117
Palhares and Pezzopane [13]	2015	BRA (Conv)	935	858	73	4
Palhares and Pezzopane [13]	2015	BRA (Org)	785	682	94	9
Zonderland-Thomassen and Ledgard [23]	2012	AUS	945	676	260	9
Zonderland-Thomassen and Ledgard [23]	2012	AUS	1084	499	336	249

* WF = water footprint; GW = green water; BW = blue water; Con = conventional farm; Org = organic farm; FPCM = fat- and protein-corrected milk; Grey = grey water; CHN = China, NTL = the Netherlands; USA = United States of America; IRE = Ireland; IND = India; BRA = Brazil. WF, GW, Grey, and BW values were transformed to L_w kg^{-1} FPCM using Equations (A1)–(A3) in Appendix A, and related fat and protein percentage values.
[1] Converted to L_w kg^{-1} FPCM using country-specific milk fat and protein percentage values [24].

Green-, blue-, and grey-water-footprint values in milk production were also calculated by Sultana et al. across 72 dairy regions from 48 countries, representing 85% of the world's milk production [22]. They identified a typical farm for each dairy region, based upon the most common farm size, farming system, production technology, and management practices used in each country. They then used the TIPI-CAL (Technology Impact Policy Impact Calculation) model, based on the principles of the farm-level impact of policy simulations, to estimate water footprints across grazing, intensive, and small-scale dairy farming within each region. The global average calculated by Sultana et al. was equal to 1643 L_w kg^{-1} FPCM, composed of 87% green-water use, 7% grey-water use, and 6% blue-water use. Western Europe was found the have the lowest green- and blue-water footprints with averages of 721 and 59 L_w kg^{-1} FPCM, respectively. On the other hand, typical small-scale African farming had the highest green-water footprint (4417 L_w kg^{-1} FPCM), while the highest blue-water footprint was found in typical Middle East feed-lot farming (295 L_w kg^{-1} FPCM) [22,24].

Utilizing data from 24 Irish dairy farms in 2013, Murphy et al. [15] calculated an overall water footprint of 690 L_w kg^{-1} FPCM, whereby 85% was attributed to green-water use for the production of pasture, 10% was required for imported forage production (hay and silage), 4% for the production of concentrated feed, and ~1% for on-farm blue-water use. They also calculated an average stress-weighted water footprint of 0.4 L_w kg^{-1} FPCM, implying that each liter of milk produced in Ireland contributed 0.4 L of freshwater consumption by an average world citizen.

Water footprinting is a popular method for attributing water consumption to specific processes throughout the milk-production cycle, as shown in the studies presented in Table 1. It is clear that on average globally, green water is the largest water consumption category in dairy farming, followed by blue-water usage. This was shown by global water-footprinting studies carried out by Mekonnen and Hoekstra [12] and Sultana et al. [22], which found that green water was responsible for 91% and 87% of the total water footprint, respectively. Although these studies showed blue-water use to be responsible for between 4% and 7% of the total water footprint, depending on the meteorological conditions of a particular region, the overall stress on localized freshwater withdrawals may be considerably high, particularly during periods of little rainfall. Moreover, the potential harm caused by blue-water withdrawals from surface and groundwater sources outweighs the potential harm associated with the consumption of green water sourced from precipitation over agricultural lands, particularly when that blue water is drawn from large underground fossil-water aquifers that have a very low recharge rate. Thus, the following section focuses on aggregating studies in the literature which focused on direct on-farm blue-water consumption.

2.1. Direct Water Consumption

Direct water consumption on dairy farms is primarily required for washing and sanitation purposes within the milking parlor, or for dairy-cow drinking water (free water intake (FWI)). Some studies have focused on calculating total direct water use, such as Shine et al. [25], who calculated a total direct water use figure of 7.2 L_w kg^{-1} FPCM on Irish dairy farms. However, it is common practice to subcategorize dairy-farm water consumption according to whether consumption takes place within the milking parlor or as FWI. Thus, the following sections highlight and critically assess dairy water literature related to parlor water usage and FWI.

2.1.1. Parlor Water Consumption

A description of parlor water consumption figures found in the literature is presented in Table 2. Shine et al. [25] found that 33% (2.4 L_w kg^{-1} milk) of total direct water consumption was used within the milking parlor. Within the milking parlor, the precooling of milk through a plate heat exchanger (PHE) required an average of 1.8 L_w kg^{-1} milk, hot-water consumption required 0.15 L_w kg^{-1} milk, and water for parlor washdown purposes required 1.3 L_w kg^{-1} milk. In the United States, Brugger and Dorsey [26] analyzed the parlor water consumption of a single Ohio dairy farm (herd size = 988 cows) in 2005. They calculated an average consumption of 0.8 L_w kg^{-1} milk with no seasonal variation trend. However, they did note an increase of 0.1 L_w kg^{-1} milk during June, July, and August when providing misting to cows to keep their body temperatures down. In Denmark, Brøgger Rasmussen and Pedersen [27] analyzed the parlor water (hot water, cold water, floor washing) consumption of 14 dairy farms which employed an AMS, and three dairy farms with traditional milking systems, between June 2003 and February 2004. Across four AMS brands, Brøgger Rasmussen and Pedersen [27] calculated an average parlor water consumption of 0.4 L_w kg^{-1} milk, equal to that consumed by the traditional milking systems. Higham et al. [28] also monitored and analyzed 35 pasture-based dairy farms (herd size range = 160–1150 cows) in New Zealand over a period of two years (June 2013 to May 2015) and calculated a parlor water usage value of 3.9 L_w kg^{-1} milk. In Germany, Krauß et al. [29] analyzed the cleaning water use within a dairy barn (herd size = 176 cows), autonomously monitoring hourly water usage across two milking systems (a herringbone and an AMS) over 806 days. Krauß et al. [29] estimated a barn-cleaning water consumption value of 0.8 L_w kg^{-1} milk for the AMS and a mean of 1.3 L_w kg^{-1} milk for the herringbone parlor system. Krauß et al. [29], along with Palhares and Pezzopane [13], highlighted the scarcity of scientific studies of dairy-barn-cleaning water consumption, while also commenting on the lack of detail regarding how cleaning water demand is measured and estimated.

Table 2. Parlor water consumption values found in the literature.

Study	Year	Country	n	Farm Type	Measurement	L_w kg^{-1} Milk
Brugger and Dorsey [26]	2006	USA	1	Confinement	Parlor	0.8 [2]
Brøgger Rasmussen and Pedersen [27]	2004	DNK	14	Confinement (AMS)	Parlor	0.4
Brøgger Rasmussen and Pedersen [27]	2004	DNK	3	Confinement	Parlor	0.4
Higham et al. [28]	2017	NZL	31	Pasture-based	Parlor	3.9 [3]
Krauß et al. [29]	2016	DEU	1	Confinement (AMS)	Cleaning [1]	0.8
Krauß et al. [29]	2016	DEU	1	Confinement	Cleaning [1]	1.3
Shine et al. [25]	2018	IRE	33	Pasture-based	Parlor	2.4
Shine et al. [25]	2018	IRE	50	Pasture-based	Cleaning [1]	1.4
Shorthall et al. [30]	2018	IRE	7	Pasture-based (AMS)	Parlor	1.5
Shorthall et al. [30]	2018	IRE	7	Pasture-based (AMS)	Cleaning [1]	0.4

AMS = automatic milking system; USA = United States of America; DNK = Denmark; NZL = New Zealand; DEU = Germany; IRE = Ireland; FPCM = fat- and protein-corrected milk; n = number of farms assessed.
[1] Water used for cold- and hot-water washdown of milking equipment, the milking parlor, and milk-cooling tank.
[2] Converted to L_w kg^{-1} milk using water consumption figure of 28.4 L_w cow^{-1} day^{-1} and mean milk production of 80 lbs cow^{-1} day^{-1}. [3] Converted to L_w kg^{-1} milk using water consumption figure of 48.7 L_w cow^{-1} day^{-1} and mean milk production of 12.5 kg cow^{-1} day^{-1}.

2.1.2. Free Water Intake

A description of FWI figures found in the literature is presented in Table 3. FWI has been extensively covered in literature, to the extent that Appuhamy et al. [31] compiled a dataset gathering results from 69 research articles and retrieved 239 FWI treatment means of lactating cows (research articles from North America (47%), Europe (25%), and Australia (8%)), of which 10% were related to pasture-based dairy-cow diets. When processed, Appuhamy et al. [31] found 55 research articles with useable FWI data to calculate a global figure for lactating cows of 78.4 L_w cow^{-1} day^{-1} (mean milk production of 28.1 kg cow^{-1} day^{-1}). Appuhamy et al. [31] also gathered results from 19 research articles related to the FWI figures of dry dairy cows, while utilizing 10 of these to estimate a global figure of 34.0 L_w cow^{-1} day^{-1}.

Table 3. Description of FWI values found in literature.

Study	Year	Country	n	Measurement	Farm Type	L_w cow^{-1} day^{-1}
Higham et al. [28]	2017	NZL	35	Lactating and dry	Grazing	36.0 *
Higham et al. [28]	2017	NZL	35	Lactating and dry	Grazing	60.0
Appuhamy et al. [31]	2016	–	–	Lactating cows	Confinement	78.4
Appuhamy et al. [31]	2016	–	–	Dry cows	Confinement	34.0
Krauß et al. [29]	2016	DEU	1	Lactating cows	Confinement (AMS)	91.1
Krauß et al. [29]	2016	DEU	1	Lactating cows	Confinement	54.4
Cardot et al. [32]	2008	FRA	1	Lactating cows	Confinement	83.6
Brugger and Dorsey [26]	2006	USA	1	Lactating and dry	Confinement	77.6

* Corrected for leaks; FWI = free water intake; AMS = automatic milking system; NZL = New Zealand; DEU = Germany; FRA = France; USA = United States of America; n = number of farms assessed.

Krauß et al. [29] calculated a mean drinking-water consumption value of 91.1 L_w cow^{-1} day^{-1} (mean milk production of 35.5 kg cow^{-1} day^{-1}) for the AMS system, and a mean of 54.4 L_w cow^{-1} day^{-1} (mean milk production of 25.4 kg cow^{-1} day^{-1}) on a farm that employed a herringbone milking system in Germany. In Ontario, Canada, Robinson et al. [33] also found that dairy farms with AMSs consumed greater volumes of water for drinking compared to free-stall and tie-stall milking operations ($p < 0.01$). Krauß et al. [29] found that 80% of dairy-cow daily water intake was consumed between 05:00 and 21:00 h, with peak water consumption occurring between 07:00 and 08:00 h for the cows in AMSs, and between both 07:00 and 08:00 h and between 17:00 and 18:00 h for cows on the dairy farm where they were milked via the herringbone system. These peak water-intake times were expected as they coincided with milking times [29]. In France, 41 lactating Holstein cows milked via a herringbone parlor system were found to consume 83.6 L_w cow^{-1} day^{-1}, of which almost 75% of drinking water was consumed between 06:00 and 19:00 h [32].

Utilizing a single dairy farm in Ohio, United States, Brugger and Dorsey [26] estimated an overall FWI value of 77.6 L_w cow^{-1} day^{-1} (mean milk production of 36.2 kg cow^{-1} day^{-1}), reporting an increased water consumption value of 120.7 L_w cow^{-1} day^{-1} in summer due to the increased average ambient temperature. The increased ambient temperature increased the drinking-water requirements of cows and increased the need to provide misting to cows to help reduce body temperature (to promote health and milk-production benefits). This concurred with findings reported by Robinson et al. [33] in which 17 Ontario dairy farms were analyzed and found to consume greater volumes of water during the summer months than in winter ($p < 0.05$). In New Zealand, Higham et al. [28] utilized 35 pasture-based dairy farms to calculate an average drinking-water consumption of 60 L_w cow^{-1} day^{-1}, although this figure was corrected to 36 L_w cow^{-1} day^{-1} when adjusted for leakage (mean milk production of 12.8 kg cow^{-1} day^{-1}). Contradictory to Brugger and Dorsey [26] and Robinson et al. [33], Higham et al. [28] reported reduced total water usage during the summer months.

2.1.3. Direct Water Consumption Summary

The literature summarized in this section focused on quantifying both parlor water consumption and FWI on dairy farms through the installation of physical metering equipment over a specified timeframe. As this method is by far the most accurate for calculations related to dairy water consumption, water-footprinting studies often incorporate the physical metering of dairy-farm direct water consumption to measure BW demand. These actual water consumption values are useful for comparing the water footprint of milk between specific regions, cognate studies, or quantifying potential improvements in water usage efficiency over time. However, the installation of metering equipment may require high capital and maintenance costs, can be quite time-consuming, and also limits the number of farms that can be included in a particular study. Thus, literature has covered the development of prediction models for total dairy water consumption as well as both dairy-cow drinking-water consumption and parlor water use. These models provide researchers the ability to make estimations with a reasonable level of accuracy.

2.2. Dairy Water Prediction Modeling

Research regarding the prediction modeling of dairy-farm water use has largely focused on predicting the daily dairy-cow FWIs of lactating dairy cows and dry dairy cows. However, research has involved the prediction of water use within the dairy parlor and total direct water consumption, as well as predicting green- and blue-water requirements of milk production. These studies have resulted in the development of a number of regression equations, as shown in Table 4. The current state-of-art approach primarily separates FWI and parlor water usage, using monitored consumption data for either a single or a number of dairy farms.

Regards the prediction of green- and blue-water consumption, Murphy et al. [34] utilized data collected from 25 Irish dairy farms (mean herd size = 126 cows) between 2014 and 2015 to develop multiple linear regression (MLR) models, as shown in Table 4. Murphy et al. [34] utilized predictive variables related to farm area, milk production, herd size, concentrates, grass growth, imported forages, and metered on-farm freshwater consumption (direct water). Nonsignificant effects ($p > 0.05$) were removed from the MLR through backward elimination to extract variables with high predictive power using SAS software (SAS, 2015). They developed the MLR models on 20 farms and calculated their prediction accuracy on the remaining 5 farms. They found that the MLR model with the input variables of concentrates fed, quantity of grass grown, and imported forages was able to predict dairy-farm green-water consumption to within 11.3% (relative prediction error (RPE)). Through a standardized regression analysis, the quantity of grass grown was shown to have the largest impact on green-water consumption, with one SD change in grass grown resulting in a 0.92 SD change in green-water demand. Regarding blue-water demand, they found that the MLR model using the input variables of concentrates fed and on-farm metered water (direct water) was able to predict dairy-farm blue-water usage to within 3.4% (RPE). Through a standardized regression analysis, they calculated on-farm metered water as having the largest impact on blue-water usage, with one SD change in on-farm metered water resulting in a 0.95 SD change in blue-water demand.

A large number of variables have been reported to influence direct water use on dairy farms, as shown via the regression models shown in Table 4. The predictive impact of these variables, and the resulting predictive capabilities vary between studies due to differences in farm data, dairy-cow breed, farming practices, research methodologies, and differences in environmental conditions between countries.

2.2.1. Direct Water Prediction

Higham et al. [28] utilized metered water data from 35 New Zealand dairy farms (herd size range = 160–1150 cows) and data related to climatic conditions, farm characteristics, and milk production to develop regression models to predict dairy-farm total direct water demand (drinking plus parlor). They selected model input variables based upon their ability to explain the variability of the observed data. Model development was carried out utilizing 21 variables, including both continuous and categorical data. Input variables were assessed as linear and quadratic terms during model development. The variables (linear or squared terms) with the lowest influence (lowest standardized regression coefficient) on total water consumption were iteratively removed to produce a model with a similar R^2 value to the original but with considerably fewer input variables. Model accuracy was then assessed on unseen data using 50-fold cross-validation. Higham et al. [28] found that the regression model with six input variables of maximum ambient temperature, evapotranspiration, radiation, milk solids (MS), milk volume, and whether a rotary or herringbone milking parlor was employed was able to explain 90% of the overall variability ($R^2 = 0.90$). Model coefficients are shown in Table 4.

Shine et al. [35] also employed MLR modeling to predict dairy water consumption; however, they took a different approach to Higham et al. [28]. As opposed to predicting daily water consumption, they predicted monthly water consumption data remotely monitored on 51 dairy farms throughout the January 2014–May 2016 period in conjunction with farm details related to milk production, stock, farm infrastructure, managerial processes, and environmental conditions. This resulted in 12 individual regression models equations being developed (one for each month), allowing for consumption trends throughout the year to be linearly modeled. In total, 20 dairy-farm variables were assessed for their ability to predict dairy-farm water consumption. The subset of farm variables that maximized the prediction accuracy of unseen water consumption was selected through application of a univariate variable selection technique, all-subsets regression, and 10-fold cross-validation. The most accurate MLR model configuration for water consumption prediction contained six variables: herd size, milk production, the number of parlor units, automatic parlor washing, whether ground water was utilized for precooling in an open-loop system, and whether water troughs were reported to contain leaks. This MLR model was found to predict dairy direct water consumption to within 49% (RPE), within which, through a standardized regression analysis, milk production, automatic parlor washing, and whether winter building troughs were reported to be leaking were shown to have the largest impacts on water consumption.

Table 4. Regression models in literature predicting free water intake (FWI), parlor water use (Parlor), total water use (TW), green-water use (GW), and blue-water use (BW).

Study	Year	Unit	Regression Model	R^2	RMSPE
Castle and Thomas [37]	1975	FWI [a]	$= -15.3 + 2.53 \times MY + 0.45 \times DM\%$	0.73	n/a
Little and Shaw [38]	1978	FWI [a]	$= 12.3 + 2.15 \times DMI + 0.73 \times MY$	0.73	n/a
Murphy et al. [39]	1983	FWI [a]	$= 16.0 + 1.58 \times DMI + 0.90 \times MY + 0.05 \times NaI + 1.20 \times minT$	0.59	n/a
Murphy et al. [39]	1983	FWI [a]	$= 23.0 + 2.38 \times DMI + 0.64 \times MY$	0.42	n/a
Stockdale and King [40]	1983	FWI [a]	$= -9.37 + 2.30 \times DMI + 0.53 \times DM\%$	0.58	n/a
Holter and Urban [41]	1992	FWI [b]	$= -10.34 + 0.23 \times DM\% + 2.212 \times DMI + 0.03944 \times CP^2$	0.64	n/a
Holter and Urban [41]	1992	FWI [a]	$= -32.4 + 2.47 \times DMI + 0.60 \times MY + 0.62 \times DM\% + 0.091 \times JD - 0.00026 \times JD^2$	0.69	n/a
Dahlborn et al. [42]	1998	FWI [a]	$= 14.3 + 1.24 \times MY + 0.32 \times DM\%$	0.67	n/a
Meyer et al. [43]	2004	FWI [a]	$= 26.1 + 1.30 \times MY + 0.406 \times NaI + 1.516 \times meanT + 0.058 \times BodyWeight$	0.60	n/a
Cardot et al. [32]	2008	FWI [a]	$= -25.65 + 1.54 \times DMI + 1.33 \times MY + 0.89 \times DM\% + 0.57 \times minT - 0.30 \times Rain$	0.74	n/a
Khelil-Arfa et al. [44]	2012	FWI [a,b]	$= -77.6 + 3.22 \times DMI + 0.92 \times MY - 0.28 \times Con\% + 0.83 \times DM\% + 0.037 \times BodyWeight$	0.92	n/a
Khelil-Arfa et al. [44]	2012	FWI [a,b]	$= -41.1 + 1.54 \times MY - 0.29 \times Con\% + 0.97 \times DM\% + 0.039 \times BodyWeight$	0.86	n/a
Appuhamy et al. [31]	2016	FWI [a]	$= -91.1 + 2.93 \times DMI + 0.61 \times DM\% + 0.062 \times NaK + 2.49 \times CP\% + 0.76 \times meanT$	n/a	14.4%
Appuhamy et al. [31]	2016	FWI [a]	$= -60.2 + 1.43 \times MY + 0.064 \times NaK + 0.83 \times DM\% + 0.54 \times meanT + 0.08 \times DIM$	n/a	17.9%
Appuhamy et al. [31]	2016	FWI [b]	$= 1.16 \times DMI + 0.23 \times DM\% + 0.44 \times meanT + 0.061 \times meanTC^2$	n/a	12.8%
Appuhamy et al. [31]	2016	FWI [b]	$= 0.010 \times BodyWeight + 0.32 \times DM\% + 0.52 \pm 0.21 \times meanT + 0.053 \times meanTC^2$	n/a	15.2%
Krauß et al. [29]	2016	FWI [a]	$= -27.94 + 0.49 \times meanT + 3.15 \times MY$	0.67	n/a
Murphy et al. [34]	2017	GW [a,b]	$= 826 \times Cn + 419 \times Gr + 498 \times lmFr$	n/a	11.3% *
Murphy et al. [34]	2017	BW [a,b]	$= -20,392 + 9.1 \times Cn + 0.92 \times MW$	n/a	3.4% *

Table 4. Cont.

Study	Year	Unit	Regression Model	R^2	RMSPE
Higham et al. [28]	2017	Log (FWI) [a,b]	$= 0.369 + 0.030 \times maxT - 0.009 \times Rain + 0.0001 \times Rain^2 - 0.117 \times Evap - 0.008 \times Evap^2 + 0.041 \times Rad + 0.261 \times MS - 0.001 \times MV^2 - 0.322 \times Milking - 0.124 \times Milking^2$	0.92	n/a
Higham et al. [28]	2017	Log(Parlor) [a,b]	$= 1.168 + 0.417 \times MS - 0.208 \times MS^2 + 0.026 \times MV + 0.315 \times Milking - 0.021 \times Milking^2 - 0.301 \times Breed - 0.029 \times CowBale$	0.95	n/a
Higham et al. [28]	2017	Log (TW) [a,b]	$= 1.104 + 0.015 \times maxT - 0.011 \times Evap^2 + 0.016 \times Rad + 0.487 \times MS - 0.265 \times MS^2 + 0.025 \times MV + 0.051 \times Rotary$	0.90	n/a
Shine et al. [35]	2018	TW_i	$= C_i + (DC_i \times \alpha_{ai}) + (MY_i \times \alpha_{ci}) + (Units_i \times \alpha_{ei}) + (AutoWash_i \times \alpha_{di}) + (PCS_i \times \alpha_{ei}) + (TroughLeaks_i \times \alpha_{fi})$	0.27	49% *

[a] denotes the water prediction of lactating cows. [b] denotes the water prediction of dry cows. TW_i = total water use in month i. RMSPE = root-mean-square prediction error as a percentage of average observed value (%). * denotes relative prediction error (RPE) (%). n/a = data unavailable. Ash % = dietary total ash content (%); AutoWash = automatic parlor washing (No = 0, Yes = 1); BodyWeight = cow weight (kg); Breed = cow breed (1 = Friesian, 0.5 = Friesian-Jersey cross, 0 = Jersey); C_i = constant value for month i; Cn = concentrates fed (kg DM year^{-1}); Con% = proportion of concentrate in diet (% of DM); CowBale = cow to cluster ratio; CP = crude protein percentage of dry matter (%); DC = number of dairy cows; DIM = day in milk; DM = dry matter (kg); DM % = dietary dry matter (%); DMI = dry matter intake (kg day^{-1}); Evap = Priestly Taylor potential evapotranspiration (mm); FWI = free water intake (L_w cow^{-1} day^{-1}); Gr = grass grown (kg DM year^{-1}); ImFr = imported forages (kg DM year^{-1}); maxT = maximum temperature (°C); meanT = mean temperature (°C); meanTC^2 = (meanT-16.4)^2; Milking = number of milkings in a day; minT = minimum temperature (°C); MS = milk solids (milk fat + milk protein, kg); MV = milk volume (L day^{-1}); MW = metered direct water (m^3 year^{-1}); MY = milk yield (kg day^{-1}); MY$_i$ = milk yield (liters month^{-1} $_i$); NaI = sodium intake (g day^{-1}); NaK = dietary sodium and potassium content (mEq kg^{-1} of DM); PCS = open-loop milk pre-cooling system (Yes = 0, No = 1); Rain = precipitation (mm day^{-1}); Rad = solar radiation (MJ m^2); Rotary = milking parlor type (rotary = 1, herringbone = 0); TroughLeaks = leaking troughs (No = 0, Yes = 1); Units = number of parlor units; α_{ai} = regression coefficient for variable a and month i.

Shine et al. [36] also looked at the potential to improve the prediction accuracy of Irish dairy-farm direct water consumption by employing various machine-learning algorithms. Using identical data to Shine et al. [25], Shine et al. [36] assessed the applicability of a classification and regression tree (CART) algorithm, a random forest ensemble algorithm, an artificial neural network and a support vector machine algorithm to predict dairy-farm water consumption. They found that the random forest algorithm maximized the prediction accuracy of dairy-farm water consumption, resulting in an RPE value of 38% (an improvement of 23% compared to MLR modeling). Thus, machine-learning algorithms failed to reduce the prediction error to a useable level (less than 20%). As the dry matter intake (DMI), live weight, and sodium intake influence cows' drinking-water intake, it is feasible that not considering these, along with individual farm effects (leaks, location of water troughs, etc.) and factors that affected the production of fresh pasture, contributed to the poor prediction accuracy of dairy-farm water consumption using machine-learning algorithms.

2.2.2. Parlor Water Prediction

Regarding parlor water demand, Higham et al. [28] built a regression model with five input variables, including milk volume, MS, number of milkings per day, body weight, and the number of cows per milking unit, which explained 95% of the variability ($R^2 = 0.95$). The model's coefficients are shown in Table 4.

In Ontario, Canada, a balance equation for predicting parlor water consumption was developed by combining four scientific research studies and was embedded as a component of the Ontario Ministry of Agriculture, Food and Rural Affairs' Nutrient Management software program [45]. The balance equation was developed using empirical usage parameters calculated by Cuthbertson et al. [46], House et al. [47], Robinson et al. [33], and Harner et al. [48] to linearly predict the water requirements of six dairy-parlor processes [49]. House et al. [47] collected data on 29 dairy farms over three years (2011–2013) and Robinson et al. [33] monitored the parlor water consumption of 17 dairy farms throughout a 20 month period (May 2013–Dec 2014), while Cuthbertson et al. [46] surveyed 308 Ontario dairy farms and washing water volumes recorded based on supplier calculations. The Ontario dairy-farm parlor water balance equation comprised six components, as detailed by Piquette [49]: (1) a mandatory base washing water usage value reflecting the type of milking system (tie stall, parlor, robotic with brush teat cleaning or robotic with water teat cleaning) multiplied by the number of lactating cows, as developed from Cuthbertson et al. [46]; (2) a plate-cooler volume figure calculated based on a milk-production value of 30 L_m cow^{-1} day^{-1} and 1.5 L_w L_m^{-1} consumed for milk precooling, for water not recycled within the milking parlor; (3) a bulk tank-cleaning water value based on milk production of 30 L_m cow^{-1} day^{-1} and 0.05 L_w L_m^{-1} consumed for bulk tank-washing purposes reflective of whether the bulk tank system did or did not have a washing system (no = 0 L_w; yes = 1.5 L_w × no. of lactating cows); (4) a parlor-wash water usage for conventional parlor milking systems (17 L_w day^{-1} × no. of lactating cows) and robotic farms (11 or 17 L_w day^{-1} (depending upon teat cleaning system) × no. of lactating cows); (5) a miscellaneous water usage value of 4 L_w cow^{-1} day^{-1}, as recommended by House et al. [47]; and (6) a heat abatement value of 18.5 L_w multiplied by the number of abatement days, as calculated by Harner et al. [48]. Piquette [49] employed the Ontario dairy-farm parlor water balance equation as part of a study aimed at calculating the water footprint of dairy farming in Eastern Ontario. However, this system had limitations regarding its inability to estimate parlor water usage on non-milking days and its simplistic nature, as each water usage process was included independently. More specifically, each model component of the Ontario dairy-farm parlor water balance equation is a separately calculated coefficient, meaning that potential relationships between the number of lactating cows, farm processes and system infrastructure are not accounted for.

In Australia, a similar approach to that of Piquette [49] was taken by the Department of Primary Industries (DPI) [50] using a mechanistic approach combined with empirical statistical data, as summarized by Callinan et al. [51]. The model was developed to estimate dairy barn water consumption, including yard cleaning, milk-cooling, parlor cleaning, cluster and platform sprays, milking-machine cleaning, and miscellaneous (yard sprinklers, fly mist sprays, washing ancillary milking equipment after milking, calf-feeding equipment). This method for estimating barn water consumption required a large amount of infrastructure data such as water pump flow rates, washdown water-container volume, the number of days the yard was cleaned annually, the average length of time the plate-cooler water was running per day at lactation and the number of days the plate cooler is used annually. Although DPI [50] failed to highlight model predication accuracy figures, this approach is an inefficient method of water consumption prediction on a large scale due to the magnitude of input variables required for predictions.

2.2.3. Free Water Intake Prediction

To explain variances in daily FWI, variables related to environmental conditions (temperature (°C), precipitation (mm), evapotranspiration (mm)), dietary consumption (dietary ash content (%), dry matter intake (DMI, kg day^{-1}), percentage of dietary dry matter (DM%, %), protein percentage of diet (%), sodium intake (g day^{-1}) etc.), dairy-cow characteristics (body weight, breed, days in milk), and milking infrastructure (rotary or herringbone milking parlor), as well as factors such as milk yield, the number of cows per milking cluster, the number of milkings per day, grass growth, and Julian day number have been employed. Of the literature contained in Table 4, 18 of the 22 studies developed prediction models for dairy-cow FWI with a mean R^2 value of 0.70 (range; 0.42–0.92), i.e., capable of explaining 70% of the variability of FWI of dairy cows. These studies were focused on explaining the variability of FWI of dairy cows as opposed to developing prediction models. Thus, the accuracy (R^2) values of these MLR models were calculated using data used for the development of the MLR coefficients (seen data). Of these, 61% (ten models) included DMI as part of the final developed MLR model, a result linked to the very strong influence of DMI on FWI for both lactating and dry dairy cows [31,41,44]. These ten studies had a mean DMI multivariate regression coefficient value of 2.19 (range = 1.16–3.22) suggesting that on average, a 1 kg increase in DMI results in an increase in FWI of 2.19 L.

As highlighted by Appuhamy et al. [31], DMI may not be routinely available on dairy farms for use in predicting drinking-water consumption. Although regression models have been developed without DMI [37,42–44], Appuhamy et al. [31] investigated the impact of excluding DMI as an independent variable for FWI prediction for lactating and dry cows using the same set of dairy farms, as accurate DMI data are often unavailable. Variables considered included DMI, milk production, body weight, number of days in milk, DM%, and concentrations of sodium and potassium in diet. Regarding the FWI of lactating cows, they developed a regression model (with DMI included) capable of predicting FWI to within 14.4% (root-mean-square prediction error (RMSPE)). When DMI was excluded, a different subset of prediction variables were selected (to minimize error), whereby the prediction error increased to 17.9% (RMSPE). Regarding the FWI of dry cows, they developed a regression model (with DMI included) capable of predicting FWI to within 12.8% (RMSPE), increasing to 15.2% when DMI was excluded, and new subset of variables utilized. They highlighted that the regression models developed without DMI as an input variable resulted in an increased impact of the DM% consumed, and that DM% represents an easier metric for dairy farmers to calculate.

Of the 18 models developed for predicting the FWI of dairy cows, 12 models included milk production (kg cow^{-1} day^{-1}) as a predictive variable, whereby on average, an increase in milk production of 1 kg cow^{-1} day^{-1} resulted in an increased FWI of 1.36 L (range = 0.60–3.15 L). Similarly, 11 models included DM% as an input variable with an average regression coefficient value of 0.53 (range = 0.23–0.97), suggesting that a 1% increase in the DM% of a cow's diet results in an increase in FWI of 0.53 L. These regression coefficients were calculated across multiple different time frames ranging from less than one week [38,41] to 806 days [29], using different cow breeds across both lactating and drying-off periods.

Higham et al. [28] utilized data related to climatic conditions, farm characteristics, and milk production to develop regression models to predict dairy-farm drinking-water use, as shown in Table 4. They calculated that 26% of stock drinking water was lost through leakage throughout the water distribution system; thus, stock drinking water was adjusted using the minimum night flow leak estimation method [52,53]. This method subtracted an average 15 min resolution night flow water usage (between 00:00 and 03:00 h) from the daytime stock drinking-water volumes. They suggested that due to similar uncorrected drinking-water values being calculated in both New Zealand and Ireland, water leakage in Ireland may be in a similar range. They found that a regression model with seven input variables, including maximum ambient temperature, rainfall, evapotranspiration, radiation, MS, milk volume, and number of milkings carried out per day, explained 92% of the variability ($R^2 = 0.92$) of corrected daily stock drinking water.

2.2.4. Dairy Water Prediction Summary

Research regarding water utilization on dairy farms has largely focused to date on modeling the FWI of dairy cows, although research has also involved the modeling of parlor water usage and totalized dairy water consumption. The prediction modeling methods used vary from MLR and polynomial regression to machine-learning algorithms to quantify water consumption. The aim of modeling the FWI of dairy cows in literature is to quantify the variability of FWI that can be explained by the variables considered in each study, as opposed to developing prediction models. Thus, recent studies have focused on developing models to predict dairy total, FWI, and parlor water consumption by assessing model prediction accuracy on unseen data [28,35]. For example, Higham et al. [28] developed prediction models using variables related the environmental conditions, milk production, and farm practices (that could be attained on a large scale). However, their models have yet to be employed to conduct a water analysis (i.e., to assess the impact of future weather conditions or varying farm practices on dairy-cow FWI and/or total or parlor water consumption). However, statistical analyses have been carried out to assess the dairy-farm variables, which have the greatest impact on dairy-farm water consumption.

2.3. Dairy Water Analyses

Literature reports indicate that direct water consumption depends upon climatic conditions such as precipitation, air temperature, and exposure to sunlight; factors such as milk production (drinking water for cows in milk), percentage of dry matter intake (DMI), PHE water use and re-use; and cleaning procedures and miscellaneous use throughout the farm [47,54,55]. Concurrently, depending upon the farming strategies employed, concerns regarding water usage may include the sizing of boreholes, the sizing of water pumps and piping, the electricity costs of pumping, the size of the manure storage and the potential volume of manure to handle (wastewater used for the washing of parlor equipment going to manure storage), the potential for the pollution of waterways, and effects on neighbors' wells [26].

Precooling milk via water through a plate cooler may reduce milk-cooling energy requirements by up to 50% [56,57]. However, careful consideration must be given to the financial and environmental impact of this utilization of water, as discussed by Murphy et al. [58]. Teagasc [7] noted the imbalance between average parlor water consumption and the cumulative consumption of precooling, hot-water, and washdown-water usage due to different farms employing different precooling strategies. More specifically, many dairy farmers recycle water utilized for milk precooling for yard washdown purposes or for animal drinking water (provided the water was at an adequate bacterial level). Concurrently, some farmers precool milk without recycling water, while others may not employ any milk precooling strategies. As a result, the milk precooling methodology employed on a dairy farm will impact overall parlor water consumption (as well as milk-cooling-related energy consumption) depending upon factors such as plate-cooler water-to-milk ratio, yard wash-down hose type, the size of the parlor, plate-cooler size, washdown water tank volume, etc.

Shine et al. [25] also conducted a detailed statistical analysis of dairy-farm direct water consumption to determine key relationships between dairy-farm characteristics as well as potential differences between categories of dairy farms. A correlation analysis found water consumption to be largely correlated with milk production, and moderately correlated with herd size and the number of lactating cows. In comparison to correlations found with electricity consumption, Shine et al. [25] found reduced correlation strengths for water, which suggested that water was less dependent on milk production and stock numbers and more dependent on managerial processes, environmental conditions, and farm infrastructure. They found that employing groundwater for milk precooling in an open-loop system increased parlor water usage by 41% on average, compared to farms which did not precool. More specifically, dairy farms which recycled milk precooling water had a 119% ($p = 0.02$) greater washdown water consumption than those without precooling facilities. Similarly, farms which recycled milk precooling water consumed 137% ($p = 0.08$) more water for washing purposes than those which did not recycle. This suggested that although farms recycled PHE water throughout the farm, use of this magnitude of water may be unwarranted for washdown purposes. They also suggested potential benefits to closed-loop milk precooling systems, allowing for a greater water-to-milk ratio as water conservation is not an issue. This would reduce milk-cooling energy use as well as reducing parlor water consumption, as only the volume of water required for washing purposes is used.

Callinan et al. [51] conducted a statistical analysis regarding daily parlor water consumption using surveyed parlor water consumption volumes and farm-practices information from over 1500 dairy farms. These farms were located in Victoria, Australia (regions: Gippsland, Northern Victoria, and Southwest Victoria) and were monitored from January 2001 to February 2009. Victoria is a relatively intensive dairy-farming region, using 53% of Australia's irrigated land and comprising 71% of the Australian dairy industry, putting major stress on the region's water supplies [59]. A significance level of 0.05 ($p < 0.05$ unless otherwise stated) was used throughout the statistical report to analyze water consumption according to region, herd size, day number, dairy-parlor type, (rotary (rotating circular milking platform), double-up herringbone and swing-over herringbone), dairy size, and yard wash type (flood, hose, or hydrant). Rotary milking parlors are rotating circular milking platforms which attach and detach milking cluster at each full rotation, meaning that rotary parlors can allow for farmers to manage a large number of dairy cows at a time [60]. A double-up herringbone parlor has a milking cluster available at each cow space with a milking line available on each side of the parlor. Swing-over herringbone milking parlors utilize one milk-harvesting cluster between a pair of adjacent cows [60]. Callinan et al. [51] found a statistically significant positive relationship between daily water use and herd size ($p < 0.001$). Concurrently, a near-significant ($p = 0.07$) reduction in water consumption per cow (L_w cow^{-1}) throughout the analysis period was calculated, highlighting the potential requirement of day number as an explanatory variable in statistical models. Regarding the multiple Australian regions analyzed, Callinan et al. [51] found that water use per day and water use per day per cow (L_w day^{-1} cow^{-1}) were significantly (both $p < 0.001$) associated with the region (Northern Victoria > Gippsland > Southwest Victoria). Concurrently, water use per day^{-1} and water use per day per cow were also significantly (both $p < 0.01$) associated with dairy type (rotary > swing-over herringbone > double-up herringbone). This result was similar to that found by Higham et al. [28] whereby parlor water usage was greater for rotary milking parlor systems when compared to herringbone systems. Regarding the recycling of used plate-cooler water, a significantly greater proportion of farms with swing-over parlors recycled plate-cooler water compared to farms with rotary and double-up parlors. Finally, regarding yard washdown, Gippsland had significantly more farms that employed hoses for yard washdown compared to the Northern or Southwest farms, while farms which had rotary parlors had significantly more flood yard washdown systems than double-up or swing-over.

Khan et al. [59] reported on historical trends and future prospects of the water use on Australian dairy farms. Khan et al. [59] highlighted growing pastures, dairy shed operations, and cattle drinking water as the three main components of dairy-farm water use in Australia; however, the proportions may vary according to precipitation influencing the requirement for irrigation. Khan et al. [59] reported a value between 2.5 and 8.5 million liters of water required to sustain pasture growth and dairy-farm operations per cow per year (assuming industrial milk production of 5231 L per cow per year), highlighting the need to plan for a future with less water availability due to climate change.

It is clear that an ample body of literature focusing on the analysis of water consumption on dairy farms is available. Both studies of Callinan et al. [51] and Shine et al. [25] found statistically significant relationships between dairy-farm water consumption and infrastructural equipment. However, as dairy-farming practices and environmental conditions vary from country to country, a similar research methodology statistically assessing water utilization on dairy farms internationally is required to further understand the factors that impact dairy-farm water consumption. These analyses are highly beneficial to quantifying the volumetric impact associated with particular water-system technologies.

3. Discussion and Perspective

The monitoring of dairy-farm water consumption is well documented throughout the literature, as presented in this review article. However, comparisons between cognate studies can be difficult when equivalent KPIs and prediction accuracy metrics are not used. For example, when reporting dairy water monitoring research results, the most common KPI is L_w cow^{-1} day^{-1}. However, results in L_w cow^{-1} day^{-1} cannot easily be compared to results of other studies that report consumption in terms of water usage per liter of water, kg^{-1} FPCM, kg^{-1} ECM, and kg milk. Concurrently, it is difficult to compare the latter set of usage consumption metrics with each other without the reporting of fat and protein percentage values of the milk production to allow for an effective comparison between international studies. It is clear that the reporting of dairy water consumption in literature should be done so in a comprehensive manner to allow for the easy comparison between studies in the domain. In the absence of studies of specific milk fat and protein values, country-/region-specific values reported by the FAO may be a suitable replacement [24].

Some studies, such as Murphy et al. [15], have calculated water-footprint figures by utilizing a relatively small sample of dairy farms compared to cognate studies (24 dairy farms). For example, Ridoutt et al. [61] calculated the water footprint of dairy farming in Australia using a sample of 75 dairy farms, while and Zonderland-Thomassen et al. [23] utilized 167 dairy farms to calculate the water footprint of dairy farms in New Zealand. However, both Ridoutt et al. [61] and Zonderland-Thomassen et al. [23] calculated stock drinking- and milking-shed water requirements based upon averaged values calculated in benchmark studies [51,62], as opposed to physical metering of on-farm direct water consumption as was carried out by Murphy et al. [15]. This underscores the importance and potential usefulness of developing dairy water prediction models, as these models complement water-footprinting studies by offering a means to estimate direct on-farm water consumption, mitigating the need of high capital investment for the purchasing, installation, and maintenance of metering equipment. However, when establishing research methodologies, researchers must consider balancing the number of farms to be included in a particular study and the accuracy of the data collected for analysis. More specifically, researchers must consider either analyzing a large number of farms using estimated data (generated using modeling methods) or analyzing a small number of farms using accurate data (collected using metering equipment). Researchers may also consider the option of acquiring accurate consumption data for a small number of farms representative of a larger dairy farm population in order to develop a model which can in turn be used to predict on-farm direct dairy water consumption. This model may then be utilized for estimating direct on-farm water usage values, and inputted to water-footprinting studies of the larger population.

Numerous prediction and analysis methodologies have been employed to predict dairy-farm direct water consumption. The development of prediction models to predict dairy-farm water consumption has generally separated FWI and parlor water usage, using monitored consumption data for either a single or a number of dairy farms. However, with the exception of Higham et al. [28], studies using the calculation of MLR equations to predict FWI have focused on understanding the variances in FWI as opposed to developing models for predicting later FWI. Thus, R^2 values are calculated based on observed data (i.e., the same data used to derive model coefficients). Although useful for determining the variance of FWI due to particular variables (such as DMI, MY, etc.), presenting R^2 values of predictions made using observed data may lead to greatly overestimating future prediction accuracy.

More recently, the development of models to predict dairy water consumption has moved towards developing prediction models utilizing "easily attainable" farm variables that may be collected on a large scale without the use of specialized equipment [28,31,35,36]. Careful consideration must be given when developing empirical models, as researchers run the risk of overestimating their prediction accuracy. To reduce the risk of overestimating prediction accuracy, techniques such as k-fold cross-validation are generally employed throughout the literature to estimate models' prediction capability on unseen data.

Balancing coarse input variables with acceptable prediction accuracy is difficult when using standard regression methods such as MLR. Thus, current work in this domain has applied machine-learning algorithms to improve the prediction accuracy when compared to state-of-the-art MLR modeling methods. However, further improvement is required to achieve greater prediction accuracies that would ensure confidence in and usefulness of model outputs prior to their adoption. Further work may look towards applying a larger range of machine-learning algorithms to dairy-farm direct water consumption data collected internationally, potentially using deep-learning algorithms if/when applicable (large numbers of data points would be required). Concurrently, the development of a global database containing international dairy water consumption figures and descriptive variables could be developed, and models built to generate a global dairy water model. If such a model were to be developed, each country would need to select a cohort of dairy farms that are representative of the country's dairy-farm demographic. These farms' water consumption levels could then be monitored in conjunction with farm characteristics, managerial strategies, and environmental data, and all collected data shared to a central database for model development. This would greatly reduce any financial cost associated with calculating the water consumption of dairy farms globally, as well as offering a means for international comparison. In conjunction with national surveys carried out throughout each country (to collect data required for water models), such a model would offer countries the opportunity to continually monitor dairy water consumption per liter of milk while also assessing the continual impact of various strategies aimed at reducing water use on dairy farms.

4. Conclusions

Scientific studies related to the use of water on dairy farms were identified and reviewed with respect to water monitoring, modeling, and analysis, placing a specific focus on studies related to on-farm total water usage, parlor usage, and free water intake (FWI) (e.g., cow drinking water). From reviewing the literature, it is clear that further studies focusing on water consumption within the milking parlor are required internationally. Additionally, future studies should consider the following three points to ensure best practice:

1. When monitoring of on-farm water use is carried out, monitoring equipment should be installed on each of the main water-consuming processes (i.e., sub-metered) to allow each component to be analyzed independently, in order to provide a greater understanding of the processes that have the greatest impact on direct on-farm water use.
2. When calculating key performance indicators, multiple units should be reported to improve their interpretation and ensure they can easily be included in cognate studies. This would reduce rounding and/or transformation errors when calculating key performance indicators across the literature. At a minimum, study-specific milk fat and protein percentage values should be reported to allow for their conversion to any key performance indicator.
3. When developing empirical prediction models, a cohort of study farms should be selected to represent the overall population of a region/country (based on which future predictions will be made). As much as possible, this cohort should cover the range and distribution of farm sizes, different infrastructural equipment employed, and managerial procedures used while being equally distributed throughout the region.

Considering these three points will help to futureproof studies in the dairy water domain and to ensure more robust analyses, with easily transferrable results that may be compared with those of future cognate studies. Additionally, increasing our understanding of dairy water consumption through statistical analyses and empirical modeling will yield an increased confidence in predictions, improving the attractiveness of empirical models as an alternative to physical metering. As well as removing time and monetary requirements associated with the physical metering of water use, these models may also offer a virtual environment whereby nonlinear impacts of changes to farm practices and equipment on dairy water use may be quantified and used by researchers, farmers, and/or policy-makers when making decisions related to dairy water consumption internationally.

Author Contributions: Conceptualization, P.S. and M.D.M.; methodology, P.S. and J.U.; data curation, P.S., J.U.; writing—original draft preparation, P.S.; writing—review and editing, P.S., J.U., M.D.M.; visualization, P.S.; project administration, M.D.M.; funding acquisition, P.S. and M.D.M. All authors have read and agreed to the published version of the manuscript.

Funding: This work has been funded by the Sustainable Energy Authority of Ireland under the SEAI Research, Development & Demonstration Funding Program 2018, Grant number 18/RDD/317.

Conflicts of Interest: The authors declare no conflict of interest.

Appendix A

A range of usage metrics relating to milk analyses are utilized throughout the dairy water consumption literature. These include: liters of milk (L_m), kilograms of milk (kg_m) (Equation (A1)), fat- and protein-corrected milk (FPCM) (Equation (A2)) [63], and energy-corrected milk (ECM) (Equation (A3)) [64]. The units of FPCM and ECM are commonly utilized for international comparisons as this ensures a fair evaluation between farms with different breeds or feed regimes [63]. However, studies utilizing different usage metrics may be compared through reporting of milk data variables such as the mean percentage of fat (% or g/kg), the percentage of protein (% or g/kg), and/or the amount of lactose (g/kg), depending on the key performance indicator required.

$$kg_m = L_m \; x \; (milk\;density) \tag{A1}$$

$$FPCM = kg_m \; x \; ((0.1226 \; x \; \%Fat) + (0.0776 \; x \; \%Protein) + 0.2534) \; kg \tag{A2}$$

$$ECM = L_m \; x \; \frac{((0.383 \; x \; \%Fat) + (0.242 \; x \; \%Protein) + 0.7832)}{3.1138} \; kg \tag{A3}$$

where *milk density* equals 1.03 kg per liter of milk, L_m is the volume of milk in liters, *%Fat* is the percentage of fat in milk, and *%Protein* is the percentage of protein in milk.

References

1. Koehler, A. Water use in LCA: Managing the planet's freshwater resources. *Int. J. Life Cycle Assess.* **2008**, *13*, 451–455. [CrossRef]
2. UN Environment. *GEO-6—Healthy Planet Healthy People*; Cambridge University Press: Cambridge, UK, 2019. [CrossRef]
3. European Union. Directive 2000/60/Ec Of The European Parliament and of the Council of 23 October 2000 Establishing a Framework for Community Action in the Field of Water Policy. *Off. J. Eur. Commun.* **2000**, *L269*, 1–93. Available online: https://eur-lex.europa.eu/legal-content/EN/TXT/?uri=celex:32000L0060 (accessed on 2 September 2020).
4. European Union. Directive 2006/118/EC of the European Parliament and of the Council of 12 December 2006 on the Protection of Groundwater against Pollution and Deterioration. *Off. J. Eur. Commun.* **2006**, *19*, 19–31. Available online: http://eur-lex.europa.eu/legal-content/EN/TXT/?uri=CELEX:32006L0118 (accessed on 2 September 2020).

5. Bruinsma, J.; Alexandratos, N. World Agriculture Towards 2030/2050: The 2012 Revision. ESA Work Pap No 12-03 2012:44–44. Available online: http://www.fao.org/docrep/016/ap106e/ap106e.pdf (accessed on 2 September 2020).
6. Mekonnen, M.M.; Hoekstra, A.Y. The green, blue and grey water footprint of crops and derived crop products. *Hydrol. Earth Syst. Sci. Discuss.* **2011**, *8*, 763–809. [CrossRef]
7. Teagasc. Dairy Farm Infrastructure Handbook 2017. Available online: https://www.teagasc.ie/media/website/publications/2017/Dairy-Farm-Infrastructure-Handbook-Moorepark2017-(V3).pdf (accessed on 2 September 2020).
8. O'Connor, D.; Kean, M. Future Expansion of the Dairy Industry in Cork: Economic Benefits and Infrastructural Requirements 2014. pp. 53–55. Available online: http://mathematics.cit.ie/contentfiles/DairyIndustry_InfrastructureReportJan27w.pdf (accessed on 2 September 2020).
9. Falkenmark, M.; Rockström, J.; Karlberg, L. Present and future water requirements for feeding humanity. *Food Secur.* **2009**, *1*, 59–69. [CrossRef]
10. Falkenmark, M.; Rockström, J. The new blue and green water paradigm: Breaking new ground for water resources planning and management. *J. Water Resour Plan Manag.* **2006**, *132*, 129–132. [CrossRef]
11. Hoekstra, A.Y.; Chapagain, A.K.; Aldaya, M.M.; Mekonnen, M.M. *The Water Footprint Assessment Manual—Setting the Global Standard*, 2nd ed.; Earthscan Ltd.: London, UK, 2011.
12. Mekonnen, M.M.; Hoekstra, A.Y. *The Green, Blue and Grey Water Footprint of Farm Animals and Animal Products. Value of Water Research Report Series No. 48. vol. 2*; UNESCO-IHE Institute for Water Education: Delft, The Netherlands, 2010.
13. Palhares, J.C.P.; Pezzopane, J.R.M. Water footprint accounting and scarcity indicators of conventional and organic dairy production systems. *J. Clean Prod.* **2015**, *93*, 299–307. [CrossRef]
14. Ercin, A.E.; Aldaya, M.M.; Hoekstra, A.Y. The water footprint of soy milk and soy burger and equivalent animal products. *Ecol. Indic.* **2012**, *18*, 392–402. [CrossRef]
15. Murphy, E.; De Boer, I.J.M.; van Middelaar, C.E.; Holden, N.M.; Shalloo, L.; Curran, T.P.; Upton, J. Water footprinting of dairy farming in Ireland. *J. Clean Prod.* **2017**, *140*, 547–555. [CrossRef]
16. De Boer, I.J.M.M.; Hoving, I.E.; Vellinga, T.V.; Van De Ven, G.W.J.J.; Leffelaar, P.A.; Gerber, P.J. Assessing environmental impacts associated with freshwater consumption along the life cycle of animal products: The case of Dutch milk production in Noord-Brabant. *Int. J. Life Cycle Assess.* **2013**, *18*, 193–203. [CrossRef]
17. Hoekstra, A.Y.; Chapagain, A.K.; Aldaya, M.M.; Mekonnen, M.M. *Water Footprint Manual State of the Art 2009*; Water Footprint Network: Enschede, The Netherlands, 2009.
18. Ran, Y.; Lannerstad, M.; Herrero, M.; Van Middelaar, C.E.; De Boer, I.J.M. Assessing water resource use in livestock production: A review of methods. *Livest. Sci.* **2016**, *187*, 68–79. [CrossRef]
19. Pfister, S.; Koehler, A.; Hellweg, S. Assessing the environmental impacts of freshwater consumption in LCA. *Environ. Sci. Technol.* **2009**, *43*, 4098–4104. [CrossRef] [PubMed]
20. Mekonnen, M.M.; Hoekstra, A.Y. A global assessment of the water footprint of farm animal products. *Ecosystems* **2012**, *15*, 401–415. [CrossRef]
21. Ridoutt, B.G.; Williams, S.R.O.; Baud, S.; Fraval, S.; Marks, N. Short communication: The water footprint of dairy products: Case study involving skim milk powder. *J. Dairy Sci.* **2010**, *93*, 5114–5117. [CrossRef] [PubMed]
22. Sultana, M.N.; Uddin, M.M.; Ridoutt, B.G.; Peters, K.J. Comparison of water use in global milk production for different typical farms. *Agric. Syst.* **2014**, *129*, 9–21. [CrossRef]
23. Zonderland-Thomassen, M.A.; Ledgard, S.F. Water footprinting—A comparison of methods using New Zealand dairy farming as a case study. *Agric. Syst.* **2012**, *110*, 30–40. [CrossRef]
24. FAO; FAOSTAT. New Food Balanc 2020. Available online: http://www.fao.org/faostat/en/#data/FBS (accessed on 22 July 2020).
25. Shine, P.; Scully, T.; Upton, J.; Shalloo, L.; Murphy, M.D. Electricity & direct water consumption on Irish pasture based dairy farms: A statistical analysis. *Appl. Energy* **2018**, *210*, 529–537. [CrossRef]
26. Brugger, M.; Dorsey, B. Water Use and Savings on a Dairy Farm. In Proceedings of the American Society of Agricultural and Biological Engineers (ASABE), St. Joseph, MI, USA, 9–12 July 2006. [CrossRef]
27. Brøgger Rasmussen, J.; Pedersen, J. *Electricity and Water Consumption at Milking*; Danish Agricultural Advisory Service: Aarhus, Denmark, 2004.

28. Higham, C.D.D.; Horne, D.; Singh, R.; Scarsbrook, M.R.R.; Kuhn-Sherlock, B.; Scarsbrook, M.R.R. Water use on nonirrigated pasture-based dairy farms:Combining detailed monitoring and modeling to set benchmarks. *J. Dairy Sci.* **2017**, *100*, 828–840. [CrossRef]
29. Krauß, M.; Drastig, K.; Prochnow, A.; Rose-Meierhöfer, S.; Kraatz, S. Drinking and cleaning water use in a dairy cow barn. *Water* **2016**, *8*, 302. [CrossRef]
30. Shortall, J.; O'Brien, B.; Sleator, R.D.; Upton, J. Daily and seasonal trends of electricity and water use on pasture-based automatic milking dairy farms. *J. Dairy Sci.* **2018**, *101*, 1565–1578. [CrossRef]
31. Appuhamy, J.A.D.R.N.; Judy, J.V.V.; Kebreab, E.; Kononoff, P.J.J. Prediction of drinking water intake by dairy cows. *J. Dairy Sci.* **2016**, *99*, 1–15. [CrossRef] [PubMed]
32. Cardot, V.; Le Roux, Y.; Jurjanz, S. Drinking behavior of lactating dairy cows and prediction of their water intake. *J. Dairy Sci.* **2008**, *91*, 2257–2264. [CrossRef] [PubMed]
33. Robinson, A.D.; Gordon, R.J.; VanderZaag, A.C.; Rennie, T.J.; Osborne, V.R. Usage and attitudes of water conservation on Ontario dairy farms. *Prof. Anim. Sci.* **2016**, *32*, 236–242. [CrossRef]
34. Murphy, E.; De Boer, I.J.M.; van Middelaar, C.; Holden, N.; Curran, P.; Upton, J. Predicting fresh water demand on Irish dairy farms using farm data. *Clean. Prod.* **2017**, *166*, 58–65. [CrossRef]
35. Shine, P.; Scully, T.; Upton, J.; Murphy, M.D. Multiple linear regression modelling of on-farm direct water and electricity consumption on pasture based dairy farms. *Comput. Electron. Agric.* **2018**, *148*, 337–343. [CrossRef]
36. Shine, P.; Murphy, M.D.; Upton, J.; Scully, T. Machine-learning algorithms for predicting on-farm direct water and electricity consumption on pasture based dairy farms. *Comput. Electron. Agric.* **2018**, *150*, 74–87. [CrossRef]
37. Castle, M.E.; Thomas, T.P. The water intake of British Friesian cows on rations containing various forages. *Anim. Prod.* **1975**, *20*, 181–189. [CrossRef]
38. Little, W.; Shaw, S.R. A note on the individuality of the intake of drinking water by dairy cows. *Anim. Prod.* **1978**, *26*, 225–227. [CrossRef]
39. Murphy, M.R.; Davis, C.L.; McCoy, G.C. Factors affecting water consumption by holstein cows in early lactation. *J. Dairy Sci.* **1983**, *66*, 35–38. [CrossRef]
40. Stockdale, C.R.; King, K.R. A note on some of the factors that affect the water consumption of lactating dairy cows at pasture. *Anim. Prod.* **1983**, *36*, 303–306. [CrossRef]
41. Holter, J.B.; Urban, W.E. Water partitioning and intake prediction in dry and lactating Holstein cows. *J. Dairy Sci.* **1992**, *75*, 1472–1479. [CrossRef]
42. Dahlborn, K.; Akerlind, M.; Gustafson, G. Water intake by dairy cows selected for high or low milk-fat percentage when fed two forage to concentrate ratios with hay or silage. *Swedish J. Agric. Res.* **1998**, *28*, 167–176.
43. Meyer, U.; Everinghoff, M.; Gädeken, D.; Flachowsky, G. Investigations on the water intake of lactating dairy cows. *Livest Prod. Sci.* **2004**, *90*, 117–121. [CrossRef]
44. Khelil-Arfa, H.; Boudon, A.; Maxin, G.; Faverdin, P. Prediction of water intake and excretion flows in Holstein dairy cows under thermoneutral conditions. *Animal* **2012**, *6*, 1662–1676. [CrossRef]
45. OMAFRA. Ontario Ministry of Agriculture, Food and Rural Affairs' (OMAFRA) Nutrient Management Software Program. Using NMAN Best Manag Pract 2016. Available online: http://www.omafra.gov.on.ca/english/nm/nman/nman3.htm (accessed on 21 May 2018).
46. Cuthbertson, E.; Senyshyn, L.; Koppen-Train, S. Milking centre waste management in Ontario. *Can. Agric. Eng.* **1995**, *37*, 258.
47. House, H.K.; Hawkins, B.C.; Barkes, B.C. Measuring and Characterizing On-Farm Milking Centre Washwater Volumes. *Am. Soc. Agric. Biol. Eng.* **2014**, *7004*, 1–16. [CrossRef]
48. Harner, J.P.; Brouk, M.J.; Potts, J.; Bradford, B.; Smith, J.F.; Hall, S. Scientific data for developing water budgets on a dairy. In Proceedings of the Western Dairy Management Conference, Reno, NV, USA, 6–8 March 2001; pp. 90–104. Available online: http://wdmc.org/2013/Scientific%20Data%20for%20Developing%20Water%20Budgets%20on%20a%20Dairy.pdf (accessed on 2 September 2020).
49. Piquette, S. *The Water Use Efficiency of Dairy Farming in Eastern Ontario: A Case Study*; Carleton University: Ottawa, ON, Canada, 2015.

50. DPI Department of Primary Industries. Dairy Shed Water—How Much do You Use? A Comprehensive Guide to Calculating Water Use in the Dairy Shed. 2009; pp. 1–17. Available online: http://agriculture.vic.gov.au/__data/assets/pdf_file/0007/197080/Dairy-shed-water.pdf (accessed on 17 May 2018).
51. Callinan, L.; Ward, R.; Kelstall, B.; McDonald, S.; Eldridge, R.; Williams, J. Dairy Shed Water Use in Victoria: 2009 Analysis 2010. Available online: http://agriculture.vic.gov.au/__data/assets/pdf_file/0003/197085/Dairy-Shed-Water-Use-in-Victoria-2009-Analysis.pdf (accessed on 17 May 2018).
52. Cheung, P.B.; Guilherme, V.; Abe, N.; Propato, M. Night flow analysis and modeling for leakage estimation in a water distribution system. In *Integrating Water Systems (CCWI 2010)*; Taylor & Francis Group: London, UK, 2010; pp. 509–513.
53. Tabesh, M.; Yekta, A.; Hossein, A. Assessment of real losses in potable water distribution systems: Some recent developments. *Water Sci. Technol. Water Supply* **2005**, *5*, 33–40. [CrossRef]
54. Clarke, J.; David, B.; Erdman, R. *Nutrient Requirements of Dairy Cattle*; National Academy Press: Washington, DC, USA, 2001; pp. 178–183. [CrossRef]
55. Teagasc. National Agri-Environment Conference 2014—Environment Conference 2014 2014:24. Available online: https://www.teagasc.ie/media/website/publications/2014/AgriEnvironment_Conference2014.pdf (accessed on 2 September 2020).
56. Shine, P.; Upton, J.; Sefeedpari, P.; Murphy, M.D. Energy consumption on dairy farms: A review of monitoring, prediction modelling and analyses. *Energies* **2020**, *13*, 1288. [CrossRef]
57. Karlsson, A.E.; Hörndahl, T.; Nordman, R. Energy recover from milk cooling. In *Report 401*; Agriculture & Industry; JTI-Swedish Institute of Agricultural and Environmental Engineering: Uppsala, Sweden, 2012.
58. Murphy, M.D.; Upton, J.; O'Mahony, M.J. Rapid milk cooling control with varying water and energy consumption. *Biosyst. Eng.* **2013**, *116*, 15–22. [CrossRef]
59. Khan, S.; Abbas, A.; Rana, T.; Jason, C. *Dairy Water Use in Australian Dairy Farms: Past Trends and Future Prospects*; CSIRO Water for a Healthy Country National Research Flagship: Canberra, Australia, 2010.
60. Ryan, T.; Donworth, J. Teagasc—Milking Facilities. 2012, pp. 125–138. Available online: https://www.teagasc.ie/media/website/animals/dairy/MilkingFacilities.pdf (accessed on 17 May 2018).
61. Ridoutt, B.; Hodges, D. From ISO14046 to water footprint labeling: A case study of indicators applied to milk production in south-eastern Australia. *Sci. Total Environ.* **2017**, *599–600*, 14–19. [CrossRef]
62. Stewart, G.; Rout, R. Reasonable Stock Water Requirements: Guidelines for Resource Consent Applications 2007. Available online: https://www.boprc.govt.nz/media/470831/reasonable-stock-water-requirements-guidelines-horizons.pdf (accessed on 2 September 2020).
63. IDF. Bulletin of the International Dairy Federation 479/2015 —A common carbon footprint approach for dairy. *Bull. Int. Dairy Fed.* 2015, pp. 1–64. Available online: https://www.fil-idf.org/wp-content/uploads/2016/09/Bulletin479-2015_A-common-carbon-footprint-approach-for-the-dairy-sector.CAT.pdf (accessed on 2 September 2020).
64. Sjaunja, L.O.; Baevre, L.; Junkkarinen, L.; Pedersen, J.; Setälä, J. A Nordic proposal for an energy corrected milk (ECM) formula. In Proceedings of the 27th Session of International Committee for Recording the Productivity of Milk Animals, Paris, France, 2–6 July 1990; pp. 156–157.

© 2020 by the authors. Licensee MDPI, Basel, Switzerland. This article is an open access article distributed under the terms and conditions of the Creative Commons Attribution (CC BY) license (http://creativecommons.org/licenses/by/4.0/).

MDPI
St. Alban-Anlage 66
4052 Basel
Switzerland
Tel. +41 61 683 77 34
Fax +41 61 302 89 18
www.mdpi.com

Sustainability Editorial Office
E-mail: sustainability@mdpi.com
www.mdpi.com/journal/sustainability

www.ingramcontent.com/pod-product-compliance
Lightning Source LLC
LaVergne TN
LVHW070449100526
838202LV00014B/1692